April 2012
Publication No. FHWA-HIF-12-004

Hydraulic Engineering Circular No. 20

# Stream Stability at Highway Structures
## Fourth Edition

U.S. Department of Transportation
**Federal Highway Administration**

Published by Books Express Publishing
Copyright © Books Express, 2012
ISBN 978-1-78266-123-8

Books Express publications are available from all good retail and online booksellers. For publishing proposals and direct ordering please contact us at: info@books-express.com

## TABLE OF CONTENTS

LIST OF FIGURES ............................................................................................................. vii

LIST OF TABLES ............................................................................................................... xi

LIST OF SYMBOLS ........................................................................................................... xiii

ACKNOWLEDGMENTS ..................................................................................................... xvii

GLOSSARY ........................................................................................................................ xix

**CHAPTER 1 - INTRODUCTION** ..................................................................................... 1.1

1.1 PURPOSE .................................................................................................................. 1.1
1.2 BACKGROUND .......................................................................................................... 1.1
1.3 COMPREHENSIVE ANALYSIS .................................................................................. 1.1
1.4 PROCEDURAL GUIDANCE ....................................................................................... 1.4
1.4.1 Stream Stability and the NBIS ................................................................................. 1.4
1.4.2 USACE Nationwide Permit Issues ........................................................................... 1.5
1.5 FACTORS THAT AFFECT STREAM STABILITY ...................................................... 1.5
1.6 IDENTIFICATION AND ANALYSIS OF STREAM STABILITY PROBLEMS .............. 1.6
1.7 ANALYSIS METHODOLOGY ..................................................................................... 1.7
1.8 MANUAL ORGANIZATION ........................................................................................ 1.8
1.9 DUAL SYSTEM OF UNITS ........................................................................................ 1.8

**CHAPTER 2 - GEOMORPHIC FACTORS AND PRINCIPLES** ...................................... 2.1

2.1 INTRODUCTION ........................................................................................................ 2.1
2.2 LANDFORM EVOLUTION .......................................................................................... 2.1
2.3 GEOMORPHIC FACTORS AFFECTING STREAM STABILITY ................................ 2.4
2.3.1 Overview .................................................................................................................. 2.4
2.3.2 Stream Size .............................................................................................................. 2.8
2.3.3 Flow Habit ................................................................................................................ 2.8
2.3.4 Bed Material ............................................................................................................. 2.9
2.3.5 Valley Setting ........................................................................................................... 2.9
2.3.6 Floodplains ............................................................................................................... 2.10
2.3.7 Natural Levees ......................................................................................................... 2.11
2.3.8 Apparent Incision ..................................................................................................... 2.11
2.3.9 Channel Boundaries and Vegetation ....................................................................... 2.11
2.3.10 Sinuosity ................................................................................................................ 2.16
2.3.11 Braided Streams .................................................................................................... 2.20
2.3.12 Anabranched Streams ........................................................................................... 2.21
2.3.13 Variability of Width and Development of Bars ....................................................... 2.22

2.4 AGGRADATION/DEGRADATION AND SEDIMENT CONTINUITY IN
    SAND-BED CHANNELS ............................................................................................ 2.22

2.4.1 Aggradation/Degradation ......................................................................................... 2.22
2.4.2 Overview of the Sediment Continuity Concept ........................................................ 2.22
2.4.3 Factors Initiating Bed Elevation Changes ................................................................ 2.23

2.5    CHANNEL STABILITY CONCEPTS FOR COHESIVE BOUNDARY CHANNELS ..2.25

2.5.1 Cohesive Streambeds ......................................................................................2.26
2.5.2 Cohesive Streambanks....................................................................................2.27

2.6    GRAVEL-BED RIVERS ........................................................................................2.27

## CHAPTER 3 - HYDRAULIC FACTORS AND PRINCIPLES .................................................3.1

3.1    INTRODUCTION ....................................................................................................3.1
3.2    HYDRAULIC DESIGN FOR SAFE BRIDGES........................................................3.1

3.2.1 Overview ............................................................................................................3.1
3.2.2 Hydraulic Modeling Criteria and Selection ........................................................3.2

3.3    BASIC HYDRAULIC PRINCIPLES .......................................................................3.2

3.3.1 Continuity Equation ...........................................................................................3.3
3.3.2 Energy Equation ................................................................................................3.4
3.3.3 Manning Equation..............................................................................................3.4

3.4    HYDRAULIC FACTORS AFFECTING STREAM STABILITY...............................3.6

3.4.1 Overview ............................................................................................................3.6
3.4.2 Magnitude and Frequency of Floods.................................................................3.6
3.4.3 Bed Configurations in Sand-Bed Streams........................................................3.8
3.4.4 Resistance to Flow ..........................................................................................3.10
3.4.5 Water Surface Profiles.....................................................................................3.15

3.5    GEOMETRY AND LOCATION OF HIGHWAY STREAM CROSSINGS.................3.17

3.5.1 Problems at Bends ..........................................................................................3.17
3.5.2 Problems at Confluences................................................................................3.17
3.5.3 Backwater Effects of Alignment and Location ................................................3.18
3.5.4 Effects of Highway Profile................................................................................3.18

3.6    BRIDGE DESIGN ................................................................................................3.20

3.6.1 Scour at Bridges ..............................................................................................3.20
3.6.2 Abutments .......................................................................................................3.22
3.6.3 Piers.................................................................................................................3.22
3.6.4 Bridge Foundations .........................................................................................3.23
3.6.5 Superstructures...............................................................................................3.23

## CHAPTER 4 - ANALYSIS PROCEDURES FOR STREAM INSTABILITY ........................4.1

4.1    INTRODUCTION ....................................................................................................4.1
4.2    GENERAL SOLUTION PROCEDURE..................................................................4.1
4.3    DATA NEEDS.........................................................................................................4.2

4.3.1 Data Needs for Level 1 Qualitative and Other Geomorphic Analyses .............4.2
4.3.2 Data Needs for Level 2 Basic Engineering Analyses .......................................4.3
4.3.3 Data Needs for Level 3 Mathematical and Physical Model Studies.................4.4

| | | |
|---|---|---|
| 4.4 | DATA SOURCES | 4.4 |
| 4.5 | LEVEL 1: QUALITATIVE GEOMORPHIC ANALYSES | 4.4 |

4.5.1 Step 1. Define Stream Characteristics ..................................................................4.4
4.5.2 Step 2. Evaluate Land Use Changes ....................................................................4.11
4.5.3 Step 3. Assess Overall Stream Stability................................................................4.11
4.5.4 Step 4. Evaluate Lateral Stability .........................................................................4.14
4.5.5 Step 5. Evaluate Vertical Stability .......................................................................4.15
4.5.6 Step 6. Evaluate Channel Response to Change ....................................................4.15

| | | |
|---|---|---|
| 4.6 | LEVEL 2: BASIC ENGINEERING ANALYSES | 4.16 |

4.6.1 Step 1. Evaluate Flood History and Rainfall-Runoff Relations.............................4.16
4.6.2 Step 2. Evaluate Hydraulic Conditions .................................................................4.18
4.6.3 Step 3. Bed and Bank Material Analysis ..............................................................4.18
4.6.4 Step 4. Evaluate Watershed Sediment Yield.........................................................4.19
4.6.5 Step 5. Incipient Motion Analysis ........................................................................4.20
4.6.6 Step 6. Evaluate Armoring Potential ....................................................................4.20
4.6.7 Step 7. Evaluation of Rating Curve Shifts ............................................................4.21
4.6.8 Step 8. Evaluate Scour Conditions .......................................................................4.22

| | | |
|---|---|---|
| 4.7 | LEVEL 3: MATHEMATICAL AND PHYSICAL MODEL STUDIES | 4.22 |
| 4.8 | ILLUSTRATIVE EXAMPLES | 4.23 |

**CHAPTER 5 - RECONNAISSANCE, CLASSIFICATION, ASSESSMENT, AND RESPONSE ........................................................................................5.1**

| | | |
|---|---|---|
| 5.1 | INTRODUCTION | 5.1 |
| 5.2 | STREAM RECONNAISSANCE | 5.1 |

5.2.1 Stream Reconnaissance Techniques .....................................................................5.2
5.2.2 Specific Applications............................................................................................5.3
5.2.3 Assessment of Drift Accumulation Potential........................................................5.4

| | | |
|---|---|---|
| 5.3 | STREAM CHANNEL CLASSIFICATION | 5.11 |

5.3.1 Overview ..............................................................................................................5.11
5.3.2 Channel Classification Concepts ..........................................................................5.12
5.3.3 Channel Classification and Stream Stability.........................................................5.19

| | | |
|---|---|---|
| 5.4 | RAPID ASSESSMENT OF CHANNEL STABILITY | 5.23 |

5.4.1 Overview ..............................................................................................................5.23
5.4.2 Rapid Assessment Method....................................................................................5.25
5.4.3 Stability Indicators................................................................................................5.26
5.4.4 Lateral and Vertical Stability ................................................................................5.26
5.4.5 Examples ..............................................................................................................5.26

| | | |
|---|---|---|
| 5.5 | QUALITATIVE EVALUATION OF CHANNEL RESPONSE | 5.34 |

5.5.1 Overview ..............................................................................................................5.34
5.5.2 Lane Relation and Other Geomorphic Concepts ..................................................5.34
5.5.3 Stream System Response ......................................................................................5.38
5.5.4 Regime Equations for Sand-Bed Channels ..........................................................5.40
5.5.5 Complex Response ...............................................................................................5.41

**CHAPTER 6 - QUANTITATIVE TECHNIQUES FOR STREAM STABILITY ANALYSIS ... 6.1**

6.1 INTRODUCTION ................................................................................................. 6.1
6.2 LATERAL CHANNEL STABILITY ....................................................................... 6.2

6.2.1 Meander Migration ..................................................................................... 6.3
6.2.2 Bank Failure ............................................................................................... 6.9
6.2.3 Channel Width Adjustments ....................................................................... 6.9

6.3 PREDICTING MEANDER MIGRATION ............................................................ 6.11

6.3.1 Map and Aerial Photograph Comparison ................................................. 6.11
6.3.2 An Overlay Comparison Technique .......................................................... 6.12

6.4 VERTICAL CHANNEL STABILITY ................................................................... 6.25

6.4.1 Overview .................................................................................................. 6.25
6.4.2 Aggradation and Degradation Analysis ................................................... 6.26
6.4.3 Sediment Continuity Analysis ................................................................... 6.34

6.5 EXAMPLE PROBLEMS - VERTICAL CHANNEL STABILITY ......................... 6.36

6.5.1 Example Problem 1 - Incipient Motion and Armoring Analysis ............... 6.36
6.5.2 Example Problem 2 - Equilibrium Slope Analysis ................................... 6.39
6.5.3 Example Problem 3 - Base Level Control ............................................... 6.41
6.5.4 Example Problem 4 - Sediment Continuity ............................................. 6.42
6.5.5 Example Problem 5 - Sediment Continuity ............................................. 6.46

**CHAPTER 7 - SEDIMENT TRANSPORT CONCEPTS ............................................ 7.1**

7.1 OVERVIEW .......................................................................................................... 7.1
7.2 SEDIMENT CONTINUITY ................................................................................... 7.2
7.3 SEDIMENT PROPERTIES .................................................................................. 7.2

7.3.1 Particle Size ............................................................................................... 7.2
7.3.2 Particle Shape ........................................................................................... 7.3
7.3.3 Fall Velocity ............................................................................................... 7.3
7.3.4 Sediment Size Distribution ........................................................................ 7.3
7.3.5 Specific Weight .......................................................................................... 7.4
7.3.6 Porosity ...................................................................................................... 7.4
7.3.7 Angle of Repose ........................................................................................ 7.4

7.4 SEDIMENT TRANSPORT CONCEPTS .............................................................. 7.4

7.4.1 Initiation of Motion ..................................................................................... 7.4
7.4.2 Modes of Sediment Transport ................................................................... 7.6
7.4.3 Effects of Bed Forms at Stream Crossings ............................................... 7.7

7.5 SEDIMENT TRANSPORT EQUATIONS ............................................................ 7.8
7.6 VARIABILITY IN SEDIMENT TRANSPORT ESTIMATES ............................... 7.10

**CHAPTER 8 - CHANNEL STABILITY IN GRAVEL-BED RIVERS ........................... 8.1**

8.1 OVERVIEW .......................................................................................................... 8.1
8.2 FLUVIAL PROCESSES IN GRAVEL-BED RIVERS ........................................... 8.1

|   |   |
|---|---|
| 8.2.1 Velocity and Flow Resistance in Gravel-Bed Rivers | 8.1 |
| 8.2.2 Sediment Characteristics and Armoring | 8.4 |
| 8.2.3 Sediment Transport | 8.6 |
| 8.2.4 Channel Pattern and Channel-Scale Bedforms | 8.8 |
| 8.2.5 Bank Erosion | 8.10 |
| 8.3 MANAGEMENT OF GRAVEL-BED RIVERS | 8.10 |
| 8.3.1 Regime Equations | 8.11 |
| 8.3.2 Channel Change and River Response | 8.14 |
| 8.3.3 River Stabilization and Training | 8.15 |
| **CHAPTER 9 - CHANNEL RESTORATION CONCEPTS** | **9.1** |
| 9.1 INTRODUCTION | 9.1 |
| 9.2 CHANNEL RESTORATION AND REHABILITATION | 9.1 |
| 9.3 DESIGN CONSIDERATIONS FOR CHANNEL RESTORATION | 9.3 |
| 9.3.1 USACE Design Methodology | 9.6 |
| 9.3.2 Natural Channel Design | 9.8 |
| 9.3.3 Restoration Project Uncertainties | 9.12 |
| 9.3.4 In-Stream Flow Control Structures | 9.12 |
| 9.4 MANAGING ROADWAY IMPACTS ON STREAM ECOSYSTEMS | 9.14 |
| **CHAPTER 10 - LITERATURE CITED** | **10.1** |
| **APPENDIX A - METRIC SYSTEM, CONVERSION FACTORS, AND WATER PROPERTIES** | **A.1** |
| **APPENDIX B - BANK EROSION AND FAILURE MECHANISMS** | **B.1** |
| **APPENDIX C - STREAM RECONNAISSANCE RECORD SHEETS** | **C.1** |
| **APPENDIX D - DATA ITEMS FROM APPENDIX C RECONNAISSANCE SHEETS RELATED TO STREAM STABILITY INDICATORS** | **D.1** |
| **APPENDIX E - SIMPLIFIED AND REVISED DATA COLLECTION SHEETS BASED ON THORNE (1998) AND JOHNSON (2006)** | **E.1** |

(page intentionally left blank)

# LIST OF FIGURES

| | | |
|---|---|---|
| Figure 1.1. | Flow chart for scour and stream stability analysis and evaluation | 1.2 |
| Figure 2.1. | The cycle of erosion, proposed by W.M. Davis, drawn by E. Raisz | 2.2 |
| Figure 2.2. | Evolution of incised channel from initial incision and widening to aggradation and eventual relative stability | 2.3 |
| Figure 2.3. | Sediment loads following channel incision | 2.5 |
| Figure 2.4. | Three possible stages in the development of a meandering reach | 2.5 |
| Figure 2.5. | Surveys showing changes of course for two meandering rivers | 2.6 |
| Figure 2.6. | Geomorphic factors that affect stream stability | 2.7 |
| Figure 2.7. | Diverse morphology of alluvial fans | 2.10 |
| Figure 2.8. | Active bank erosion illustrated by vertical cut banks, slump blocks, and falling vegetation | 2.14 |
| Figure 2.9. | Typical bank failure surfaces | 2.15 |
| Figure 2.10. | Plan view of a meandering stream | 2.18 |
| Figure 2.11. | Modes of meander loop development | 2.19 |
| Figure 2.12. | Definition sketch of sediment continuity concept applied to a given channel reach over a given time period | 2.24 |
| Figure 2.13. | Stream channel with erodible rock bed and semi-cohesive banks | 2.26 |
| Figure 3.1. | Sketch of the continuity concept | 3.3 |
| Figure 3.2. | Sketch of the energy concept for open channel flow | 3.4 |
| Figure 3.3. | Hydraulic, location, and design factors that affect stream stability | 3.7 |
| Figure 3.4(a). | Forms of bed roughness in sand channels | 3.9 |
| Figure 3.4(b). | Relation between water surface and bed configuration | 3.9 |
| Figure 3.5. | Relative resistance to flow in sand-bed channels | 3.13 |
| Figure 3.6. | Types of water surface profiles through bridge openings | 3.16 |
| Figure 3.7. | Superelevation of water surface in a bend | 3.17 |
| Figure 3.8. | Backwater effect associated with three types of stream crossings | 3.19 |

| | | |
|---|---|---|
| Figure 3.9. | Various highway profiles | 3.21 |
| Figure 4.1. | Flow chart for Level 1: Qualitative Geomorphic Analyses | 4.10 |
| Figure 4.2. | Channel classification and relative stability as hydraulic factors are varied | 4.13 |
| Figure 4.3. | Hydraulic problems at bridges attributed to erosion at a bend or to lateral migration of the channel | 4.13 |
| Figure 4.4. | Flow chart for Level 2: Basic Engineering Analyses | 4.17 |
| Figure 4.5. | Specific gage data for Cache Creek, California | 4.22 |
| Figure 4.6. | Local scour and contraction scour related hydraulic problems at bridges | 4.23 |
| Figure 5.1. | Increased scour at bridge piers as a result of debris | 5.4 |
| Figure 5.2. | Flow chart for evaluating debris production potential | 5.6 |
| Figure 5.3. | Flow chart for determining the potential for debris transport and delivery | 5.8 |
| Figure 5.4. | Hypothetical debris and flood flow paths during in-bank and out-of-bank flood flows for a low sinuosity channel | 5.9 |
| Figure 5.5. | Schematic of design (or key) log length, butt diameter, and root mass extension | 5.10 |
| Figure 5.6. | Alluvial channel pattern classification devised by Brice | 5.13 |
| Figure 5.7. | The range of alluvial channel patterns | 5.15 |
| Figure 5.8. | Idealized long profile from hillslopes and unchanneled hollows downslope through the channel network showing the general distribution of alluvial channel types | 5.16 |
| Figure 5.9. | Key to classification of rivers in Rosgen¡s method | 5.18 |
| Figure 5.10. | Longitudinal, cross-sectional, and planform views of major stream types in Rosgen¡s method | 5.19 |
| Figure 5.11. | Significant cattle activity, agricultural activity; poorly maintained engineered channel | 5.30 |
| Figure 5.12. | Cattle and farming activity; flashy flows; braiding; small loose bed material; minor obstructions | 5.31 |
| Figure 5.13. | Development in the watershed; perennial stream; low levees; well-packed bed material | 5.32 |
| Figure 5.14. | Stable watershed; perennial stream; mild meanders; no entrenchment; large, packed bed material | 5.33 |

| Figure 5.15. | Sinuosity vs. slope with constant discharge | 5.34 |
| Figure 5.16. | Slope-discharge relationship for braiding or meandering in sand-bed streams | 5.36 |
| Figure 5.17. | Changes in channel slope in response to a decrease in sediment supply at point C | 5.39 |
| Figure 5.18. | Use of geomorphic relationships of Figures 5.15 and 5.16 in a qualitative analysis | 5.39 |
| Figure 6.1. | Types of lateral activity and typical associated floodplain features | 6.2 |
| Figure 6.2. | Models of flow structure and associated bed forms in straight alluvial channels | 6.4 |
| Figure 6.3. | Flow patterns in meanders | 6.5 |
| Figure 6.4. | Schematic of an idealized meander bend illustrating geometric variables | 6.6 |
| Figure 6.5. | Modified Brice classification of meandering channels | 6.17 |
| Figure 6.6. | Aerial photograph of a site on the White River in Indiana showing four registration points common to the 1966 aerial photo | 6.19 |
| Figure 6.7. | Aerial photo of a site on the White River in Indiana showing the four registration points common to the 1937 aerial photo | 6.19 |
| Figure 6.8. | The 1966 aerial photo of the White River in Indiana with the 1937 bankline tracing and registration points | 6.20 |
| Figure 6.9. | Meander loop evolution and classification scheme proposed by Brice | 6.21 |
| Figure 6.10. | Circles that define the average outer banklines from the 1937 aerial photo of the White River site in Indiana | 6.21 |
| Figure 6.11. | Depiction of the bends from 1937 and 1966 outer banklines as defined by best-fit circles | 6.22 |
| Figure 6.12. | Depiction of the bends from the 1937 and 1966 outer banklines, as defined by best-fit circles, and the predicted location and radius of the 1998 outer bankline circle | 6.23 |
| Figure 6.13. | Aerial photo of the White River in 1966 showing the actual 1937 banklines and the predicted 1998 bankline positions | 6.23 |
| Figure 6.14. | Aerial photograph of the White River site in Indiana in 1998 comparing the predicted bankline positions with the actual banklines | 6.24 |
| Figure 6.15. | Channel armoring | 6.28 |
| Figure 6.16. | Base level control and degradation due to changes in slope | 6.32 |

| | | |
|---|---|---|
| Figure 6.17. | Headcuts and nickpoints | 6.33 |
| Figure 6.18. | Headcut downstream of bridge | 6.33 |
| Figure 6.19. | Definition of sediment load components | 6.34 |
| Figure 7.1. | Suspended sediment concentration profiles | 7.7 |
| Figure 7.2. | Velocity and sediment concentration profiles | 7.9 |
| Figure 7.3. | Bed-material size effects on bed material transport | 7.11 |
| Figure 7.4. | Effect of slope on bed material transport | 7.12 |
| Figure 7.5. | Effect of kinematic viscosity (temperature) on bed material transport | 7.12 |
| Figure 7.6. | Variation of bed material load with depth of flow | 7.12 |
| Figure 8.1. | Distinction between bed surface armor layer and subsurface sediment common in many gravel-bed rivers and streams | 8.5 |
| Figure 8.2. | Classification of channel pattern and overlapping pool-bar units in gravel-bed rivers of different channel pattern | 8.9 |
| Figure 8.3. | Definition sketch for a riffle-pool channel | 8.9 |
| Figure 9.1. | Flow chart depicting the sequence of implementation of Rosgen's eight sequence phases associated with natural channel design using a geomorphic approach | 9.10 |
| Figure 9.2. | Typical in-stream deflector and sill | 9.12 |

# LIST OF TABLES

| | | |
|---|---|---|
| Table 1.1. | Commonly Used Engineering Terms in English and SI Units | 1.8 |
| Table 2.1. | Sediment Grade Scale | 2.13 |
| Table 3.1. | Base Values of Manning n | 3.11 |
| Table 3.2. | Adjustment Factors for the Determination of n Values for Channels | 3.12 |
| Table 3.3. | Manning n ($n_b$) Roughness Coefficients for Alluvial Sand-bed Channels | 3.13 |
| Table 4.1. | List of Data Sources | 4.5 |
| Table 4.2. | List of Internet Data Sources | 4.9 |
| Table 4.3. | Interpretation of Observed Data | 4.12 |
| Table 5.1. | Major Phases, Tasks, and Subtasks for Assessing the Potential for Debris Production and Accumulation at a Bridge | 5.5 |
| Table 5.2. | Classification of Alluvial Channels | 5.14 |
| Table 5.3. | Classification of Channel-Reach Morphology in Mountain Drainage Basins of the Pacific Northwest | 5.17 |
| Table 5.4. | Summary of Common Indicators Used in Channel Stability Assessment Methods | 5.24 |
| Table 5.5. | Stability Indicators, Descriptions, and Ratings | 5.27 |
| Table 5.6. | Overall Scores for Three Classifications of Channels | 5.29 |
| Table 5.7. | Stability Ratings for Streams in Figures 5.11 – 5.14 | 5.29 |
| Table 5.8. | Lateral and Vertical Stability for Streams in Figures 5.11 – 5.14 | 5.29 |
| Table 6.1. | Sources of Contemporary and Historical Aerial Photographs and Maps | 6.14 |
| Table 6.2. | Range of Parameters | 6.30 |
| Table 8.1. | Generalized Relative Differences of Sand-bed vs. Gravel/Cobble-bed Streams | 8.2 |
| Table 8.2. | Examples of River Metamorphosis | 8.15 |
| Table 9.1. | Hierarchical List and Classification of Environmentally Sensitive Channel and Bank Protection Techniques | 9.5 |

(page intentionally left blank)

# LIST OF SYMBOLS

| | | |
|---|---|---|
| $A$ | = | Cross-sectional flow area, ft$^2$ (m$^2$) |
| $A_m$ | = | Meander amplitude, ft (m) |
| a, b, c | = | Coefficients or exponents |
| b | = | Length of the bridge opening, ft (m) |
| C | = | Conveyance, ft$^3$/s (m$^3$/s) |
| $C_d$ | = | Coefficient of drag |
| $C_t$ | = | Sediment concentration in parts per million by weight |
| c | = | Sediment concentration at height y above bed |
| $c_a$ | = | Sediment concentration at height a above bed |
| $c_d$ | = | Bank material cohesion, lbs/ft$^2$ (Pa or N/m$^2$) |
| d | = | Flow depth (also y), ft (m) |
| $d_m$ | = | Bankfull maximum depth, ft (m) |
| $D_c$ | = | Diameter of sediment particle at incipient motion conditions, in or ft (mm or m) |
| $D_i$ | = | The i$^{th}$ percentile size of bed material finer than a given size, in or ft (mm or m) |
| $D_{50}$ | = | Median sediment size, in or ft (mm or m) |
| Fr | = | Froude Number |
| FS | = | Factor of safety |
| g | = | Acceleration of gravity, ft/s$^2$ (m/s$^2$) |
| $h_L$ | = | Energy loss ft (m) |
| H | = | Depth of submergence, ft (m) |
| K | = | Conveyance, ft$^3$/s (m$^3$/s) |
| $K_u$ | = | Conversion constant (English/SI units) |
| $K_s$ | = | Shield's Parameter |
| $k_s$ | = | Grain roughness |
| L | = | Distance upstream of base level control, ft (m) |
| L | = | Reach length, ft (m) |
| M | = | Mass of the debris slugs or lbs (kg) |
| M | = | Weighted silt-clay index or percent silt-clay |
| MR | = | Migration rate, ft/yr (m/yr) |
| $MD_5$ | = | 5-year migration distance, ft (m) |

## LIST OF SYMBOLS (continued)

| | | |
|---|---|---|
| $m$ | = | Roughness correction factor for sinuosity of the channel |
| $n$ | = | Manning roughness coefficient |
| $n_b$ | = | Base value for straight, uniform channel |
| $n_1$ | = | Value for surface irregularities in the cross section |
| $n_2$ | = | Value for variations in shape and size of the channel |
| $n_3$ | = | Value for obstructions |
| $n_4$ | = | Value for vegetation and flow conditions |
| $P$ | = | Planform sinuosity |
| $P$ | = | Wetted perimeter, boundary, ft (m) |
| $P_d$ | = | Bankfull pool depth, ft (m) |
| $P_c$ | = | Decimal fraction of material coarser than the armoring size |
| $P_w$ | = | Bankfull pool width, ft (m) |
| $Q$ | = | Discharge, total discharge, ft$^3$/s (m$^3$/s) |
| $Q_r$ | = | Radial stress, lbs/ft$^2$ (N/m$^2$) |
| $Q_s$ | = | Sediment discharge, ft$^3$/s (m$^3$/s) |
| $q$ | = | Discharge per unit width, ft$^3$/s/ft (m$^3$/s/m or m$^2$/s) |
| $q_b$ | = | Bedload discharge per unit width, ft$^2$/sec (m$^2$/sec) |
| $q_s$ | = | Sediment discharge per unit width, ft$^3$/s/ft (m$^3$/s/m or m$^2$/s) |
| $R$ | = | Hydraulic radius (ratio of flow area to wetted perimeter), ft (m) |
| $R_c$ | = | Radius of the center of the stream, ft (m) |
| $R_d$ | = | Bankfull riffle depth, ft (m) |
| $R_i$ | = | Radius of the inside bank, ft (m) |
| $R_o$ | = | Radius of the outside bank at the bend, ft (m) |
| $R_w$ | = | Bankfull riffle width, ft (m) |
| $S$ | = | Stopping distance, ft (m) |
| $S$ | = | Energy slope or channel slope, ft/ft (m/m) |
| $S_{eq}$ | = | Equilibrium channel slope, ft/ft (m/m) |
| $S_{ex}$ | = | Existing channel slope, ft/ft (m/m) |
| $s$ | = | Shear strength, lbs/ft$^2$ (Pa or N/m$^2$) |
| $V_*$ | = | Shear velocity, ft/sec (m/s) |

## LIST OF SYMBOLS (continued)

| | | |
|---|---|---|
| $V$ | = | Velocity or average velocity of flow, ft/s (m/s) |
| $V_{cr}$ | = | Critical velocity, ft/s (m/s) |
| $V_{s\,(inflow)}$ | = | Volume of sediment supplied, ft³ (m³) |
| $V_{s\,(outflow)}$ | = | Volume of sediment transport, ft³ (m³) |
| $W$ | = | Width, ft (m) |
| $W_D$ | = | Width under dominant discharge, ft (m) |
| $x$ | = | Sinusoidal function of distance |
| $Y$ | = | Flow depth (also d or y) ft (m) |
| $Y$ | = | Depth of tension cracking, ft (m) |
| $Y_a$ | = | Thickness of the armoring layer, ft (m) |
| $Y_s$ | = | Depth of scour, ft (m) |
| $Y_s$ | = | Ultimate degradation amount, ft (m)) |
| $Z$ | = | Bed elevation referenced to a common datum, ft (m) |
| $z$ | = | Rouse Number |
| $Z_o$ | = | Depth of tensile stress, ft (m) |
| $\beta$ | = | Angle of bank failure plane relative to horizontal |
| $\gamma$ | = | Specific weight of water, lbs/ft³ (N/m³) |
| $\gamma_s$ | = | Specific weight of sediment, lbs/ft³ (N/m³) |
| $\Delta t$ | = | Time increment, s |
| $\Delta V$ | = | Volume of sediment stored or eroded, ft³ (m³) |
| $\Delta W$ | = | Lateral erosion distance, ft (m) |
| $\Delta Z$ | = | Change in bed elevation, ft (m) |
| $\Delta Z$ | = | Difference in water surface elevation between concave and convex banks, ft (m) |
| $\Delta_{max}$ | = | Maximum lateral (migration) erosion distance, ft (m) |
| $\eta$ | = | Porosity of bed material |
| $\theta$ | = | Bank slope angle |
| $\theta$ | = | Channel direction |
| $\lambda$ | = | Meander wavelength, ft (m) |

## LIST OF SYMBOLS (continued)

| | | |
|---|---|---|
| $\mu$ | = | Pore water pressure, lbs/ft$^2$ (Pa or N/m$^2$) |
| $\nu$ | = | Kinematic viscosity, ft$^2$/s (m$^2$/s) |
| $\rho$ | = | Density of water, slugs/ft$^3$ (kg/m$^3$) |
| $\rho_s$ | = | Sediment density, slugs/ft$^3$ (kg/m$^3$) |
| $\sigma$ | = | Normal stress, lbs/ft$^2$ (Pa or N/m$^2$) |
| $\tau_o$ | = | Average boundary shear stress, lbs/ft$^2$ (Pa or N/m$^2$) |
| $\tau_c$ | = | Critical shear stress, lbs/ft$^2$ (Pa or N/m$^2$) |
| $\tau_e$ | = | Shear stress ratio, $\tau_o/\tau_c$ |
| $\phi$ | = | Bank material friction angle |
| $\phi'$ | = | Apparent angle of internal friction |
| $\omega$ | = | Maximum angle between a channel segment and the mean downvalley axis |
| $\omega$ | = | Fall velocity of sediment, ft/s (m/s) |

## ACKNOWLEDGMENTS

This document is a major revision of the 2001 Third Edition of HEC-20. The writers wish to acknowledge the contributions made by F. Johnson, F. Chang, and J.R. Richardson, as co-authors of the 1991 First Edition and J.D. Schall and E.V. Richardson as co-authors of the 1995 Second Edition. Dr. Peggy Johnson (Pennsylvania State University) provided input and review of selected sections of this fourth edition, including update of the channel classification discussion, her rapid channel stability methodology and example problems, and a review of the channel restoration (Chapter 9) discussion. Her contributions are gratefully acknowledged.

The writers also wish to recognize the National Cooperative Highway Research Program (NCHRP) and its researchers for the projects completed that contributed to this document. We also recognize State Departments of Transportation for the research they have sponsored and completed as well as research completed by other federal agencies, universities, and private researchers.

All of this work will enable improved bridge design practice and bridge maintenance and inspection procedures resulting in greater safety for the users of the nation's bridges.

## DISCLAIMER

*Mention of a manufacturer, registered or trade name does not constitute a guarantee or warranty of the product by the U.S. Department of Transportation or the Federal Highway Administration and does not imply their approval and/or endorsement to the exclusion of other products and/or manufacturers that may also be suitable.*

# GLOSSARY

abrasion: Removal of streambank material due to entrained sediment, ice, or debris rubbing against the bank.

aggradation: General and progressive buildup of the longitudinal profile of a channel bed due to sediment deposition.

alluvial channel: Channel wholly in alluvium; no bedrock is exposed in channel at low flow or likely to be exposed by erosion.

alluvial fan: A fan-shaped deposit of material at the place where a stream issues from a narrow valley of high slope onto a plain or broad valley of low slope. An alluvial cone is made up of the finer materials suspended in flow while a debris cone is a mixture of all sizes and kinds of materials.

alluvial stream: A stream which has formed its channel in cohesive or noncohesive materials that have been and can be transported by the stream.

alluvium: Unconsolidated material deposited by a stream in a channel, floodplain, alluvial fan, or delta.

alternating bars: Elongated deposits found alternately near the right and left banks of a channel.

anabranch: Individual channel of an anabranched stream.

anabranched stream: A stream whose flow is divided at normal and lower stages by large islands or, more rarely, by large bars; individual islands or bars are wider than about three times water width; channels are more widely and distinctly separated than in a braided stream.

anastomosing stream: An anabranched stream.

angle of repose: The maximum angle (as measured from the horizontal) at which gravel or sand particles can stand.

annual flood: The maximum flow in one year (may be daily or instantaneous).

apron: Protective material placed on a streambed to resist scour.

apron, launching: An apron designed to settle and protect the side slopes of a scour hole after settlement.

# GLOSSARY (continued)

**armor (armoring):** Surfacing of channel bed, banks, or embankment slope to resist erosion and scour. (a) Natural process whereby an erosion-resistant layer of relatively large particles is formed on a streambed due to the removal of finer particles by streamflow; (b) placement of a covering to resist erosion.

**articulated concrete mattress:** Rigid concrete slabs which can move without separating as scour: occurs; usually hinged together with corrosion-resistant cable fasteners; primarily placed for lower bank protection.

**average velocity:** Velocity at a given cross section determined by dividing discharge by cross sectional area.

**avulsion:** A sudden change in the channel course that usually occurs when a stream breaks through its banks; usually associated with a flood or a catastrophic event.

**backfill:** The material used to refill a ditch or other excavation, or the process of doing so.

**backwater:** The increase in water surface elevation relative to the elevation occurring under natural channel and floodplain conditions. It is induced by a bridge or other structure that obstructs or constricts the free flow of water in a channel.

**backwater area:** The low-lying lands adjacent to a stream that may become flooded due to backwater.

**bank:** The sides of a channel between which the flow is normally confined.

**bank, left (right):** The side of a channel as viewed in a downstream direction.

**bankfull discharge:** Discharge that, on the average, fills a channel to the point of overflowing.

**bank protection:** Engineering works for the purpose of protecting streambanks from erosion.

**bank revetment:** Erosion-resistant materials placed directly on a streambank to protect the bank from erosion.

**bar:** An elongated deposit of alluvium within a channel, not permanently vegetated.

**base floodplain:** The floodplain associated with the flood with a 100-year recurrence interval.

## GLOSSARY (continued)

bed:
: The bottom of a channel bounded by banks.

bed form:
: A recognizable relief feature on the bed of a channel, such as a ripple, dune, plane bed, antidune, or bar. Bed forms are a consequence of the interaction between hydraulic forces (boundary shear stress) and the bed sediment.

bed layer:
: A flow layer, several grain diameters thick (usually two) immediately above the bed.

bed load:
: Sediment that is transported in a stream by rolling, sliding, or skipping along the bed or very close to it; considered to be within the bed layer (contact load).

bed load discharge (or bed load):
: The quantity of bed load passing a cross section of a stream in a unit of time.

bed material:
: Material found in and on the bed of a stream (May be transported as bed load or in suspension).

bedrock:
: The solid rock exposed at the surface of the earth or overlain by soils and unconsolidated material.

bed material discharge:
: The part of the total sediment discharge that is composed of grain sizes found in the bed and is equal to the transport capability of the flow.

bed shear (tractive force):
: The force per unit area exerted by a fluid flowing past a stationary boundary.

bed slope:
: The inclination of the channel bottom.

blanket:
: Material covering all or a portion of a streambank to prevent erosion.

boulder:
: A rock fragment whose diameter is greater than 250 mm.

braid:
: A subordinate channel of a braided stream.

braided stream:
: A stream whose flow is divided at normal stage by small mid-channel bars or small islands; the individual width of bars and islands is less than about three times water width; a braided stream has the aspect of a single large channel within which are subordinate channels.

bridge opening:
: The cross-sectional area beneath a bridge that is available for conveyance of water.

## GLOSSARY (continued)

bridge waterway:
: The area of a bridge opening available for flow, as measured below a specified stage and normal to the principal direction of flow.

bulk density:
: Density of the water sediment mixture (mass per unit volume), including both water and sediment.

bulkhead:
: A vertical, or near vertical, wall that supports a bank or an embankment; also may serve to protect against erosion.

bulking:
: Increasing the water discharge to account for high concentrations of sediment in the flow.

catchment:
: See drainage basin.

causeway:
: Rock or earth embankment carrying a roadway across water.

caving:
: The collapse of a bank caused by undermining due to the action of flowing water.

cellular-block mattress:
: Interconnected concrete blocks with regular cavities placed directly on a streambank or filter to resist erosion. The cavities can permit bank drainage and the growth of vegetation where synthetic filter fabric is not used between the bank and mattress.

channel:
: The bed and banks that confine the surface flow of a stream.

channel classification:
: Classifying a stream according to a set of observations or typical characteristics (e.g., straight, meandering, braided).

channel diversion:
: The removal of flows by natural or artificial means from a natural length of channel.

channel pattern:
: The aspect of a stream channel in plan view, with particular reference to the degree of sinuosity, braiding, and anabranching.

channel process:
: Behavior of a channel with respect to shifting, erosion and sedimentation.

channelization:
: Straightening or deepening of a natural channel by artificial cutoffs, grading, flow-control measures, or diversion of flow into an engineered channel.

check dam:
: A low dam or weir across a channel used to control stage or degradation.

choking (of flow):
: Excessive constriction of flow which may cause severe backwater effect.

# GLOSSARY (continued)

clay (mineral): A particle whose diameter is in the range of 0.00024 to 0.004 mm.

clay plug: A cutoff meander bend filled with fine grained cohesive sediments.

clear-water scour: Scour at a pier or abutment (or contraction scour) when there is no movement of the bed material upstream of the bridge crossing at the flow causing bridge scour.

cobble: A fragment of rock whose diameter is in the range of 64 to 250 mm.

cohesive streambed: Cohesive bed material can include caliche, hardpan, loess, highly compact and dense clays, and in the broader sense, erodible rock.

concrete revetment: Unreinforced or reinforced concrete slabs placed on the channel bed or banks to protect it from erosion.

confluence: The junction of two or more streams.

constriction: A natural or artificial control section, such as a bridge crossing, channel reach or dam, with limited flow capacity in which the upstream water surface elevation is related to discharge.

contact load: Sediment particles that roll or slide along in almost continuous contact with the streambed (bed load).

contraction: The effect of channel or bridge constriction on flow streamlines.

contraction scour: Contraction scour, in a natural channel or at a bridge crossing, involves the removal of material from the bed and banks across all or most of the channel width. This component of scour results from a contraction of the flow area at the bridge which causes an increase in velocity and shear stress on the bed at the bridge. The contraction can be caused by the bridge or from a natural narrowing of the stream channel.

countermeasure: A measure intended to prevent, delay or reduce the severity of hydraulic problems.

crib: A frame structure filled with earth or stone ballast, designed to reduce energy and to deflect streamflow away from a bank or embankment.

## GLOSSARY (continued)

**critical shear stress:** The minimum amount of shear stress required to initiate soil particle motion (i.e., the point of incipient motion).

**crossing:** The relatively short and shallow reach of a stream between bends; also crossover or riffle.

**cross section:** A section normal to the trend of a channel or flow.

**current:** Water flowing through a channel.

**current meter:** An instrument used to measure flow velocity.

**cut bank:** The concave wall of a meandering stream.

**cutoff:** (a) A direct channel, either natural or artificial, connecting two points on a stream, thereby shortening the original length of the channel and increasing its slope; (b) A natural or artificial channel which develops across the neck of a meander loop (neck cutoff) or across a point bar (chute cutoff).

**cutoff wall:** A wall, usually of sheet piling or concrete, that extends down to scour-resistant material or below the expected scour depth.

**daily discharge:** Discharge averaged over one day (24 hours).

**debris:** Floating or submerged material, such as logs, vegetation, or trash, transported by a stream.

**degradation (bed):** A general and progressive (long-term) lowering of the channel bed due to erosion, over a relatively long channel length.

**depth of scour:** The vertical distance a streambed is lowered by scour below a reference elevation.

**design flow (design flood):** The discharge that is selected as the basis for the design or evaluation of a hydraulic structure.

**dike:** An impermeable linear structure for the control or containment of overbank flow. A dike-trending parallel with a streambank differs from a levee in that it extends for a much shorter distance along the bank, and it may be surrounded by water during floods.

## GLOSSARY (continued)

**dike (groin, spur, jetty):** A structure extending from a bank into a channel that is designed to: (a) reduce the stream velocity as the current passes through the dike, thus encouraging sediment deposition along the bank (permeable dike); or (b) deflect erosive current away from the streambank (impermeable dike).

**discharge:** Volume of water passing through a channel during a given time.

**dominant discharge:** (a) The discharge of water which is of sufficient magnitude and frequency to have a dominating effect in determining the characteristics and size of the stream course, channel, and bed; (b) That discharge which determines the principal dimensions and characteristics of a natural channel. The dominant formative discharge depends on the maximum and mean discharge, duration of flow, and flood frequency. For hydraulic geometry relationships, it is taken to be the bankfull discharge which has a return period of approximately 1.5 years in many natural channels.

**drainage basin:** An area confined by drainage divides, often having only one outlet for discharge (catchment, watershed).

**drift:** Alternative term for vegetative "debris."

**eddy current:** A vortex-type motion of a fluid flowing contrary to the main current, such as the circular water movement that occurs when the main flow becomes separated from the bank.

**entrenched stream:** Stream cut into bedrock or consolidated deposits.

**ephemeral stream:** A stream or reach of stream that does not flow for parts of the year. As used here, the term includes intermittent streams with flow less than perennial.

**equilibrium scour:** Scour depth in sand-bed stream with dune bed about which live bed pier scour level fluctuates due to variability in bed material transport in the approach flow.

**equilibrium slope:** Channel slope at which alluvial particles on the channel bed will no longer move.

**erosion:** Displacement of soil particles due to water or wind action.

**erosion control matting:** Fibrous matting (e.g., jute, paper, etc.) placed or sprayed on a stream-bank for the purpose of resisting erosion or providing temporary stabilization until vegetation is established.

## GLOSSARY (continued)

**fabric mattress:** Grout-filled mattress used for streambank protection.

**fall velocity:** Velocity at which a sediment particle falls through a column of still water.

**fascine:** Matrix of willow or other natural material woven in bundles and used as a filter. Also, a streambank protection technique consisting of wire mesh or timber attached to a series of posts, sometimes in double rows; space between rows may be filled with rock, brush, or other materials.

**fill slope:** Side or end slope of an earth-fill embankment. Where a fill-slope forms the streamward face of a spill-through abutment, it is regarded as part of the abutment.

**filter:** Layer of fabric (geotextile) or granular material (sand, gravel, or graded rock) placed between bank revetment (or bed protection) and soil for the following purposes: (1) to prevent the soil from moving through the revetment by piping, extrusion, or erosion; (2) to prevent the revetment from sinking into the soil; and (3) to permit natural seepage from the streambank, thus preventing the buildup of excessive hydrostatic pressure.

**filter blanket:** A layer of graded sand and gravel laid between fine-grained material and riprap to serve as a filter.

**filter fabric (cloth):** Geosynthetic fabric that serves the same purpose as a granular filter blanket.

**fine sediment load:** That part of total sediment load that is composed of particle sizes finer than those represented in the bed (wash load). Normally, fine-sediment load is finer than 0.062 mm for sand-bed channels. Silts, clays and sand could be considered wash load in coarse gravel and cobble-bed channels.

**flanking:** Erosion around the landward end of a stream stabilization countermeasure.

**flashy stream:** Stream characterized by rapidly rising and falling stages, as indicated by a sharply peaked hydrograph. Typically associated with mountain streams or highly disturbed urbanized catchments. Most flashy streams are ephemeral, but some are perennial.

**flood-frequency curve:** A graph indicating the probability that the annual flood discharge will exceed a given magnitude, or the recurrence interval corresponding to a given magnitude.

## GLOSSARY (continued)

floodplain: A nearly flat, alluvial lowland bordering a stream, that is subject to frequent inundation by floods.

flow-control structure: A structure either within or outside a channel that acts as a countermeasure by controlling the direction, depth, or velocity of flowing water.

flow habit: The general characteristics of river flow: ephemeral, perennial, or flashing.

flow hazard: Flow characteristics (discharge, stage, velocity, or duration) that are associated with a hydraulic problem or that can reasonably be considered of sufficient magnitude to cause a hydraulic problem or to test the effectiveness of a countermeasure.

flow slide: Saturated soil materials which behave more like a liquid than a solid. A flow slide on a channel bank can result in a bank failure.

fluvial geomorphology: The science dealing with the morphology (form) and dynamics of streams and rivers.

fluvial system: The natural river system consisting of (1) the drainage basin, watershed, or sediment source area, (2) tributary and mainstem river channels or sediment transfer zone, and (3) alluvial fans, valley fills and deltas, or the sediment deposition zone.

freeboard: The vertical distance above a design stage that is allowed for waves, surges, drift, and other contingencies.

Froude Number: A dimensionless number that represents the ratio of inertial to gravitational forces in open channel flow.

gabion: A basket or compartmented rectangular container made of wire mesh. When filled with cobbles or other rock of suitable size, the gabion becomes a flexible and permeable unit with which flow- and erosion-control structures can be built.

geomorphology/morphology: That science that deals with the form of the Earth, the general configuration of its surface, and the changes that take place due to erosion and deposition.

grade-control structure (sill, check dam): Structure placed bank to bank across a stream channel (usually with its central axis perpendicular to flow) for the purpose of controlling bed slope and preventing scour or headcutting.

## GLOSSARY (continued)

**graded stream:** A geomorphic term used for streams that have apparently achieved a state of equilibrium between the rate of sediment transport and the rate of sediment supply throughout long reaches.

**gravel:** A rock fragment whose diameter ranges from 2 to 64 mm.

**groin:** A structure built from the bank of a stream in a direction transverse to the current to redirect the flow or reduce flow velocity. Many names are given to this structure, the most common being "spur," "spur dike," "transverse dike," "jetty," etc. Groins may be permeable, semi-permeable, or impermeable.

**grout:** A fluid mixture of cement and water or of cement, sand, and water used to fill joints and voids.

**guide bank:** A dike extending upstream from the approach embankment at either or both sides of the bridge opening to direct the flow through the opening. Some guide banks extend downstream from the bridge (also spur dike).

**hardpoint:** A streambank protection structure whereby "soft" or erodible materials are removed from a bank and replaced by stone or compacted clay. Some hard points protrude a short distance into the channel to direct erosive currents away from the bank. Hard points also occur naturally along streambanks as passing currents remove erodible materials leaving nonerodible materials exposed.

**headcutting:** Channel degradation associated with abrupt changes in the bed elevation (headcut) that generally migrates in an upstream direction.

**helical flow:** Three-dimensional movement of water particles along a spiral path in the general direction of flow. These secondary-type currents are of most significance as flow passes through a bend; their net effect is to remove soil particles from the cut bank and deposit this material on a point bar.

**hydraulics:** The applied science concerned with the behavior and flow of liquids, especially in pipes, channels, structures, and the ground.

**hydraulic geometry:** General term applied to alluvial channels to denote relationships between discharge Q and channel morphology, hydraulics, and sediment transport.

## GLOSSARY (continued)

| | |
|---|---|
| hydraulic model: | A small-scale physical or mathematical representation of a flow situation. |
| hydraulic radius: | The cross-sectional area of a stream divided by its wetted perimeter. |
| hydraulic structures: | The facilities used to impound, accommodate, convey or control the flow of water, such as dams, weirs, intakes, culverts, channels, and bridges. |
| hydrograph: | The graph of stage or discharge against time. |
| hydrology: | The science concerned with the occurrence, distribution, and circulation of water on the earth. |
| imbricated: | In reference to stream bed sediment particles, having an overlapping or shingled pattern. |
| icing: | Masses or sheets of ice formed on the frozen surface of a river or floodplain. When shoals in the river are frozen to the bottom or otherwise dammed, water under hydrostatic pressure is forced to the surface where it freezes. |
| incised reach: | A stretch of stream with an incised channel that only rarely overflows its banks. |
| incised stream: | A stream which has deepened its channel through the bed of the valley floor, so that the floodplain is a terrace. |
| invert: | The lowest point in the channel cross section or at flow control devices such as weirs, culverts, or dams. |
| island: | A permanently vegetated area, emergent at normal stage, that divides the flow of a stream. Islands originate by establishment of vegetation on a bar, by channel avulsion, or at the junction of minor tributary with a larger stream. |
| jack: | A device for flow control and protection of banks against lateral erosion consisting of three mutually perpendicular arms rigidly fixed at the center. Kellner jacks are made of steel struts strung with wire, and concrete jacks are made of reinforced concrete beams. |
| jack field: | Rows of jacks tied together with cables, some rows generally parallel with the banks and some perpendicular thereto or at an angle. Jack fields may be placed outside or within a channel. |

## GLOSSARY (continued)

jetty:
: (a) An obstruction built of piles, rock, or other material extending from a bank into a stream, so placed as to induce bank building, or to protect against erosion; (b) A similar obstruction to influence stream, lake, or tidal currents, or to protect a harbor (also spur).

knick point:
: Head cut in non-cohesive alluvial material.

lateral erosion:
: Erosion in which the removal of material is extended horizontally as contrasted with degradation and scour in a vertical direction.

launching:
: Release of undercut material (stone riprap, rubble, slag, etc.) downslope or into a scoured area.

levee:
: An embankment, generally landward of top bank, that confines flow during high-water periods, thus preventing overflow into lowlands.

live-bed scour:
: Scour at a pier or abutment (or contraction scour) when the bed material in the channel upstream of the bridge is moving at the flow causing bridge scour.

load (or sediment load):
: Amount of sediment being moved by a stream.

local scour:
: Removal of material from around piers, abutments, spurs, and embankments caused by an acceleration of flow and resulting vortices induced by obstructions to the flow.

longitudinal profile:
: The profile of a stream or channel drawn along the length of its centerline. In drawing the profile, elevations of the water surface or the thalweg are plotted against distance as measured from the mouth or from an arbitrary initial point.

lower bank:
: That portion of a streambank having an elevation less than the mean water level of the stream.

mathematical model:
: A numerical representation of a flow situation using mathematical equations (also computer model).

mattress:
: A blanket or revetment of materials interwoven or otherwise lashed together and placed to cover an area subject to scour.

meander or full meander:
: A meander in a river consists of two consecutive loops, one flowing clockwise and the other counter-clockwise.

## GLOSSARY (continued)

**meander amplitude:** The distance between points of maximum curvature of successive meanders of opposite phase in a direction normal to the general course of the meander belt, measured between center lines of channels.

**meander belt:** The distance between lines drawn tangent to the extreme limits of successive fully developed meanders.

**meander length:** The distance along a stream between corresponding points of successive meanders.

**meander loop:** An individual loop of a meandering or sinuous stream lying between inflection points with adjoining loops.

**meander ratio:** The ratio of meander width to meander length.

**meander radius of curvature:** The radius of a circle inscribed on the centerline of a meander loop.

**meander scrolls:** Low, concentric ridges and swales on a floodplain, marking the successive positions of former meander loops.

**meander width:** The amplitude of a fully developed meander measured from midstream to midstream.

**meander stream:** A stream having a sinuosity greater than some arbitrary value. The term also implies a moderate degree of pattern symmetry, imparted by regularity of size and repetition of meander loops. The channel generally exhibits a characteristic process of bank erosion and point bar deposition associated with systematically shifting meanders.

**median diameter:** The particle diameter of the 50th percentile point on a size distribution curve such that half of the particles (by weight, number, or volume) are larger and half are smaller ($D_{50}$).

**mid-channel bar:** A bar lacking permanent vegetal cover that divides the flow in a channel at normal stage.

**middle bank:** The portion of a streambank having an elevation approximately the same as that of the mean water level of the stream.

**migration:** Change in position of a channel by lateral erosion of one bank and simultaneous accretion of the opposite bank.

**mud:** A soft, saturated mixture mainly of silt and clay.

## GLOSSARY (continued)

natural levee: A low ridge that slopes gently away from the channel banks that is formed along streambanks during floods by deposition.

nominal diameter: Equivalent spherical diameter of a hypothetical sphere of the same volume as a given sediment particle.

nonalluvial channel: A channel whose boundary is in bedrock or non-erodible material.

normal stage: The water stage prevailing during the greater part of the year.

overbank flow: Water movement that overtops the bank either due to stream stage or to overland surface water runoff.

oxbow: The abandoned former meander loop that remains after a stream cuts a new, shorter channel across the narrow neck of a meander. Often bow-shaped or horseshoe-shaped.

pavement: Streambank surface covering, usually impermeable, designed to serve as protection against erosion. Common pavements used on streambanks are concrete, compacted asphalt, and soil-cement.

paving: Covering of stones on a channel bed or bank (used with reference to natural covering).

peaked stone dike: Riprap placed parallel to the toe of a streambank (at the natural angle of repose of the stone) to prevent erosion of the toe and induce sediment deposition behind the dike.

pebble count: Method used to determine size distribution of coarse bed materials which are too large to be sieved.

perennial stream: A stream or reach of a stream that flows continuously for all or most of the year.

phreatic line: The upper boundary of the seepage water surface landward of a streambank.

pile: An elongated member, usually made of timber, concrete, or steel, that serves as a structural component of a river-training structure or bridge.

pile dike: A type of permeable structure for the protection of banks against caving; consists of a cluster of piles driven into the stream, braced and lashed together.

## GLOSSARY (continued)

piping:
: Removal of soil material through subsurface flow of seepage water that develops channels or "pipes" within the soil bank.

planform:
: River characteristics as viewed from above (e.g., on a map or vertical aerial photograph).

point bar:
: An alluvial deposit of sand or gravel lacking permanent vegetal cover occurring in a channel at the inside of a meander loop, usually somewhat downstream from the apex of the loop.

poised stream:
: A stream which, as a whole, maintains its slope, depths, and channel dimensions without any noticeable raising or lowering of its bed (stable stream). Such condition may be temporary from a geological point of view, but for practical engineering purposes, the stream may be considered stable.

probable maximum flood:
: A very rare flood discharge value computed by hydro-meteorological methods, usually in connection with major hydraulic structures.

quarry-run stone:
: Stone as received from a quarry without regard to gradation requirements.

railbank protection:
: A type of countermeasure composed of rock-filled wire fabric supported by steel rails or posts driven into streambed.

rapid drawdown:
: Lowering the water against a bank more quickly than the bank can drain without becoming unstable.

reach:
: A segment of stream length that is arbitrarily bounded for purposes of study.

recurrence interval:
: The reciprocal of the annual probability of exceedance of a hydrologic event (also return period, exceedance interval).

regime:
: The condition of a stream or its channel with regard to stability. A stream is in regime if its channel has reached an equilibrium form as a result of its flow characteristics. Also, the general pattern of variation around a mean condition, as in flow regime, tidal regime, channel regime, sediment regime, etc. (used also to mean a set of physical characteristics of a river).

regime change:
: A change in channel characteristics resulting from such things as changes in imposed flows, sediment loads, or slope.

## GLOSSARY (continued)

**regime channel:** Alluvial channel that has attained, more or less, a state of equilibrium with respect to erosion and deposition.

**regime formula:** A formula relating stable alluvial channel dimensions or slope to discharge and sediment characteristics.

**reinforced-earth bulkhead:** A retaining structure consisting of vertical panels and attached to reinforcing elements embedded in compacted backfill for supporting a streambank.

**reinforced revetment:** A streambank protection method consisting of a continuous stone toe-fill along the base of a bank slope with intermittent fillets of stone placed perpendicular to the toe and extending back into the natural bank.

**rehabilitation:** Making the land (or river) useful again after a disturbance.

**restoration:** The process of repairing damage to the diversity and dynamics of ecosystems.

**relief bridge:** An opening in an embankment on a floodplain to permit passage of overbank flow.

**retard (retarder structure):** A permeable or impermeable linear structure in a channel parallel with the bank and usually at the toe of the bank, intended to reduce flow velocity, induce deposition, or deflect flow from the bank.

**revetment:** Rigid or flexible armor placed to inhibit scour and lateral erosion. (See bank revetment).

**riffle:** A natural, shallow flow area extending across a streambed in which the surface of flowing water is broken by waves or ripples. Typically, riffles alternate with pools along the length of a stream channel (e.g., in a gravel-bed channel).

**riparian:** Pertaining to anything connected with or adjacent to the banks of a stream (corridor, vegetation, zone, etc.).

**riprap:** Layer or facing of rock or broken concrete dumped or placed to protect a structure or embankment from erosion; also the rock or broken concrete suitable for such use. Riprap has also been applied to almost all kinds of armor, including wire-enclosed riprap, grouted riprap, sacked concrete, and concrete slabs.

# GLOSSARY (continued)

**river training:** Engineering works with or without the construction of embankment, built along a stream or reach of stream to direct or to lead the flow into a prescribed channel. Also, any structure configuration constructed in a stream or placed on, adjacent to, or in the vicinity of a streambank that is intended to deflect currents, induce sediment deposition, induce scour, or in some other way alter the flow and sediment regimes of the stream.

**rock and wire mattress:** A flat wire cage or basket filled with stone or other suitable material and placed as protection against erosion.

**roughness coefficient:** Numerical measure of the frictional resistance to flow in a channel, as in the Manning or Chezy's formulas.

**rubble:** Rough, irregular fragments of materials of random size used to retard erosion. The fragments may consist of broken concrete slabs, masonry, or other suitable refuse.

**runoff:** That part of precipitation which appears in surface streams of either perennial or intermittent form.

**sack revetment:** Sacks (e.g., burlap, paper, or nylon) filled with mortar, concrete, sand, stone or other available material used as protection against erosion.

**saltation load:** Sediment bounced along the streambed by energy and turbulence of flow, and by other moving particles.

**sand:** A rock fragment whose diameter is in the range of 0.062 to 2.0 mm.

**scour:** Erosion of streambed or bank material due to flowing water; often considered as being localized (see local scour, contraction scour, total scour).

**sediment or fluvial sediment:** Fragmental material transported, suspended, or deposited by water.

**sediment concentration:** Weight or volume of sediment relative to the quantity of transporting (or suspending) fluid.

**sediment discharge:** The quantity of sediment that is carried past any cross section of a stream in a unit of time. Discharge may be limited to certain sizes of sediment or to a specific part of the cross section.

**sediment load:** Amount of sediment being moved by a stream.

## GLOSSARY (continued)

sediment yield:
: The total sediment outflow from a watershed or a drainage area at a point of reference and in a specified time period. This outflow is equal to the sediment discharge from the drainage area.

seepage:
: The slow movement of water through small cracks and pores of the bank material.

shear stress:
: See unit shear force.

shoal:
: A relatively shallow submerged bank or bar in a body of water.

sill:
: (a) A structure built under water, across the deep pools of a stream with the aim of changing the depth of the stream; (b) A low structure built across an effluent stream, diversion channel or outlet to reduce flow or prevent flow until the main stream stage reaches the crest of the structure.

silt:
: A particle whose diameter is in the range of 0.004 to 0.062 mm.

sinuosity:
: The ratio between the thalweg length and the valley length of a stream.

slope (of channel or stream):
: Fall per unit length along the channel centerline or thalweg.

slope protection:
: Any measure such as riprap, paving, vegetation, revetment, brush or other material intended to protect a slope from erosion, slipping or caving, or to withstand external hydraulic pressure.

sloughing:
: Sliding or collapse of overlying material; same ultimate effect as caving, but usually occurs when a bank or an underlying stratum is saturated.

slope-area method:
: A method of estimating unmeasured flood discharges in a uniform channel reach using observed high-water levels.

slump:
: A sudden slip or collapse of a bank, generally in the vertical direction and confined to a short distance, probably due to the substratum being washed out or having become unable to bear the weight above it.

soil-cement:
: A designed mixture of soil and Portland cement compacted at a proper water content to form a blanket or structure that can resist erosion.

## GLOSSARY (continued)

**sorting:** Progressive reduction of size (or weight) of particles of the sediment load carried down a stream.

**spill-through abutment:** A bridge abutment having a fill slope on the streamward side. The term originally referred to the "spill-through" of fill at an open abutment but is now applied to any abutment having such a slope.

**spread footing:** A pier or abutment footing that transfers load directly to the earth.

**spur:** A permeable or impermeable linear structure that projects into a channel from the bank to alter flow direction, induce deposition, or reduce flow velocity along the bank.

**spur dike:** See guide bank.

**stability:** A condition of a channel when, though it may change slightly at different times of the year as the result of varying conditions of flow and sediment charge, there is no appreciable change from year to year; that is, accretion balances erosion over the years.

**stable channel:** A condition that exists when a stream has a bed slope and cross section which allows its channel to transport the water and sediment delivered from the upstream watershed without aggradation, degradation, or bank erosion (a graded stream).

**stage:** Water-surface elevation of a stream with respect to a reference elevation.

**stone riprap:** Natural cobbles, boulders, or rock dumped or placed as protection against erosion.

**stream:** A body of water that may range in size from a large river to a small rill flowing in a channel. By extension, the term is sometimes applied to a natural channel or drainage course formed by flowing water whether it is occupied by water or not.

**streambank erosion:** Removal of soil particles or a mass of particles from a bank surface due primarily to water action. Other factors such as weathering, ice and debris abrasion, chemical reactions, and land use changes may also directly or indirectly lead to bank erosion.

**streambank failure:** Sudden collapse of a bank due to an unstable condition such as removal of material at the toe of the bank by scour.

## GLOSSARY (continued)

**streambed mining:** Removal of alluvial streambed material (generally sand and gravel) by mechanical or hydraulic methods.

**streambank protection:** Any technique used to prevent erosion or failure of a streambank.

**suspended sediment discharge:** The quantity of sediment passing through a stream cross section above the bed layer in a unit of time suspended by the turbulence of flow (suspended load).

**sub-bed material:** Material underlying that portion of the streambed which is subject to direct action of the flow. Also, substrate.

**subcritical, supercritical flow:** Open channel flow conditions with Froude Number less than and greater than unity, respectively.

**tetrahedron:** Component of river-training works made of six steel or concrete struts fabricated in the shape of a pyramid.

**tetrapod:** Bank protection component of precast concrete consisting of four legs joined at a central joint, with each leg making an angle of 109.5E with the other three.

**thalweg:** The line extending down a channel that follows the lowest elevation of the bed.

**tieback:** Structure placed between revetment and bank to prevent flanking.

**timber or brush mattress:** A revetment made of brush, poles, logs, or lumber interwoven or otherwise lashed together. The completed mattress is then placed on the bank of a stream and weighted with ballast.

**toe of bank:** That portion of a stream cross section where the lower bank terminates and the channel bottom or the opposite lower bank begins.

**toe protection:** Loose stones laid or dumped at the toe of an embankment, groin, etc., or masonry or concrete wall built at the junction of the bank and the bed in channels or at extremities of hydraulic structures to counteract erosion.

**total scour:** The sum of long-term degradation, general (contraction) scour, and local scour.

**total sediment load:** The sum of suspended load and bed load or the sum of bed material load and wash load of a stream (total load).

# GLOSSARY (continued)

**tractive force:** The drag or shear on a streambed or bank caused by passing water which tends to move soil particles along with the streamflow.

**trench-fill revetment:** Stone, concrete, or masonry material placed in a trench dug behind and parallel to an eroding streambank. When the erosive action of the stream reaches the trench, the material placed in the trench armors the bank and thus retards further erosion.

**turbulence:** Motion of fluids in which local velocities and pressures fluctuate irregularly in a random manner as opposed to laminar flow where all particles of the fluid move in distinct and separate lines.

**ultimate scour:** The maximum depth of scour attained for a given flow condition. May require multiple flow events and in cemented or cohesive soils may be achieved over a long time period.

**uniform flow:** Flow of constant cross section and velocity through a reach of channel at a given time. Both the energy slope and the water slope are equal to the bed slope under conditions of uniform flow.

**unit discharge:** Discharge per unit width (may be average over a cross section, or local at a point).

**unit shear force (shear stress):** The force or drag developed at the channel bed by flowing water. For uniform flow, this force is equal to a component of the gravity force acting in a direction parallel to the channel bed on a unit wetted area. Usually in units of stress, Pa ($N/m^2$) or ($lb/ft^2$).

**unsteady flow:** Flow of variable discharge and velocity through a cross section with respect to time.

**upper bank:** The portion of a streambank having an elevation greater than the average water level of the stream.

**velocity:** The time rate of flow usually expressed in m/s (ft/sec). The average velocity is the velocity at a given cross section determined by dividing discharge by cross-sectional area.

**vertical abutment:** An abutment, usually with wingwalls, that has no fill slope on its streamward side.

## GLOSSARY (continued)

**vortex:** Turbulent eddy in the flow generally caused by an obstruction such as a bridge pier or abutment (e.g., horseshoe vortex).

**wandering channel:** A channel exhibiting a more or less non-systematic process of channel shifting, erosion and deposition, with no definite meanders or braided pattern.

**wandering thalweg:** A thalweg whose position in the channel shifts during floods and typically serves as an inset channel that conveys all or most of the stream flow at normal or lower stages.

**wash load:** Suspended material of very small size (generally clays and colloids) originating primarily from erosion on the land slopes of the drainage area and present to a negligible degree in the bed itself.

**watershed:** See drainage basin.

**waterway opening width (area):** Width (area) of bridge opening at (below) a specified stage, measured normal to the principal direction of flow.

**weephole:** A hole in an impermeable wall or revetment to relieve the neutral stress or pore pressure in the soil.

**windrow revetment:** A row of stone placed landward of the top of an eroding streambank. As the windrow is undercut, the stone is launched downslope, thus armoring the bank.

**wire mesh:** Wire woven to form a mesh; where used as an integral part of a countermeasure, openings are of suitable size and shape to enclose rock or broken concrete or to function on fence-like spurs and retards.

# CHAPTER 1

# INTRODUCTION

## 1.1 PURPOSE

The purpose of this document is to provide guidelines for identifying stream instability problems at highway stream crossings. Techniques for stream channel classification and reconnaissance, as well as rapid assessment methods for channel instability are summarized. Qualitative and quantitative geomorphic and engineering techniques useful in stream channel stability analysis are presented.

## 1.2 BACKGROUND

Approximately 500,000 bridges in the National Bridge Inventory (NBI) are built over streams. A large proportion of these bridges span alluvial streams that are continually adjusting their beds and banks. Many, especially those on more active streams, will experience problems with aggradation, degradation, bank erosion, and lateral channel shift during their useful life. The magnitude of these problems is demonstrated by the average annual flood damage repair costs of approximately $50 million for highways on the Federal-aid system (Rhodes and Trent 1993).

## 1.3 COMPREHENSIVE ANALYSIS

This manual is part of a set of Hydraulic Engineering Circulars (HEC) issued by the Federal Highway Administration (FHWA) to provide guidance for bridge scour and stream stability analyses. The three manuals in this set are:

    HEC-18    Evaluating Scour at Bridges (FHWA 2012b)
    HEC-20    Stream Stability at Highway Structures
    HEC-23    Bridge Scour and Stream Instability Countermeasures (FHWA 2009)

The Flow Chart shown in Figure 1.1 illustrates the interrelationship between these three documents and emphasizes that they should be used as a set. A comprehensive scour analysis, stability evaluation, or countermeasure design must be based on information presented in all three documents.

While the flow chart does not attempt to present every detail of a complete stream stability and scour evaluation, it has sufficient detail to show the major elements in a complete analysis, the logical flow of a typical analysis or evaluation, and the most common decision points and feedback loops. It clearly shows how the three documents tie together and recognizes the differences between design of a new bridge and evaluation of an existing bridge.

The HEC-20 block of the flow chart outlines initial data collection and site reconnaissance activities leading to an understanding of the problem, evaluation of river system stability and potential future response. The HEC-20 procedures include both qualitative and quantitative geomorphic and engineering analysis techniques which help establish the level of analysis necessary to solve the stream instability and scour problem for design of a new bridge, or for the evaluation of an existing bridge that may require rehabilitation or countermeasures.

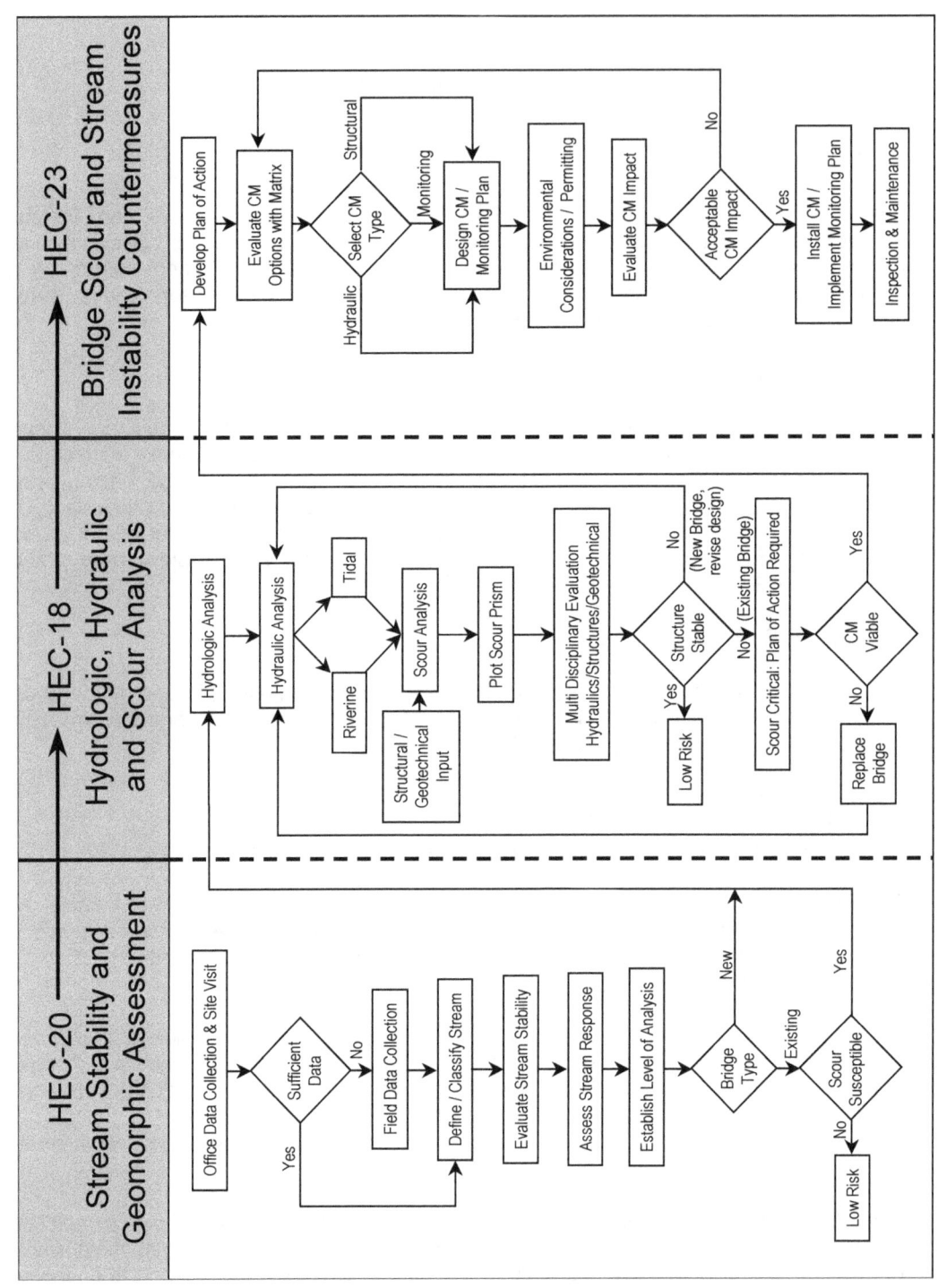

Figure 1.1. Flow chart for scour and stream stability analysis and evaluation.

The "Classify Stream," "Evaluate Stream Stability," and "Assess Stream Response" portions of the HEC-20 block are expanded in Chapter 4 into a six-step Level 1 and an eight-step Level 2 analysis procedure. In some cases, the HEC-20 analysis may be sufficient to determine that stream instability and/or scour problems do not exist, i.e., the bridge has a "low risk" of failure regarding scour susceptibility.

In most cases, the analysis or evaluation will progress to the HEC-18 block of the flow chart. Here more detailed hydrologic and hydraulic data are developed, with the specific approach determined by the level of complexity of the problem and waterway characteristics (e.g., tidal or riverine). The "Scour Analysis" portion of the HEC-18 block encompasses a seven-step specific design approach which includes evaluation of the components of total scour.

Since bridge scour evaluation requires multidisciplinary inputs, it is often advisable for the hydraulic engineer to involve structural and geotechnical engineers at this stage of the analysis. **Once the total scour prism is plotted, then all three disciplines must be involved in a determination of the structural stability of the bridge foundation**.

For a new bridge design, if the structure is stable the design process can proceed to consideration of environmental impacts, cost, constructability, and maintainability or if the bridge is unstable, revise the design and repeat the analysis. For an existing bridge, a finding of structural stability at this stage will result in a "low risk" evaluation, with no further action required. However, a Plan of Action must be developed for an unstable existing bridge (scour critical) to correct the problem as referenced in HEC-18 (FHWA 2012b) and HEC-23 (FHWA 2009).

The scour problem may be so serious that installing countermeasures would not provide a viable solution and a replacement or substantial bridge rehabilitation would be required. If countermeasures would correct the stream instability or scour problem at a reasonable cost and with acceptable environmental impacts, the analysis would progress to the HEC-23 block of the flow chart.

HEC-23 provides a range of resources to support bridge scour and stream instability countermeasure selection and design. A countermeasure matrix in HEC-23 presents a variety of countermeasures that have been used by State Departments of Transportation (DOTs) to control scour and stream instability at bridges. The matrix is organized to highlight the various groups of countermeasures and identifies distinctive characteristics of each countermeasure. The matrix identifies most countermeasures used and lists information on their functional applicability to a particular problem, their suitability to specific river environments, the general level of maintenance resources required, and which DOTs have experience with specific countermeasures. Finally, a reference source for design guidelines is noted.

HEC-23 includes specific design guidelines for the most common (and some uncommon) countermeasures used by DOTs, or references to sources of design guidance. Inherent in the design of any countermeasure are an evaluation of potential environmental impacts, permitting for countermeasure installation, and redesign, if necessary, to meet environmental requirements. As shown in the flow chart, to be effective most countermeasures will require a monitoring plan, inspection, and maintenance.

## 1.4 PROCEDURAL GUIDANCE

### 1.4.1 Stream Stability and the NBIS

The Federal requirements for bridge inspection are set forth in the National Bridge Inspection Standards (NBIS). The NBIS require bridge owners to maintain a bridge inspection program that includes procedures for underwater inspection. This information may be found in the FHWA Federal Register, Title 23, Code of Federal Regulations, Highways, Part 650, Bridges, Structures, and Hydraulics, Subpart C, National Bridge Inspection Standards (23 CFR 650, Subpart C). The most recent ruling was enacted on January 13, 2005 (USDOT 2004).

The primary purpose of NBIS is to identify and evaluate existing bridge deficiencies to ensure the safety of the traveling public. The NBIS sets national policy regarding bridge inspection and rating procedures, frequency of inspections, inspector qualifications, report formats, and the preparation and maintenance of a State bridge inventory. Each State or Federal agency must prepare and maintain an inventory of all bridges subject to the NBIS. Certain Structure Inventory and Appraisal (SI&A) data must be collected and retained by the State or Federal agency for collection and reported to FHWA on an annual basis. A tabulation of this data is contained in the SI&A sheet which may be found in the FHWA's "Recording and Coding Guide for the Structure Inventory and Appraisal of the Nation's Bridges" (FHWA 1995). The National Bridge Inventory (NBI) is the aggregation of Structure Inventory and Appraisal data collected to fulfill the requirements of the NBIS.

A national scour evaluation program as an integral part of the NBIS was established in 1988 by Technical Advisory (TA) T5140.20 (USDOT 1988). This TA was published following the April 1987 collapse of New York's Schoharie Bridge due to scour. In 1991 T5140.20 was superseded by T5140.23, "Evaluating Scour at Bridges." This Technical Advisory provides more guidance on the development and implementation of procedures for evaluating bridge scour to meet the requirements of 23 CFR 650, Subpart C. Specifically, Technical Advisory T5140.23 provides guidance on:

1. Developing and implementing a scour evaluation for designing new bridges
2. Evaluating existing bridges for scour vulnerability
3. Using scour countermeasures
4. Improving the state-of-practice for estimating scour at bridges

While the emphasis in T5140.23 is on "bridge scour," stream stability issues are referenced as well. The Technical Advisory (TA) states that bridge inspectors should receive appropriate training and instruction in inspecting bridges for scour. The TA notes that bridge inspectors should accurately record the present condition of the bridge and the stream, and should identify conditions that are indicative of potential problems with scour and stream stability. Thus, consideration must be given to stream stability issues as they affect future conditions at a bridge site and the potential for scour.

The FHWA Recording and Coding Guide (FHWA 1995) contains guidance for the following items related to channel hydraulics and scour:

1. Item 60: Substructure
2. Item 61. Channel and Channel Protection
3. Item 71: Waterway Adequacy
4. Item 113: Scour Critical Bridges

During a bridge inspection under the NBIS, the condition of the bridge waterway opening, substructure, channel protection, and scour countermeasures should be evaluated, along with the condition of the stream.

Item 61 describes the physical conditions associated with the flow of water through the bridge "such as stream stability" and the condition of the channel. Specifically, factors that would indicate potentially serious problems include: stream bed aggradation, degradation, or lateral movement which may have changed the channel to now threaten the bridge and/or approach roadway. While the scour critical bridge coding item (Item 113) does not address stream stability issues directly, by inference stream stability problems that might lead to serious scour problems during some future flood event should be considered. The topics addressed in the following chapters have a direct bearing on the scour status of a bridge, both at present and in the future. For more detailed guidance on inspecting bridges for scour, reference to HEC-18 "Evaluating Scour at Bridges" (Chapter 10) is suggested (FHWA 2012b). Also see USDOT 2001 and 2003.

### 1.4.2 USACE Nationwide Permit Issues

The United States Army Corps of Engineers (USACE), Regulatory Program regulates the discharge of fill placed within waters of the United States, through Section 404 of the Federal Clean Water Act. There are two main types of permits the Regulatory program authorizes: a Standard Permit and a General Permit. A Standard Permit is a permit which authorizes impacts which have more than minimal impact to the waters of the United States. These permits are typically for the larger impact projects and require a more thorough review by the Regulatory Program. A General Permit is one which authorizes minimal adverse impacts to waters of the United States. There are two types of General Permits: Regional General Permits and Nationwide Permits. Regional General Permits are issued for projects which are similar in nature and are typically issued for a specific geographic region by a local USACE District. Nationwide permits are permits which authorize only minimal adverse impacts. The USACE Headquarters issues the Nationwide Permits, typically, every 4 years. Currently there are 50 Nationwide Permits, which authorize activities from utility line installation to minor fill discharge. The three Nationwide Permits which are often utilized by State DOT's for scour or stream instability repair are Nationwide Permits 3 (Maintenance of Currently Serviceable Structures), 13 (Bank Stabilization), and 14 (Linear Transportation Projects).

A stream stability or scour protection project that involves work within a water of the United States must be coordinated with the local USACE Regulatory office to determine if a permit is required. Some activities for stream stability or scour protection projects may be exempt, but many require a permit. The permitting process can take a few months to a year, so it is imperative that coordination take place early in the design process.

### 1.5 FACTORS THAT AFFECT STREAM STABILITY

Factors that affect stream stability and, potentially, bridge stability at highway stream crossings can be classified as geomorphic factors and hydraulic factors. Rapid and unexpected changes can occur in streams in response to human activities in the watershed and/or natural disturbances of the fluvial system, making it important to anticipate changes in channel geomorphology, location and behavior. Geomorphic characteristics of particular interest to the highway engineer are the alignment, geometry, and form of the stream channel. The behavior of a stream at a highway crossing depends not only on the apparent stability of the stream at the bridge, but also on the behavior of the stream system of which it is a part. Upstream and downstream changes may affect future stability at the site. Natural disturbances such as floods, drought, earthquakes, landslides, forest fires, etc., may result in large changes in sediment load in a stream and major changes in the stream channel.

These changes can be reflected in aggradation, degradation, or lateral migration of the stream channel.

Geomorphic factors that can influence stream stability include stream size, flow habit (i.e., ephemeral or perennial) and the characteristics of channel boundaries. The bed material of a stream can be a cohesive material, sand, gravel, cobbles, boulders, or bedrock. Bank material is also composed of these materials and may be dissimilar from the bed material. The stability and rate of change in a stream are dependent on material in the bed and banks. Other natural factors such as the stream's relationship to its valley, floodplain and planform characteristics, and features such as natural levees, incision, and riparian vegetation are important indicators of stream stability (or instability).

Human-induced changes in the drainage basin and the stream channel, such as alteration of vegetative cover and changes in pervious (or impervious) area can alter the hydrology of a stream, sediment yield, and channel geometry. Channelization, stream channel straightening, streamside levees and dikes, bridges and culverts, reservoirs, gravel mining, and changes in land use can have major effects on streamflow, sediment transport, and channel geometry and location. Geomorphic factors are discussed in detail in Chapter 2.

Hydraulic factors which affect stream channel and bridge stability are numerous and include bed forms and their effects on sediment transport, resistance to flow, flow velocities and flow depths. They also include the magnitude and frequency of floods; characteristics of floods, (i.e., duration, time to peak, and time of recession); flow classification (e.g., unsteady, nonuniform, turbulent, supercritical or subcritical); ice and other floating debris in the flow; and flow constrictions. Other factors are bridge length, location, orientation, span lengths, pier location and design; superstructure elevation and design; the location and design of countermeasures; and the effects of natural and human-induced changes which affect the hydrology and hydraulic flow conditions of the stream. In the bridge reach, bridge design and orientation can induce contraction scour and local scour at piers and abutments. Hydraulic factors are discussed in Chapter 3.

## 1.6 IDENTIFICATION AND ANALYSIS OF STREAM STABILITY PROBLEMS

Identification of the geomorphic factors that can affect channel stability in the bridge reach provides a useful first step in detecting existing or potential channel instability and scour problems at highway bridges. Consideration of fundamental geomorphic principles can lead to a qualitative prediction, in terms of trends, of the most likely direction of channel response to natural and human-induced change in the watershed and river system. However, more general methods of river classification can also provide insight on potential instability problems common to a given stream type.

A necessary first step in any channel classification or stability analysis is a field site visit. Geomorphologists have developed stream reconnaissance guidelines and specific techniques, including geomorphic assessment checklists, which can be useful to the highway engineer during a site visit. In addition, a rapid assessment methodology using both geomorphic and hydraulic factors could help identify the most likely sources of stability problems in a stream reach. Guidance for reconnaissance, classification, and rapid assessment techniques, as well as qualitative techniques for evaluating channel response, are presented in Chapter 5. The potential for debris accumulation which could have a significant impact on both stream stability and scour at a bridge is also addressed in Chapter 5.

In analyzing stream stability problems, it may be necessary to go beyond a qualitative analysis of trends, particularly if remedial action or countermeasures are required. Quantitative geomorphic and engineering techniques are available for analyzing bank stability and lateral and vertical channel stability. In addition, quantitative techniques may be necessary when channel stability or river restoration design are components of a highway project. Chapter 6 provides an introduction to quantitative geomorphic and engineering techniques useful for stream stability analyses. Many of these techniques can be applied to the restoration and rehabilitation of environmentally degraded stream channels. Chapter 9 provides an introduction to currently available channel restoration guidelines.

Alluvial channels are formed by sediment that has been transported and deposited by flowing water and can be transported by the channel in the future. Channel adjustments include aggradation, degradation, width adjustment, and lateral shifting. Sediment transport involves complex processes that interact to produce the existing channel form and future channel adjustments. The amount of material transported or deposited in a stream under a given set of conditions is the result of the interaction of variables related to both the source and caliber of the sediment and the hydraulic capacity of the stream to transport sediment. Sediment transport and channel stability depend not only on the specific physical processes, but also on the history of natural and human-induced factors in the watershed. Fundamental sediment transport concepts are introduced in Chapter 7.

Many of the characteristics of sand-bed channels are found in gravel-bed channels, but gravel-bed channels also have characteristics not common to sand-bed channels. Gravel-bed channels consist of gravel-sized materials, can have planforms ranging from straight to braided and, where they take the form of a meandering channel, they often have a riffle-pool profile. In general, the processes of bank erosion, lateral migration, and channel instability in gravel-bed channels are similar to those in sand-bed channels. However, flow hydraulics, flow resistance, and sediment transport in gravel-bed channels are distinctly different than those in sand-bed channels. These differences are highlighted in Chapter 8.

## 1.7 ANALYSIS METHODOLOGY

The evaluation and design of a highway stream crossing or encroachment should begin with a qualitative assessment of stream stability. This involves application of geomorphic concepts to identify potential problems and alternative solutions. This analysis should be followed with quantitative analyses using basic hydrologic, hydraulic and sediment transport engineering concepts. Such analyses could include evaluation of flood history, channel hydraulic conditions (up to and including, for example, water surface profile analysis) and basic sediment transport analyses such as evaluation of watershed sediment yield, incipient motion analysis and scour calculations. This analysis can be considered adequate for many locations if the problems are resolved and the relationships among different factors affecting stability are adequately explained. If not, a more complex quantitative analysis based on detailed mathematical modeling and/or physical hydraulic models should be considered. A step-wise methodology for analyzing stream stability problems is presented in Chapter 4.

In general, the solution procedure for analyzing stream stability could involve the following three levels of analysis:

Level 1: Application of Simple Geomorphic Concepts and other Qualitative Analyses
Level 2: Application of Basic Hydrologic, Hydraulic and Sediment Transport Engineering Concepts
Level 3: Application of Mathematical or Physical Modeling Studies

## 1.8 MANUAL ORGANIZATION

This manual is organized to:

- Familiarize the user with the important geomorphic factors which are indicators of and contributors to potential and existing stream and bridge stability problems (Chapter 2)
- Provide a summary of hydraulic factors that can affect stream stability (Chapter 3)
- Provide a procedure for the analysis of potential and existing stability problems (Chapter 4)
- Provide guidance for stream channel reconnaissance, classification, and rapid assessment techniques, and an introduction to qualitative methods for evaluating channel response (Chapter 5)
- Introduce quantitative geomorphic and engineering techniques to assess river channel stability (Chapter 6)
- Provide an overview of sediment transport concepts important to stream stability analyses (Chapter 7)
- Introduce channel stability concepts for gravel-bed rivers (Chapter 8)
- Provide an introduction to channel restoration concepts (Chapter 9)
- Provide selected references (Chapter 10)

## 1.9 DUAL SYSTEM OF UNITS

This edition of HEC-20 uses dual units (English and SI metric). The "English" system of units as used throughout this manual refers to U.S. Customary units. **In Appendix A, the metric (SI) unit of measurement is explained. The conversion factors, physical properties of water in the SI and English systems of units, sediment particle size grade scale, and some common equivalent hydraulic units are also given.** This edition uses for the unit of length the foot (ft) or meter (m); of mass the slug or kilogram (kg); of weight/force the pound (lb) or newton (N); of pressure the lb/ft$^2$ or Pascal (Pa, N/m$^2$); and of temperature degrees Fahrenheit (EF) or Centigrade (EC). The unit of time is the same in English as in SI system (seconds, s). Sediment particle size is given in millimeters (mm), but in calculations the decimal equivalent of millimeters in meters is used (1 mm = 0.001 m) or for the English system feet (ft). The value of some hydraulic engineering terms used in the text in English units and their equivalent SI units are given in Table 1.1.

| Table 1.1. Commonly Used Engineering Terms in English and SI Units. | | |
|---|---|---|
| Term | English Units | SI Units |
| Length | 3.28 ft | 1 m |
| Volume | 35.31 ft$^3$ | 1 m$^3$ |
| Discharge | 35.31 ft$^3$/s | 1 m$^3$/s |
| Acceleration of Gravity | 32.2 ft/s$^2$ | 9.81 m/s$^2$ |
| Unit Weight of Water | 62.4 lb/ft$^3$ | 9800 N/m$^3$ |
| Density of Water | 1.94 slugs/ft$^3$ | 1000 kg/m$^3$ |
| Density of Quartz | 5.14 slugs/ft$^3$ | 2647 kg/m$^3$ |
| Specific Gravity of Quartz | 2.65 | 2.65 |
| Specific Gravity of Water | 1 | 1 |
| Temperature | °F | °C = 5/9 (°F - 32) |

# CHAPTER 2

# GEOMORPHIC FACTORS AND PRINCIPLES

## 2.1 INTRODUCTION

Most streams that highways cross or encroach upon are alluvial; that is, the streams are formed in materials that have been and can be transported by the stream. In alluvial stream systems, it is the rule rather than the exception that banks will erode; sediments will be deposited; and floodplains, islands and side channels will undergo modification with time. Alluvial channels continually change position and shape as a consequence of hydraulic forces exerted on the bed and banks. These changes may be gradual or rapid and may be the result of natural causes or human activities.

Some streams are not alluvial. The bed and bank material is very coarse, and except at extreme flood events, does not erode. These streams are classified as sediment supply deficient, i.e., the transport capacity of the streamflow is greater than the availability of bed material for transport. The bed and bank material of these streams may consist of cobbles, boulders or even bedrock. In general these streams are stable, but should be carefully analyzed for stability at large flows.

A study of the plan and profile of a stream is very useful in understanding stream morphology. Plan view appearances of streams are varied and result from many interacting variables. Small changes in a variable can change the plan view and profile of a stream, adversely affecting a highway crossing or encroachment. This is particularly true for alluvial streams. Conversely, a highway crossing or encroachment can inadvertently change a variable, adversely affecting the stream.

This chapter presents an overview of general landform and channel evolutionary processes to illustrate the dynamics of alluvial channel systems. A checklist of geomorphic properties of interest to the highway engineer is presented as a framework for identifying and understanding river channel dynamics. Finally, factors affecting bed elevation changes and the sediment continuity principle provide an introduction to alluvial channel response to natural and human-induced change.

## 2.2 LANDFORM EVOLUTION

Earth scientists (geomorphologists) have historically concerned themselves with documenting and explaining the changing morphology of the landscape through time. For example, Figure 2.1 illustrates the changing character of a landscape during a million years of geologic time. Initially, this type of evolution of landforms would appear to be of no interest to the highway or bridge engineer, but it serves as an alert that change can be expected at the scale of individual landforms (hillslopes, channels), and the change can be sufficiently rapid to cause problems.

In the extreme case of incised channels (gullies, arroyos) rapid incision is followed by channel adjustment (deepening, widening) to a new condition of relative stability as erosion decreases, sediment storage increases and a floodplain develops (Figure 2.2). Simon (1989) obtained data on the sediment loads transported through incised channels in Tennessee (Figure 2.3A). The stages of channel evolution shown in Figure 2.2 are reflected in the changing sediment loads of Figure 2.3A. Note that there is an apparent increase of sediment load at stage E (Figure 2.3A) as some stored sediment is remobilized.

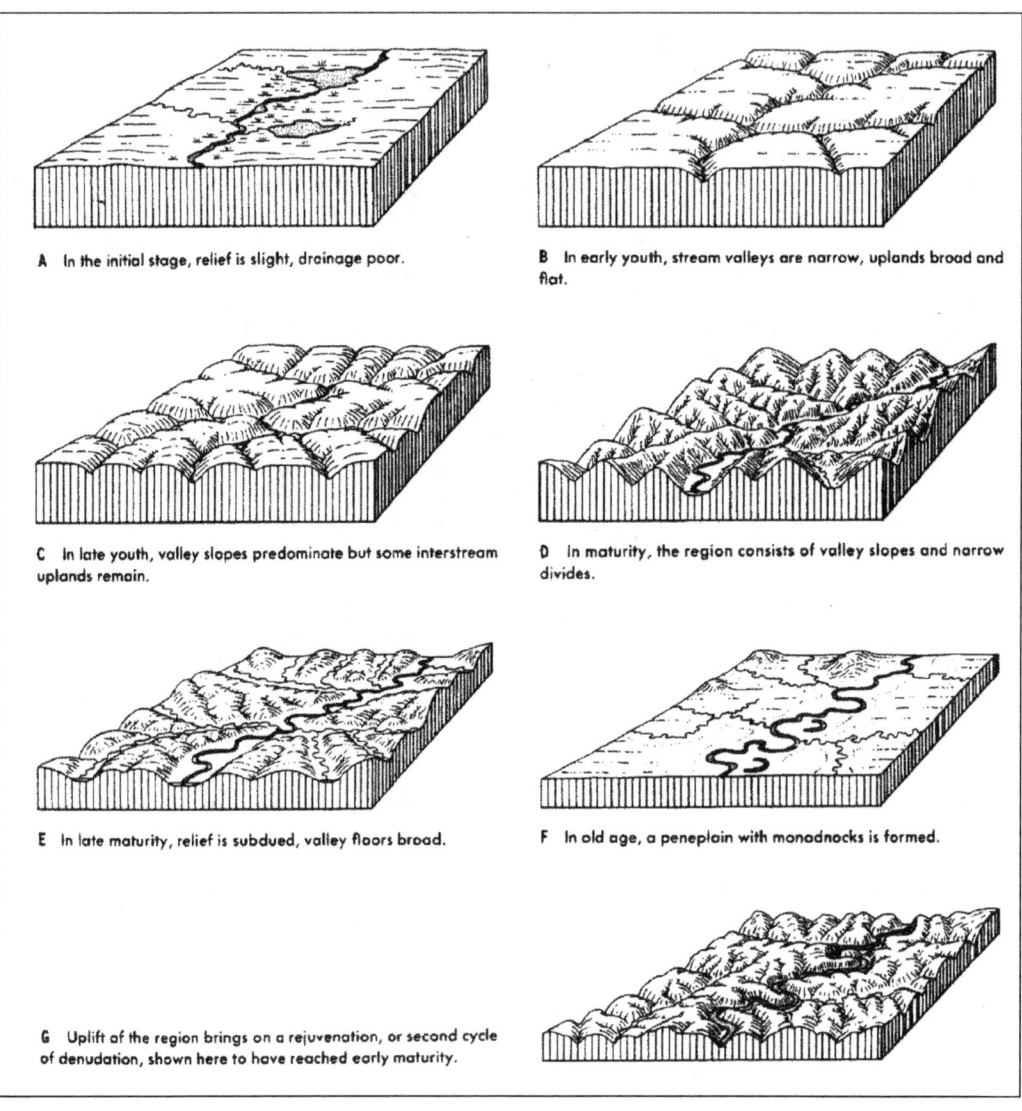

Figure 2.1. The cycle of erosion, proposed by W.M. Davis, drawn by E. Raisz (Strahler 1965).

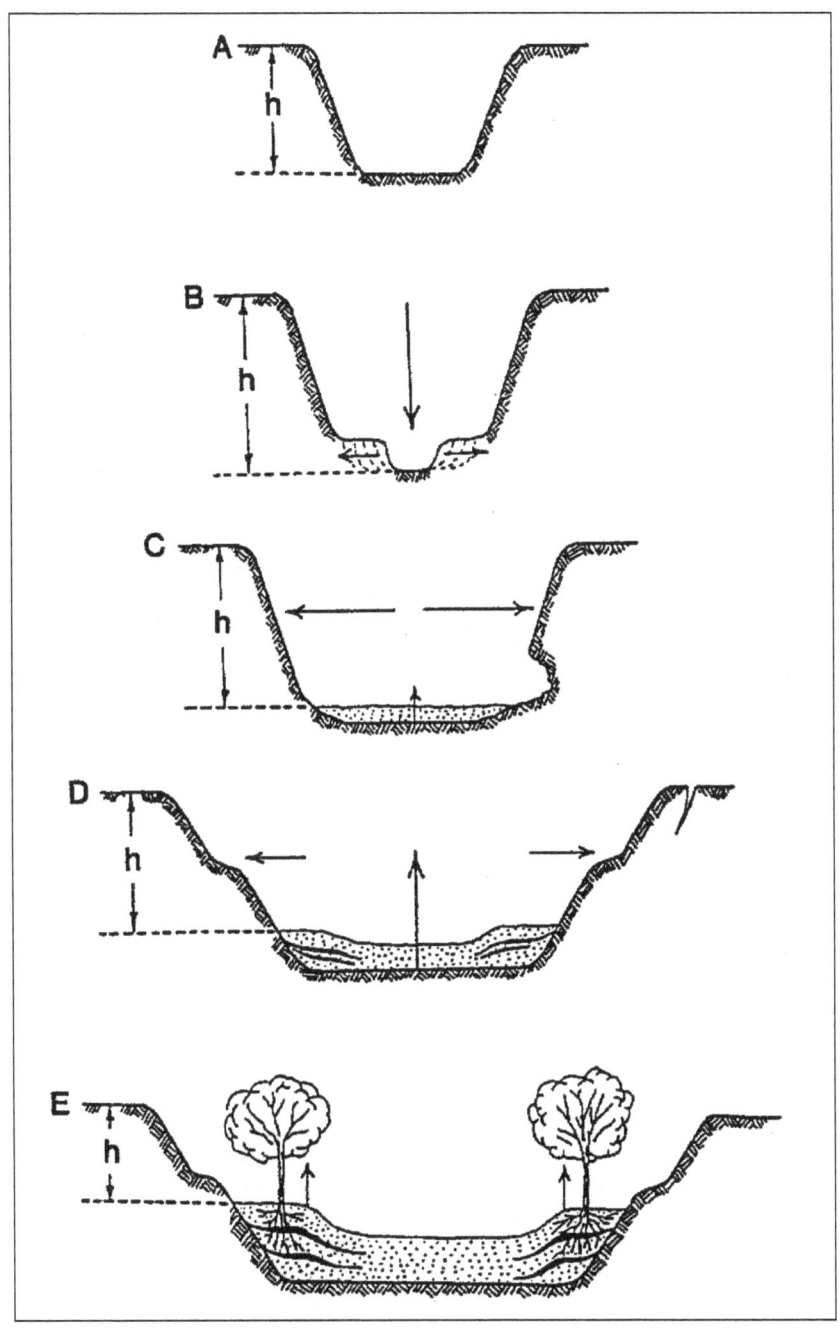

Figure 2.2. Evolution of incised channel from initial incision (A, B) and widening (C, D) to aggradation (D, E) and eventual relative stability; h is bank height (Schumm et al. 1984).

Field investigations in the upper Colorado River basin have also revealed that the large arroyos formed by incision of valley-floor alluvium in the latter part of the nineteenth century are at present storing sediment in newly developed floodplains (Gellis et al. 1991). Daily sediment-load data were collected starting in 1930. In 1963, upstream dams trapped much of sediment and the post-1963 record was not used. These incised channels are also behaving as illustrated in Figure 2.2. At the later stages of adjustment, they are eroding less sediment and storing larger amounts of sediment. As a result, sediment loads at the Grand Canyon gaging station have decreased, during the period of record, prior to closure of Glen Canyon Dam and other upstream dams in 1963 (Figure 2.3B). In addition, sediment deposition in Lake Powell between 1963 and 1986 is only 43 percent of that estimated prior to dam construction (USBR 1988), which indicates that the channel adjustment process is occurring throughout the upper Colorado River basin, in a manner similar to that in the incised channels of Tennessee (Figure 2.3A).

Because of climatic differences, the evolutionary changes involved in the complex response of Figure 2.2 require about 100 years in the southwest but only about 40 years in the southeast. Additional discussion of complex response of fluvial systems is provided in Section 4.4.

As the cross section of an incised channel (Figure 2.2) changes through time, the pattern can also evolve from straight to sinuous (Figure 2.4). In fact, a river that straightens naturally by meander cutoffs will also evolve to restore the meandering pattern (Figure 2.4, 2.5A). The downstream shift of meanders (Figure 2.5A) and the cutoff and regrowth of meanders (Figure 2.5A, B) are all part of the natural evolution of channel patterns through time.

Although the landscape as a whole may appear unchanging except over vast periods of time (Figure 2.1), components of the landscape can evolve or adjust to human activities (Figures 2.2, 2.3) and hydrologic variations (Figures 2.3, 2.4) during relative short periods of time and can pose serious problems for the highway engineer.

## 2.3 GEOMORPHIC FACTORS AFFECTING STREAM STABILITY

### 2.3.1 Overview

Figure 2.6 introduces a set of geomorphic factors that can affect stream stability. Each of the geomorphic properties listed in the left column of Figure 2.6 could be used as the basis of a valid stream characterization at a bridge site. The approach presented here is based on stream properties observed on aerial photographs and in the field. Its major purpose is to facilitate the assessment of streams for engineering purposes, particularly regarding lateral stability of a stream. Common stream types are described and their engineering significance discussed. Data and observations are derived from a study of case histories of 224 bridge sites in the United States and Canada (FHWA 1978a and b).

This section is organized according to Figure 2.6. No particular significance is assigned to the order of the figure, and association of characteristics should not be inferred with descriptions above or below in the figure. Chapter 5 contains an introduction to more general stream channel classification systems.

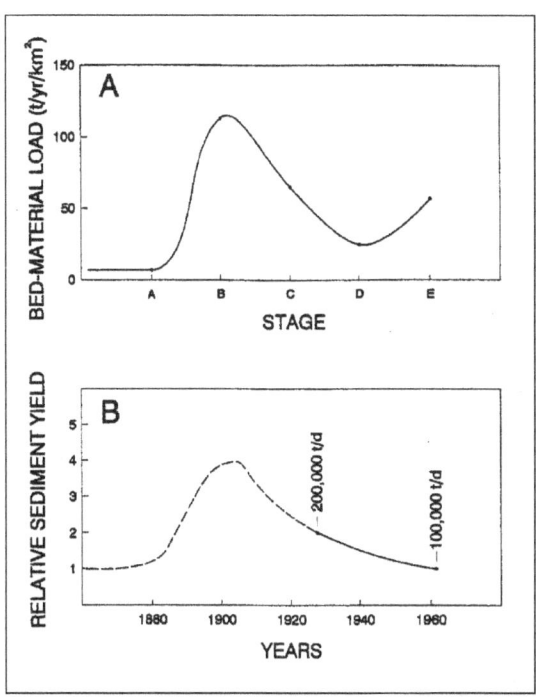

Figure 2.3. Sediment loads following channel incision: (A) Bed-material load transported by incised Tennessee streams for each stage of incised-channel evolution (Figure 2.2), (B) Hypothetical (dashed line) and measured (solid line) sediment volumes transported through Grand Canyon (Simon 1989).

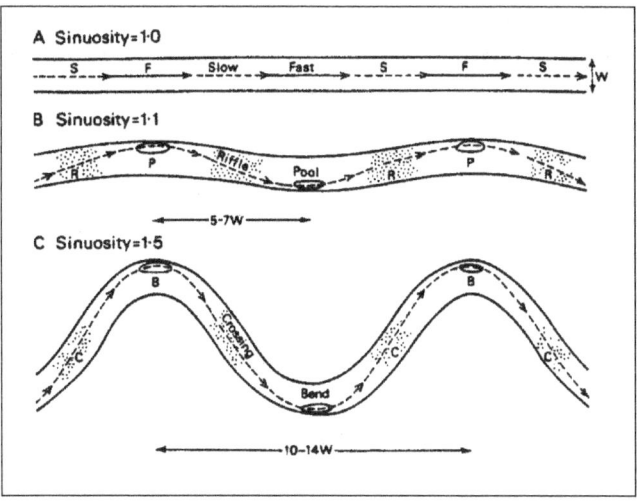

Figure 2.4. Three possible stages in the development of a meandering reach: (A) Reaches of faster and slower eddy flow at bankfull discharge, (B) Development of pools and riffles with spacing of 5-7 channel widths, (C) Development of meanders with a wavelength of 10-14 channel widths (Richards 1982).

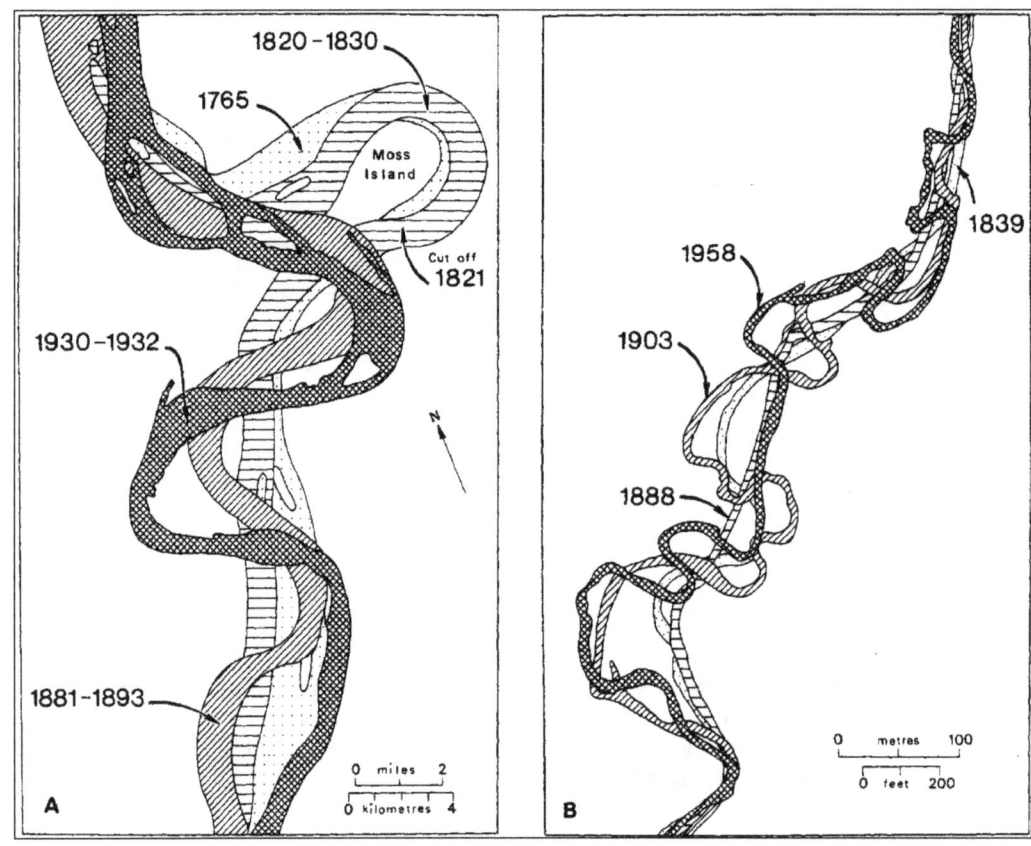

Figure 2.5. Surveys showing changes of course for two meandering rivers: (A) Mississippi River in northern Tennessee during the period 1765-1932 (Strahler 1965), (B) River Sid in east Devon during the period 1839-1958 (Hooke 1977).

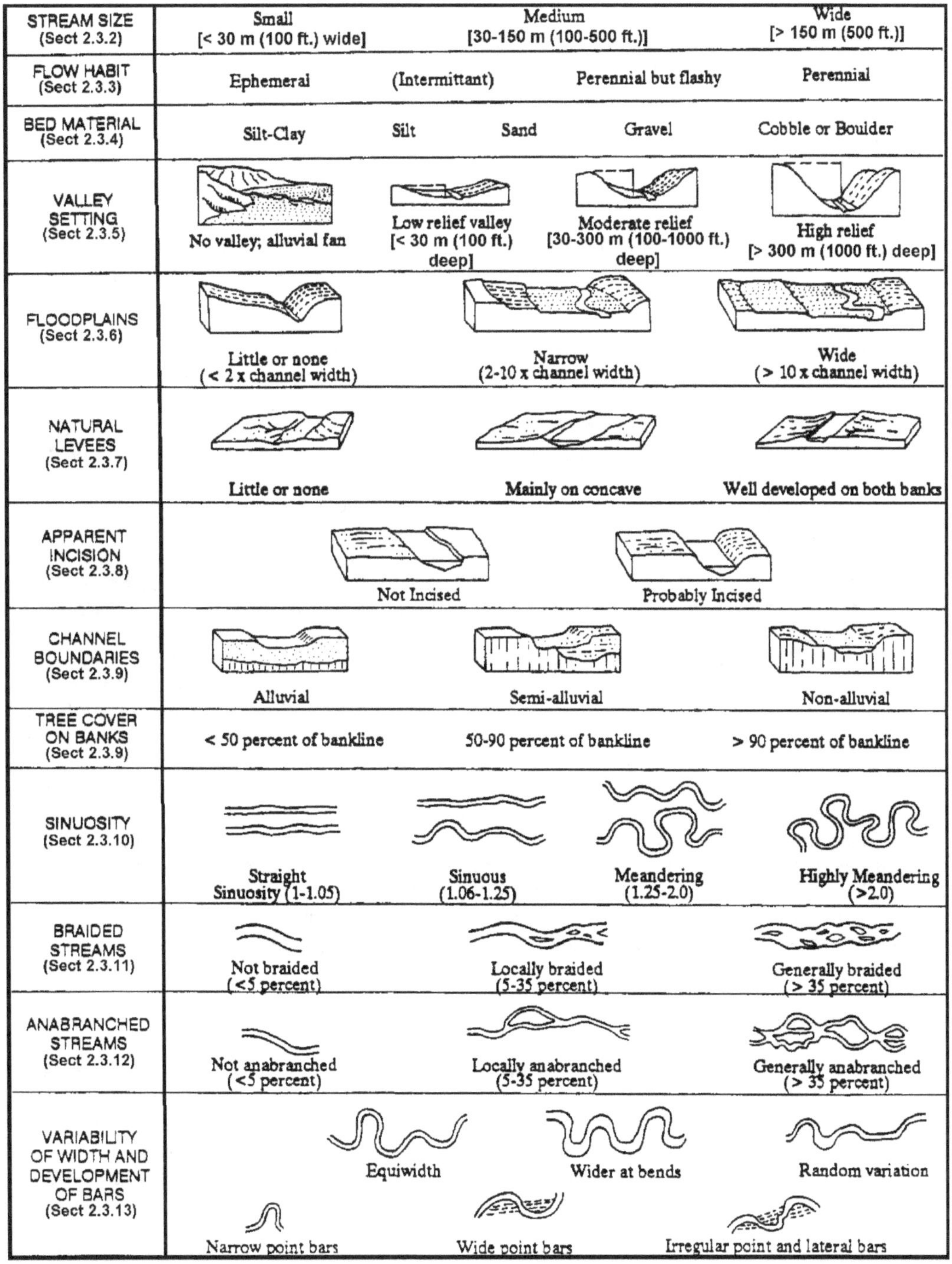

| STREAM SIZE (Sect 2.3.2) | Small [< 30 m (100 ft.) wide] | Medium [30-150 m (100-500 ft.)] | | Wide [> 150 m (500 ft.)] |
|---|---|---|---|---|
| FLOW HABIT (Sect 2.3.3) | Ephemeral | (Intermittant) | Perennial but flashy | Perennial |
| BED MATERIAL (Sect 2.3.4) | Silt-Clay | Silt | Sand | Gravel | Cobble or Boulder |
| VALLEY SETTING (Sect 2.3.5) | No valley; alluvial fan | Low relief valley [< 30 m (100 ft.) deep] | Moderate relief [30-300 m (100-1000 ft.) deep] | High relief [> 300 m (1000 ft.) deep] |
| FLOODPLAINS (Sect 2.3.6) | Little or none (< 2 x channel width) | Narrow (2-10 x channel width) | | Wide (> 10 x channel width) |
| NATURAL LEVEES (Sect 2.3.7) | Little or none | Mainly on concave | | Well developed on both banks |
| APPARENT INCISION (Sect 2.3.8) | | Not Incised | Probably Incised | |
| CHANNEL BOUNDARIES (Sect 2.3.9) | Alluvial | Semi-alluvial | | Non-alluvial |
| TREE COVER ON BANKS (Sect 2.3.9) | < 50 percent of bankline | 50-90 percent of bankline | | > 90 percent of bankline |
| SINUOSITY (Sect 2.3.10) | Straight Sinuosity (1-1.05) | Sinuous (1.06-1.25) | Meandering (1.25-2.0) | Highly Meandering (>2.0) |
| BRAIDED STREAMS (Sect 2.3.11) | Not braided (<5 percent) | Locally braided (5-35 percent) | | Generally braided (> 35 percent) |
| ANABRANCHED STREAMS (Sect 2.3.12) | Not anabranched (<5 percent) | Locally anabranched (5-35 percent) | | Generally anabranched (> 35 percent) |
| VARIABILITY OF WIDTH AND DEVELOPMENT OF BARS (Sect 2.3.13) | Equiwidth; Narrow point bars | Wider at bends; Wide point bars | | Random variation; Irregular point and lateral bars |

Figure 2.6. Geomorphic factors that affect stream stability (adapted from FHWA 1978a).

## 2.3.2 Stream Size

Stream depth tends to increase with size, and potential for scour increases with depth. Thus, potential depth of scour increases with increasing stream size.

The potential for lateral erosion also increases with stream size. This fact may be less fully appreciated than the increased potential for deep scour. Brice et al. cite as examples the lower Mississippi River, with a width of about 5,000 ft (1500 m), which may shift laterally 100 ft (30 m) or more in a single major flood; the Sacramento River, where the width is about 1,000 ft (300 m), is unlikely to shift more than 25 ft (8 m) in a single flood; and streams whose width is about 100 ft (30 m) are unlikely to shift more than 10 ft (3 m) in a single flood (FHWA 1978a). Except for the fact that the potential for lateral migration increases with stream size, no generalization is currently possible regarding migration rates.

The size of a stream can be indicated by discharge, drainage area, or some measure of channel dimensions, such as width or cross-sectional area. No single measure of size is satisfactory because of the diversity of stream types. For purposes of stream classification (Figure 2.6), bank-to-bank channel width is chosen as the most generally useful measure of size, and streams are arbitrarily divided into three size categories on the basis of width. The width of the stream does not include the width of the floodplain, but floodplain width is an important factor in bridge design if significant overbank flow occurs.

Bank-to-bank width is sometimes difficult to define for purposes of measurement when one of the banks is indefinite. This is particularly true at bends, where the outside bank is likely to be vertical and sharply defined but the inside bank slopes gradually up to floodplain level. The position of the line of permanent vegetation on the inside bank is the best available indicator of the bankline, and it tends to be rather sharply defined along many rivers in humid regions. The width of a stream is measured along a perpendicular line drawn between its opposing banks, which are defined either by their form or as the riverward edge of a line of permanent vegetation. For sinuous or meandering streams, width is measured at straight reaches or at the inflections between bends, where it tends to be most consistent. For multiple channel streams, width is the sum of the widths of individual, unvegetated channels.

The National Mapping Division of the U.S. Geological Survey (USGS) uses, insofar as possible, the so-called "normal" stage or the stage prevailing during the greater part of the year for representing streams on topographic maps. They find that the "normal" stage for a perennial river usually corresponds to the water level filling the channel to the line of permanent vegetation along its banks. Normal stage is also adopted here to define channel width.

## 2.3.3 Flow Habit

The flow habit of a stream may be ephemeral, perennial but flashy, or perennial. An ephemeral stream flows briefly in direct response to precipitation, and as used here, includes intermittent streams. A perennial stream flows all or most of the year, and a perennial but flashy stream responds to precipitation by rapid changes in stage and discharge. Perennial streams may be relatively stable or unstable, depending on other factors such as channel boundaries and bed material.

In arid regions, ephemeral streams may be relatively large and unstable. They may pose problems in determining the stage-discharge relationship and in estimating the depth of scour. A thalweg that shifts with stage and channel degradation by headcutting may also cause problems. In humid regions, ephemeral streams are likely to be small and pose few problems of instability.

## 2.3.4 Bed Material

Streams are classified, according to the dominant size of the sediment on their beds, as silt-clay bed, sand bed, gravel bed, and cobble or boulder bed. Accurate determination of the particle size distribution of bed material requires careful sampling and analysis, particularly for coarse bed material, but for most of the bed material designations, rough approximations can be derived from visual observation.

The greatest depths of scour are usually found on streams having sand or sand-silt beds. The general conclusion is that scour problems are as common on streams having coarse bed material as on streams having fine bed material. However, very deep scour is more probable in fine bed material (FHWA 1978a). In general, sand-bed alluvial streams are less stable than streams with coarse or cohesive bed and bank material (see Section 2.5).

## 2.3.5 Valley Setting

Valley relief is used as a means of indicating whether the surrounding terrain is generally flat, hilly, or mountainous. For a particular site, relief is measured (usually on a topographic map) from the valley bottom to the top of the highest adjacent divide. Relief greater than 1,000 ft (300 m) is regarded as mountainous, and relief in the range of 100 to 1,000 ft (30 to 300 m) as hilly. Streams in mountainous regions are likely to have steep slopes, coarse bed materials, narrow floodplains and be nonalluvial, i.e., supply-limited sediment transport rates. In many regions, channel slope increases as the steepness of valley side slopes increases. Brice et al. reported no specific hydraulic problems at bridges at 23 study sites in mountainous terrain, at which all have beds of gravel or cobble-boulder (FHWA 1978b). Streams in regions of lower relief are usually alluvial and exhibit more problems because of lateral erosion in the channels.

Streams on alluvial fans or on piedmont slopes in arid regions pose special problems. A piedmont slope is a broad slope along a mountain front, and streams issuing from the mountain front may have shifting courses and poorly defined channels, as on an alluvial fan. Alluvial fans are among the few naturally occurring cases of aggradation problems at transverse highway crossing. They occur wherever there is a change from a steep to a flat gradient. As the bed material and water reaches the flatter section of the stream, the coarser bed materials are deposited because of the sudden reduction in both slope and velocity. Consequently, a cone or fan builds out as the material is dropped with the steep side of the fan facing the floodplain. Although typically viewed as a depositional zone, alluvial fans are also characterized by unstable channel geometries and rapid lateral movement. Deposition tends to be episodic, being interrupted by periods of fan trenching and sediment reworking.

The occurrence of deposition versus fan trenching on an alluvial fan surface are important factors in the assessment of stream stability at bridge crossings (Figure 2.7). On an untrenched fan, the sediment depositional zone will be nearer the mountain front, possibly creating more channel instability on the upper fan surface than on the lower fan surface. In contrast, a fan that is trenched will promote sediment movement across the fan and move the depositional zone closer to the toe of the fan, suggesting that the upper fan surface will be more stable than the lower fan surface. However, the general instability of fan channels and their tendency for rapid changes during large floods, and the possible channel avulsion created by deposition near the fan head suggest that any location of an alluvial fan surface is, or could easily become, an area where channel instability is a serious concern to bridge safety (Schumm and Lagasse 1998, FHWA 2001).

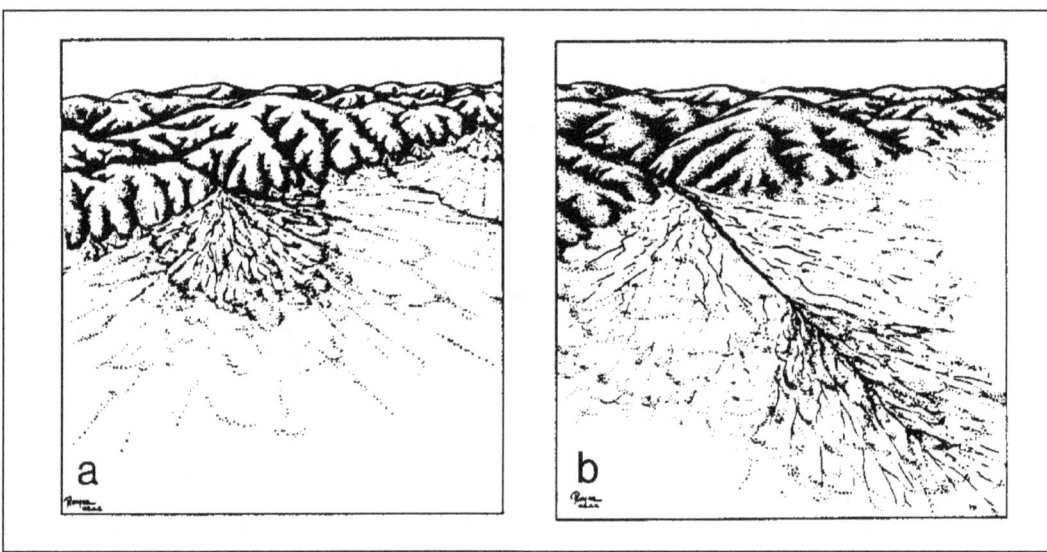

Figure 2.7. Diverse morphology of alluvial fans: (a) area of deposition at fan head, (b) fan-head trench with deposition at fan toe (after Bull 1968).

There is considerable similarity between deltas and alluvial fans. Both result from reductions in slope and velocity, have steep slopes at their outer edges and tend to reduce upstream slopes. Deposits very similar to a delta develop where a steep tributary enters a larger stream. The steep channel tends to drop part of its sediment load in the main channel building out into the main stream. In some instances, drastic changes can occur in the main stream channel as a result of deposition from the tributary stream. Channels on both alluvial fans and deltaic deposits commonly change through avulsion, a sudden change in channel course that occurs when a stream breaks through its banks.

### 2.3.6 Floodplains

Floodplains are described as the nearly flat alluvial lowlands bordering a stream that are subject to inundation by floods. Many geomorphologists prefer to define a floodplain as the surface presently under construction by the stream, which is flooded with a frequency of about 1.5 years (excluding incised channels, see Section 2.3.8). According to this definition, surfaces flooded less frequently are terraces, abandoned floodplains, or flood-prone areas. However, flood-prone areas are considered herein as part of the floodplain. Vegetative cover, land use, and flow depth on the floodplain are also significant factors in stream channel stability. In Figure 2.6, floodplains are categorized according to floodplain width relative to channel width.

Over time, the highlands of an area are worn down, streams erode their banks, and the material that is eroded is utilized farther downstream to build banks and bars. Streams move laterally, pushing the highlands back. Low, flat valley land and floodplains are formed. As streams transport sediment to areas of flatter slopes and, in particular, to bodies of water where the velocity and turbulence are too small to sustain transport of the material, the material is deposited forming deltas. As deltas build outward, the upstream portion of the channel is elevated through deposition and becomes part of the floodplain. Also, the stream channel is lengthened and the slope is further reduced. The upstream streambed is filled in and average flood elevations are increased. As the stream works across the stream valley, deposition causes the total floodplain to raise in elevation. Hence, even old streams are far from static. Old rivers meander, and they are affected by changes in sea level, influenced by movements of the earth's crust, changed by delta formations or glaciation, and subject to modifications due to climatological changes and as a consequence of man's development.

### 2.3.7 Natural Levees

Natural levees form during floods as the stream stage exceeds bankfull conditions. Sediment is then deposited on the floodplain due to the reduced velocity and transporting capacity of the flood in these overbank areas. The natural levees formed near the stream are rather steep because coarse material drops out quickly as the overbank velocity is smaller than the stream velocity. Farther from the stream, the gradients are flatter and finer materials drop out. Swamp areas are found beyond the levees.

Classification based on natural levees is illustrated in Figure 2.6. Streams with well-developed natural levees tend to be of constant width and have low rates of lateral migration. Well-developed levees usually occur along the lower courses of streams or where the floodplain is submerged for several weeks or months a year. If the levee is breached, the stream course may change through the breach. Areas between natural levees and the valley sides may drain, but slowly. Streams tributary to streams with well-developed natural levees may flow approximately parallel with the larger stream for long distances before entering the larger stream.

### 2.3.8 Apparent Incision

The apparent incision of a stream channel is judged from the height of its banks at normal stage relative to its width. For a stream whose width is about 100 ft (30 m), bank heights in the range of 6 to 10 ft (1.8 to 3.0 m) are about average, and higher banks indicate probable incision. For a stream whose width is about 1,000 ft (300 m), bank heights in the range of 10 to 15 ft (3.0 to 5.0 m) are about average, and higher banks indicate probable incision. Incised streams tend to be fixed in position and are not likely to bypass a bridge or to shift in alignment at a bridge. Lateral erosion rates are likely to be slow, except for western arroyos with high, vertical, and clearly unstable banks.

### 2.3.9 Channel Boundaries and Vegetation

Although no precise definitions can be given for alluvial, semi-alluvial, or non-alluvial streams, some distinction with regard to the erosional resistance of the earth material in channel boundaries is needed. In geology, bedrock is distinguished from alluvium and other surficial materials mainly on the basis of age, rather than on resistance to erosion. A compact alluvial clay is likely to be more resistant than a weakly cemented sandstone that is much older. Nevertheless, the term "bedrock" does carry a connotation of greater resistance to erosion, and it is used here in that sense. An alluvial channel is in alluvium, a non-alluvial channel is in bedrock or in very large material (cobbles and boulders) that do not move except at very large flows, and a semi-alluvial channel has both bedrock and alluvium in its boundaries. The bedrock of non-alluvial channels may be wholly or partly covered with sediment at low stages, but is likely to be exposed by scour during floods.

Most highway stream crossings are over alluvial streams which are susceptible to more hydraulic problems than non-alluvial streams. However, the security of a foundation in bedrock depends on the quality of the bedrock (FHWA 2012b) and the care with which foundation is set. Serious problems and failures have developed at bridges with foundations on shale, sandstone, limestone, glacial till, and other erodible rock. The New York State Thruway Schoharie Creek bridge failure is a catastrophic example of such a failure. Bed material at the bridge site was highly cemented glacial till which scoured, undermining spread footings (NTSB 1988).

Changes in channel geometry with time are particularly significant during periods when alluvial channels are subjected to high flows, and few changes occur during relatively dry periods. Erosive forces during high-flow periods may have a capacity as much as 100 times greater than those forces acting during periods of intermediate and low-flow rates. When considering the stability of alluvial streams, in most instances it can be shown that approximately 90 percent of all changes occur during that small percentage of the time when the flow equals or exceeds dominant discharge. A discussion of dominant discharge may be found in Hydraulic Design Series (HDS) No. 6, but the bankfull flow condition is recommended for use where a detailed analysis of dominant discharge is not feasible (FHWA 2001).

The most significant property of materials of which channel boundaries are comprised is particle size. It is the most readily measured property, and, in general, represents a sufficiently complete description of the sediment particle for many practical purposes. Other properties such as shape and fall velocity tend to vary with size in a roughly predictable manner.

In general, sediments have been classified into boulders, cobbles, gravel, sands, silts, and clays on the basis of their nominal or sieve diameters. The size range in each general class is given in Table 2.1. Note that even when the English system of units is used, sand size particles and smaller are typically described in millimeters. Noncohesive material generally consists of silt (0.004 - 0.062 mm), sand (0.062 - 2.0 mm), gravel (2.0 - 64 mm), or cobbles (64 - 250 mm).

The appearance of the stream bank is a good indication of relative stability. A field inspection of a channel will help to identify characteristics which are associated with erosion rates:

- Unstable banks with moderate to high erosion rates usually have slopes which exceed 30 percent, and a cover of woody vegetation is rarely present. At a bend, the point bar opposite an unstable cut bank is likely to be bare at normal stage, but it may be covered with annual vegetation and low woody vegetation, especially willows. Where very rapid erosion is occurring, the bank may have irregular indentations. Fissures, which represent the boundaries of actual or potential slump blocks along the bankline indicate the potential for very rapid bank erosion.

- Unstable banks with slow to moderate erosion rates may be partly reshaped to a stable slope. The degree of instability is difficult to assess, and reliance is placed mainly on vegetation. The reshaping of a bank typically begins with the accumulation of slumped material at the base such that a slope is formed and progresses by smoothing of the slope and the establishment of vegetation.

- Eroding banks are a source of debris when trees fall as they are undermined. Therefore, debris can be a sign of unstable banks and of great concern due to potential blockage of bridge openings.

| Table 2.1. Sediment Grade Scale. ||||||
| Size ||| Approximate Sieve Mesh Openings (per inch) || Class |
| Millimeters | Microns | Inches | Tyler | U.S. Standard | |
|---|---|---|---|---|---|
| 4000-2000 | --- | 180-160 | --- | --- | Very large boulders |
| 2000-1000 | --- | 80-40 | --- | --- | Large boulders |
| 1000-500 | --- | 40-20 | --- | --- | Medium boulders |
| 500-250 | --- | 20-10 | --- | --- | Small boulders |
| 250-130 | --- | 10-5 | --- | --- | Large cobbles |
| 130-64 | --- | 5-2.5 | --- | --- | Small cobbles |
| 64-32 | --- | 2.5-1.3 | --- | --- | Very coarse gravel |
| 32-16 | --- | 1.3-0.6 | --- | --- | Coarse gravel |
| 16-8 | --- | 0.6-0.3 | 2.5 | --- | Medium gravel |
| 8-4 | --- | 0.3-0.16 | 5 | 5 | Fine gravel |
| 4-2 | --- | 0.16-0.08 | 9 | 10 | Very fine gravel |
| 2.00-1.00 | 2000-1000 | --- | 16 | 18 | Very coarse sand |
| 1.00-0.50 | 1000-500 | --- | 32 | 35 | Coarse sand |
| 0.50-0.25 | 500-250 | --- | 60 | 60 | Medium sand |
| 0.25-0.125 | 250-125 | --- | 115 | 120 | Fine sand |
| 0.125-0.062 | 125-62 | --- | 250 | 230 | Very fine sand |
| 0.062-0.031 | 62-31 | --- | --- | --- | Coarse silt |
| 0.031-0.016 | 31-16 | --- | --- | --- | Medium silt |
| 0.016-0.008 | 16-8 | --- | --- | --- | Fine silt |
| 0.008-0.004 | 8-4 | --- | --- | --- | Very fine silt |
| 0.004-0.0020 | 4-2 | --- | --- | --- | Coarse clay |
| 0.0020-0.0010 | 2-1 | --- | --- | --- | Medium clay |
| 0.0010-0.0005 | 1-0.5 | --- | --- | --- | Fine clay |
| 0.0005-0.0002 | 0.5-0.24 | --- | --- | --- | Very fine clay |

- Stable banks with very slow erosion rates tend to be graded to a smooth slope of less than about 30 percent. Mature trees on a graded bank slope are convincing evidence of bank stability. In most regions of the United States, the upper parts of stable banks are vegetated, but the lower part may be bare at normal stage, depending on bank height and flow regime of the stream. Where banks are low, dense vegetation may extend to the water's edge at normal stage. Where banks are high, occasional slumps may occur on even the most stable graded banks. Shallow mountain streams that transport coarse bed sediment tend to have stable banks.

Active bank erosion can be recognized by falling or fallen vegetation along the bankline, cracks along the bank surface, slump blocks, deflected flow patterns adjacent to the bankline, live vegetation in the flow, increased turbidity, fresh vertical faces, newly formed bars immediately downstream of the eroding area, and, in some locations, a deep scour pool adjacent to the toe of the bank. These indications of active bank erosion can be noted in the field and on stereoscopic pairs of aerial photographs. Color infrared photography is particularly useful in detecting most of the indicators listed above, especially differences in turbidity (FHWA 1981). Figure 2.8 illustrates some of the features which indicate that a bankline is actively eroding.

Figure 2.8. Active bank erosion illustrated by vertical cut banks, slump blocks, and falling vegetation.

Bank Materials. Resistance of a stream bank to erosion is closely related to several characteristics of the bank material. Bank material deposited in the stream can be broadly classified as cohesive, noncohesive, and composite. Typical bank failure surfaces of various materials are shown in Figure 2.9 and are described as follows (FHWA 1985):

- Noncohesive bank material tends to be removed grain by grain from the bank. The rate of particle removal and, hence, the rate of bank erosion is affected by factors such as particle size, bank slope, the direction and magnitude of the velocity adjacent to the bank, turbulent velocity fluctuations, the magnitude of and fluctuations in the shear stress exerted on the banks, seepage force, piping, and wave forces. Figure 2.9(a) illustrates failure of banks of noncohesive material from flow slides resulting from a loss of shear strength because of saturation and failure from sloughing resulting from the removal of materials in the lower portion of the bank.

- Cohesive material is more resistant to surface erosion and has low permeability, which reduces the effects of seepage, piping, frost heaving, and subsurface flow on the stability of the banks. However, when undercut and/or saturated, such banks are more likely to fail due to mass wasting processes. Failure mechanisms for cohesive banks are illustrated in Figure 2.9(b).

- Composite or stratified banks consist of layers of materials of various sizes, permeability, and cohesion. The layers of noncohesive material are subject to surface erosion, but may be partly protected by adjacent layers of cohesive material. This type of bank is also vulnerable to erosion and sliding as a consequence of subsurface flows and piping. Typical failure modes are illustrated in Figure 2.9(c).

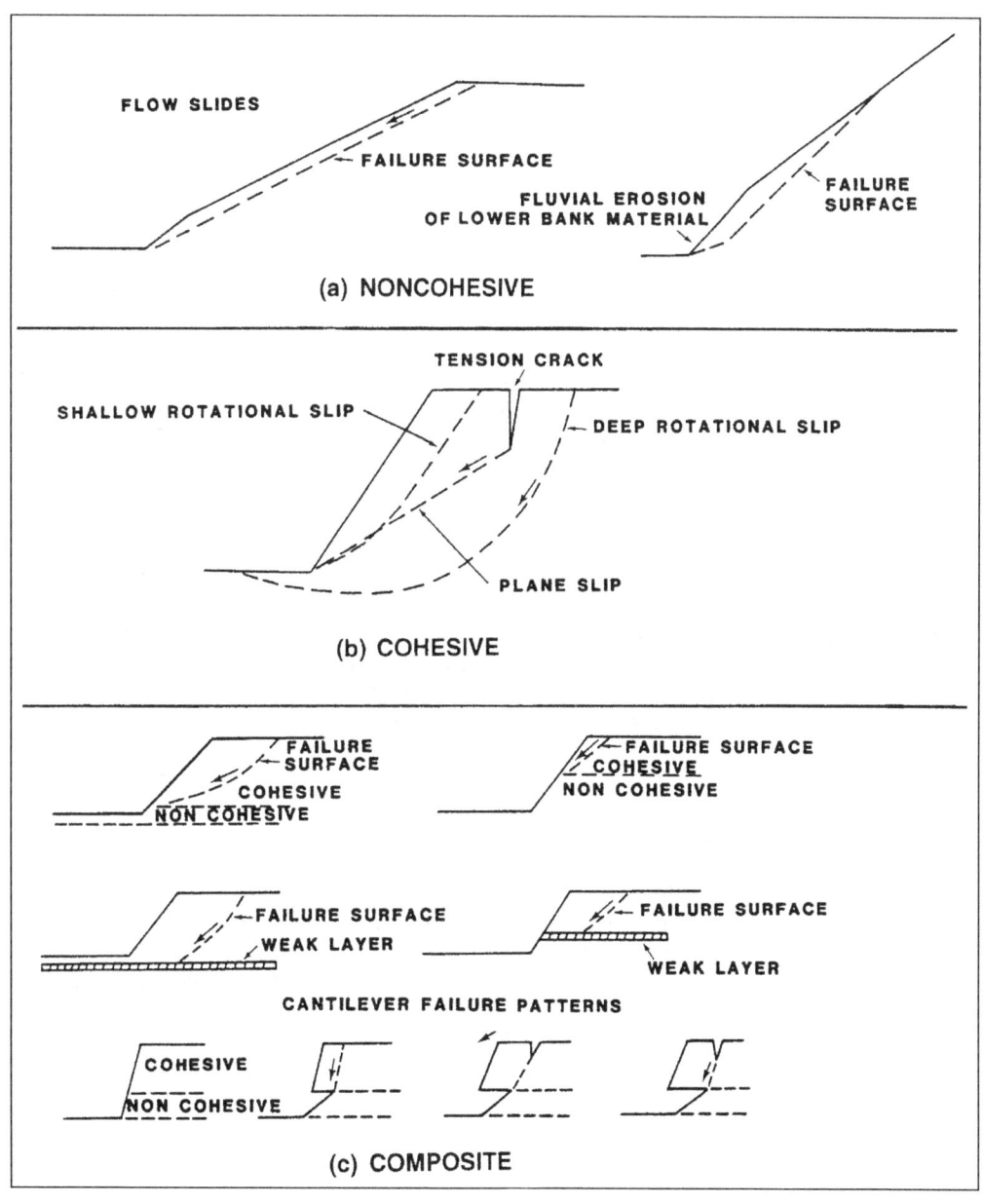

Figure 2.9. Typical bank failure surfaces: (a) noncohesive, (b) cohesive, and (c) composite (after FHWA 1985).

Piping. Piping is a phenomenon common to alluvial stream banks. With stratified banks, flow is induced in more permeable layers by changes in stream stage and by waves. If flow through the permeable lenses is capable of dislodging and transporting particles, the material is slowly removed, forming "pipes" which undermine portions of the bank. Without this foundation material to support the overlying layers, a block of bank material drops down and results in the development of tension cracks as sketched in Figure 2.9(c). These cracks allow surface flows to enter, further reducing the stability of the affected block of bank material. Bank erosion may continue on a grain-by-grain basis or the block of bank material may ultimately slide downward and outward into the channel, with bank failure resulting from a combination of seepage forces, piping, and mass wasting.

Mass Wasting. Local mass wasting is another form of bank failure. If a bank becomes saturated and possibly undercut by flowing water, blocks of the bank may slump or slide into the channel. Mass wasting may be caused or aggravated by the construction of homes on river banks, operation of equipment adjacent to the banks, added gravitational force resulting from tree growth, location of roads that cause unfavorable drainage conditions, agricultural uses on adjacent floodplain, saturation of banks by leach fields from septic tanks, and increased infiltration of water into the floodplain as a result of changing land-use practices.

Various forces are involved in mass wasting. Landslides, the downslope movement of earth and organic materials, result from an imbalance of forces. These forces are associated with the downslope gravity component of the slope mass. Resisting these downslope forces are the shear strength of the materials and any contribution from vegetation via root strength or engineered slope reinforcement activities. When the toe of a slope is removed, as by a stream, the slope materials may move downward into the void in order to establish a new equilibrium. Often, this equilibrium is a slope configuration with less than original surface gradient. The toe of the failed mass then provides a new buttress against further movements. Erosion of the toe of the slope then begins the process over again.

Bank Erosion and Failure. The erosion, instability, and/or retreat of a stream bank is dependent on the processes responsible for the erosion of material from the bank and the mechanisms of failure resulting from the instability created by those processes. Bank retreat is often a combination of these processes and mechanisms operating at various timescales. While the detailed analysis of bank stability is, primarily, a geotechnical problem (see for example, FHWA publications on soil slope stability FHWA 1994a and b), insight on the relationship between stream channel degradation and bank failure, for example, can be important to the hydraulic engineer concerned with bank instability. The processes responsible for bank erosion and bank failure mechanisms are discussed in more detail in Appendix B.

### 2.3.10 Sinuosity

Sinuosity is the ratio of the length of a stream reach measured along its centerline, to the length measured along the valley centerline or along a straight line connecting the ends of the reach. The valley centerline is preferable when the valley itself is curved. Sometimes, sinuosity is defined as the ratio of valley slope to stream slope or, more commonly, the ratio of the thalweg length to the valley length, where the thalweg is the trace of the deepest point in successive channel cross sections. Straight stream reaches have a sinuosity of one, and the maximum value of sinuosity for natural streams is about four.

A straight stream, or one that directly follows the valley centerline, sometimes has the same slope as the valley. As the sinuosity of the stream increases, its slope decreases in direct proportion. Similarly, if a sinuous channel is straightened, the slope increases in direct proportion to the change in length.

The size, form, and regularity of meander loops are aspects of sinuosity. Symmetrical meander loops are not very common, and a sequence of two or three identical symmetrical loops is even less common. In addition, meander loops are rarely of uniform size. The largest is commonly about twice the diameter of the smallest. Statistically, the size-frequency distribution of loop radii tends to have a normal distribution.

There is little relation between degree of sinuosity and lateral stream stability. A highly meandering stream may have a lower rate of lateral migration than a less sinuous stream of similar size (Figure 2.6). Stability is largely dependent on other properties, especially bar development and the variability of channel width (see Section 2.3.13).

Streams are broadly classified as straight, meandering or braided. Any change imposed on a stream system may change its planform geometry.

Straight Streams. A straight stream has small sinuosity at bankfull stage. At low stage, the channel develops alternate sandbars, and the thalweg meanders around the sandbars in a sinuous fashion. Straight streams are considered a transitional stage to meandering, since straight channels are relatively stable only where sediment size and load are small, gradient, velocities, and flow variability are low, and the channel width-depth ratio is relatively low. Straight channel reaches of more than 10 channel widths are not common in nature.

Meandering Streams. Alluvial channels of all types deviate from a straight alignment. The thalweg oscillates transversely and initiates the formation of bends. In a straight stream, alternate bars and the thalweg are continually changing; thus, the current is not uniformly distributed through the cross section, but is deflected toward one bank and then the other. Sloughing of the banks, nonuniform deposition of bed load, debris such as trees, and the Coriolis force due to the Earth⍳s rotation have been cited as causes for the meandering of streams. When the current is directed toward a bank, the bank is eroded in the area of impingement, and the current is deflected and impinges on the opposite bank farther downstream. The angle of deflection of the current is affected by the curvature formed in the eroding bank and the lateral depth of erosion. Figure 2.10 shows bars, pools, and crossings (riffles) typical of a meandering channel (see also Figures 2.4 and 2.5). A more detailed explanation of the meandering process and flow patterns through meanders is provided in Chapter 6.

Sinuous, meandering, and highly meandering streams have more or less regular inflections that are sinuous in plan view, consisting of a series of bends connected by crossings. In the bends, deep pools are carved adjacent to the concave bank by the relatively high velocities. Because velocities are lower on the inside of bends, sediments are deposited in this region, forming point bars. Also, the centrifugal force in the bend causes a transverse water surface slope and helicoidal flow with a bottom velocity away from the outer bank toward the point bar. These transverse velocities enhance point bar building by sweeping the heavier concentrations of bed load toward the convex bank where they are deposited to form the point bar. Some transverse currents have a magnitude of about 15 percent of the average channel velocity.

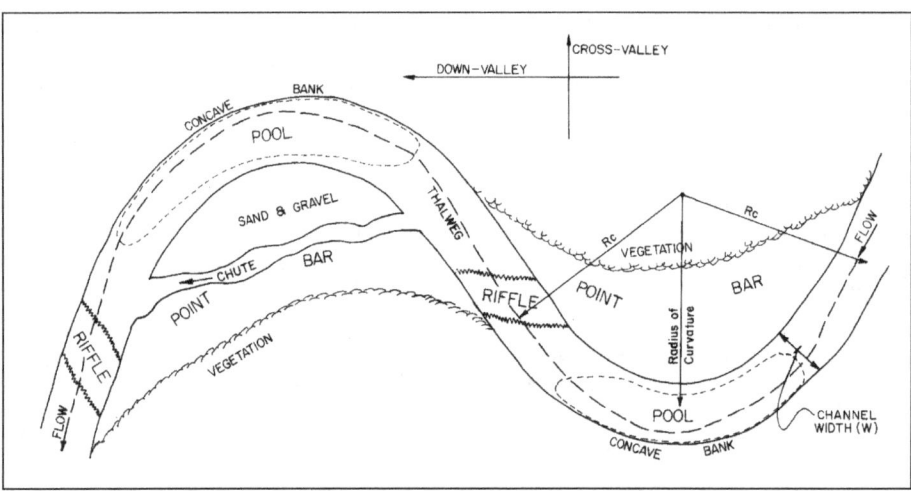

Figure 2.10. Plan view of a meandering stream.

The bends in meandering streams are connected by crossings (short straight reaches) which are quite shallow compared to the pools in the bendways. At low flow, large sandbars form in the crossings if the channel is not well confined. Scour in the bend causes the bend to migrate downstream and sometimes laterally. Lateral movements as large as 2,500 ft/yr (750 m/yr) have been observed in large alluvial rivers. Much of the sediment eroded from the outside bank is deposited in the crossing and on the point bar in the next bend downstream. The variability of bank materials and the fact that the stream encounters and produces such features as clay plugs causes a wide variety of river forms. The meander belt formed is often fifteen to twenty times the channel width.

On a laterally unstable channel, or at actively migrating bends on an otherwise stable channel, point bars are usually wide and unvegetated and the opposite bank is cut and often scalloped by erosion. The crescent-shaped scars of slumping may be visible from place to place along the bankline. The presence of a cut bank opposite a point bar is evidence of instability. Sand or gravel on the bar appears as a light tone on aerial photographs. The unvegetated condition of the point bar is attributed to a rate of outbuilding that is too rapid for vegetation to become established. However, the establishment of vegetation on a point bar is dependent on factors other than the rate of growth of the point bar, such as climate and the timing of floods. Therefore, the presence of vegetation on a point bar is not conclusive evidence of stability. If the width of an unvegetated point bar is considered as part of the channel width, the channel tends to be wider at bends.

As a meandering stream system moves laterally and longitudinally, meander loops move at unequal rates because of unequal erodibility of the banks. This causes the channel to appear as a slowly developing bulb-form. Channel geometry depends upon the local slope, bank material, and the geometry of adjacent bends. Years may be required before a configuration characteristic of average conditions in the stream is attained.

If the proposed highway or highway stream crossing is located near a meander loop, it is useful to have some insight into the probable way in which the loop will migrate or develop, as well as its rate of growth. No two meanders will behave in exactly the same way, but the meanders on a particular stream reach tend to conform to one of the several modes of behavior illustrated in Figure 2.11, which is based on a study of about 200 sinuous or meandering stream reaches (FHWA 1978a).

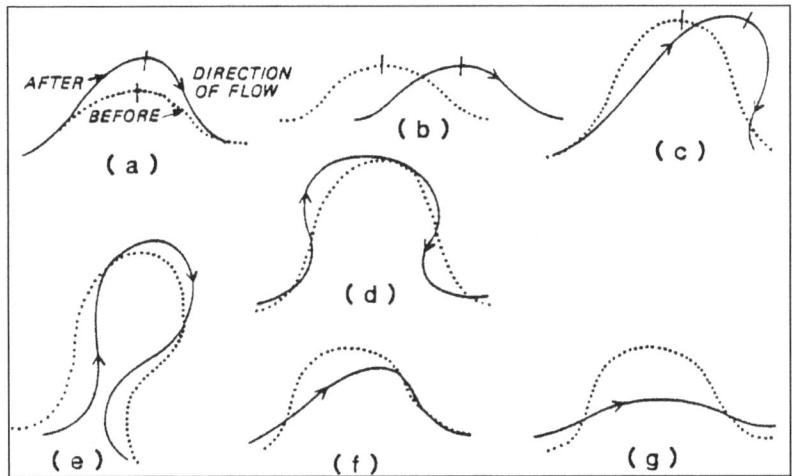

Figure 2.11. Modes of meander loop development: (a) extension, (b) translation, (c) rotation, (d) conversion to a compound loop, (e) neck cutoff by closure, (f) diagonal cutoff by chute, and (g) neck cutoff by chute (after FHWA 1978a).

Mode a (Figure 2.11) represents the typical development of a loop of low amplitude, which decreases in radius as it extends slightly in a downstream direction. Mode b rarely occurs unless meanders are confined by artificial levees or by valley sides on a narrow floodplain. Well developed meanders on streams that have moderately unstable banks are likely to follow Mode c. Mode d applies mainly to larger loops on meandering or highly meandering streams. The meander has become too large in relation to stream size and flow, and secondary meanders develop, converting it to a compound loop. Mode e also applies to meandering or highly meandering streams, usually of the equiwidth, point-bar type. The banks have been sufficiently stable for an elongated loop to form without being cut off, but the neck of the loop is gradually being closed and cutoff will eventually occur at the neck. Modes f and g apply mainly to locally braided, sinuous, or meandering streams having unstable banks. Loops are cut off by chutes that break diagonally or directly across the neck.

Oxbow lakes are formed by the cutoff of meander loops, which occurs either by gradual closure of the neck (neck cutoffs) or by a chute that cuts across the neck (chute cutoffs). Neck cutoffs are associated with relatively stable channels and chute cutoffs with relatively unstable channels. Recently formed oxbow lakes along a channel are evidence of recent lateral migration. Commonly, a new meander loop soon forms at the point of cutoff and grows in the same direction as the previous meander. Cutoffs tend to induce rapid bank erosion at adjacent meander loops. The presence of abundant oxbow lakes on a floodplain does not necessarily indicate a rapid channel migration rate because an oxbow lake may persist for hundreds of years.

Usually the upstream end of the oxbow lake fills quickly to bank height. Overflow during floods and overland flow entering the oxbow lake carry fine materials into the oxbow lake area. The lower end of the oxbow remains open and drainage entering the system can flow out from the lower end. The oxbow gradually fills with fine silts and clays which are plastic and cohesive. As the stream channel meanders, old bendways filled with cohesive materials (referred to as clay plugs) are sufficiently resistant to erosion to serve as semipermanent geologic controls which can drastically affect planform geometry.

The local increase in channel slope due to cutoff usually results in an increase in the growth rate of adjoining meanders and an increase in channel width at the point of cutoff. On a typical wide-bend point-bar stream, the effects of cutoff do not extend very far upstream or downstream. The consequences of cutoffs are an abruptly steeper stream gradient at the point of the cutoff, scour at the cutoff, and a propagation of the scour in an upstream direction. Downstream of a cutoff, the gradient of the channel is not changed and, therefore, the increased sediment load caused by upstream scour will usually be deposited at the site of the cutoff or below it, forming a large bar.

In summary, there is little relation between degree of sinuosity, as considered apart from other properties, and lateral stream stability (FHWA 1978a). A highly meandering stream may have a lower rate of lateral migration than a sinuous stream of similar size. Assessment of stability is based mainly on additional properties, especially on bar development and the variability of channel width. However, many hydraulic problems are associated with the location of highway crossings at a meander or bend. These include the shift of flow direction (angle of attack) at flood stage, shift of thalweg toward piers or abutments, development of point bars in the bridge reach, and lateral channel erosion at piers, abutments, or approaches.

In general, the most rapid bank erosion is generally at the outside of meanders, downstream from the apex of the loop. The cutoff of a meander, whether done artificially or naturally, causes a local increase in channel slope and a more rapid growth rate of adjoining meanders. Adjustment of the channel to increase in slope seems to be largely accomplished by increase in channel width (wetted perimeter) at and near the point of cutoff.

Some generalizations can be made, from knowledge of stream behavior, about the probable consequences of controlling or halting the development of a meander loop by the use of countermeasures. The most probable consequences relate to change in flow alignment (or lack of change, if the position of a naturally eroding bank is held constant). The development of a meander is affected by the alignment of the flow that enters it. Any artificial influence on flow alignment is likely to affect meander form. Downstream bank erosion rates are not likely to be increased, but the points at which bank erosion occurs are likely to be changed. In the case where flow is deflected directly at a bank, an increase in erosion rates would be expected. The failure of a major bridge on the Hatchie River near Covington, Tennessee has been attributed, in part, to lateral migration of the channel in the bridge reach (NTSB 1990).

### 2.3.11 Braided Streams

A braided stream is one that consists of multiple and interlacing channels (Figure 2.6). In general, a braided channel has a large slope, a large bed-material load in comparison with its suspended load, and relatively small amounts of silts and clays in the bed and banks. The magnitude of the bed load is more important than its size. If the flow is overloaded with sediment, deposition occurs, the bed aggrades, and the slope of the channel increases in an effort to obtain a graded (equilibrium) condition. As the channel steepens, velocity increases, and multiple channels develop. Multiple channels are generally formed as bars of sediment and deposited within the main channel, causing the overall channel system to widen.

Multiple, mid-channel islands and bars are characteristic of streams that transport large bed loads. The presence of bars obstructs flow and scour occurs, either lateral erosion of banks on both sides of the bar, scour of the channels surrounding the bar, or both. This erosion will enlarge the channel and, with reduced water levels, an island may form at the site of a gravel or sand bar. The worst case will be where major bar or island forms at a bridge site. This can produce erosion of both banks of the stream and bed scour along both sides of the island. Reduction in the flow capacity beneath the bridge can result as a vegetated island forms under the bridge. An island or bar that forms upstream or downstream of a bridge can change flow alignment and create bank erosion or scour problems at the bridge site.

Island shift is easily identified because active erosion at one location and active deposition at another on the edge of an island can be recognized in the field. Also, the development or abandonment of flood channels and the joining together of islands can be detected by observing vegetation differences and patterns of erosion and deposition.

The degree of channel braiding is indicated by the percent of reach length that is divided by bars and islands, as shown in Figure 2.6. Braided streams tend to be common in arid and semiarid parts of the western United States and regions having active glaciers.

Braided streams may present difficulties for highway construction because they are unstable, change alignment rapidly, carry large quantities of sediment, are very wide and shallow even at flood flow and are, in general, unpredictable. Deep scour holes can develop downstream of a gravel bar or island where the flow from two channels comes together.

Braided streams generally require long bridges if the full channel width is crossed or effective flow-control measures if the channel is constricted. The banks are likely to be easily erodible, and unusual care must be taken to prevent lateral erosion at or near abutments. The position of braids is likely to shift during floods, resulting in unexpected velocities, angle of attack, and depths of flow at individual piers. Lateral migration of braided streams takes place by lateral shift of a braid against the bank, but available information indicates that lateral migration rates are generally less than for meandering streams. Along braided streams, however, migration is not confined to the outside of bends but can take place at any point by the lateral shift of individual braids.

**2.3.12 Anabranched Streams**

An anabranched stream differs from a braided stream in that the flow is divided by islands rather than bars, and the islands are large relative to channel width (also called an anastomosing stream). The anabranches, or individual channels, are more widely and distinctly separated and more fixed in position than the braids of a braided stream. An anabranch does not necessarily transmit flow at normal stage, but it is an active and well-defined channel, not blocked by vegetation. The degree of anabranching is arbitrarily categorized in Figure 2.6 in the same way as the degree of braiding was described.

Although the distinction between braiding and anabranching may seem academic, it has real significance for engineering purposes. Inasmuch as anabranches are relatively permanent channels that may convey substantial flow, diversion and confinement of an anabranched stream is likely to be more difficult than for a braided stream. Problems associated with crossings on anabranched streams can be avoided if a site where the channel is not anabranched can be chosen. If not, the designer may be faced with a choice of either building more than one bridge, building a long bridge, or diverting anabranches into a single channel. Problems with flow alignment may occur if a bridge is built at or near the junction of anabranches. Where anabranches are crossed by separate bridges, the design discharge for the bridges may be difficult to estimate. If one anabranch should become partly blocked, as by floating debris or ice, an unexpected amount of flow may be diverted to the other.

## 2.3.13 Variability of Width and Development of Bars

The variability of unvegetated channel width is a useful indication of the lateral stability of a channel. The visual impression of unvegetated channel width on aerial photographs depends on the relatively dark tones of vegetation as contrasted with the lighter tones of sediment or water. A channel is considered to be of uniform width (equiwidth) if the unvegetated width at bends is not more than 1.5 times the average width at the narrowest places.

The relationship between width variability and lateral stability is based on the rate of development of point bars and alternate bars. If the concave bank at a bend is eroding slowly, the point bar will grow slowly and vegetation will become established on it. The unvegetated part of the bar will appear as a narrow crescent. If the bank is eroding rapidly, the unvegetated part of the rapidly growing point bar will be wide and conspicuous. A point bar with an unvegetated width greater than the width of flowing water at the bend is considered to be wider than average. Lateral erosion rates are probably high in stream reaches where bare point bars tend to exceed average width. In areas where vegetation is quickly established, as in rainy southern climates, cut banks at bends may be a more reliable indication of instability than the unvegetated width of point bars.

Three categories of width variability are distinguished in Figure 2.6, but the relative lateral stability of these must be assessed in connection with bar development and other properties. In general, equiwidth streams having narrow point bars are the most stable laterally, and random-width streams having wide, irregular point bars are the least stable. Vertical stability, or the tendency to scour, cannot be assessed from these properties. Scour may occur in any alluvial channel. In fact, the greatest potential for deep scour might be expected in laterally stable equiwidth channels, which tend to have relatively deep and narrow cross sections and bed material in the size range of silt and sand.

## 2.4 AGGRADATION/DEGRADATION AND SEDIMENT CONTINUITY IN SAND-BED CHANNELS

### 2.4.1 Aggradation/Degradation

Aggradation and degradation are the vertical raising and lowering, respectively, of the streambed over relatively long distances and time frames. Such changes can be the result of both natural and man-induced changes in the watershed. The sediment continuity concept is the primary principle applied in both qualitative and quantitative analyses of bed elevation changes. After an introduction to the concept of sediment continuity, some factors causing a bed elevation change are reviewed.

### 2.4.2 Overview of the Sediment Continuity Concept

The amount of material transported, eroded, or deposited in an alluvial channel is a function of sediment supply and channel transport capacity. Sediment supply is provided from the tributary watershed and from any erosion occurring in the upstream channel. Sediment transport capacity is a function of the sediment size, the discharge of the stream, and the geometric and hydraulic properties of the channel. When the transport capacity (sediment outflow) equals sediment supply (sediment inflow), a state of equilibrium exists.

Application of the sediment continuity concept to a single channel reach illustrates the relationship between sediment supply and transport capacity. Technically, the sediment continuity concept states that the sediment inflow minus the sediment outflow equals the time rate of change of sediment volume in a given reach. More simply stated, during a given time period the amount of sediment coming into the reach minus the amount leaving the downstream end of the reach equals the change in the amount of sediment stored in that reach (Figure 2.12). The sediment inflow to a given reach is defined by the sediment supply from the watershed (upstream of the study reach plus any significant lateral input directly to the study reach). The transport capacity of the channel within the given reach defines the sediment outflow. Changes in the sediment volume within the reach occur when the total input to the reach (sediment supply) is not equal to the downstream output (sediment transport capacity). When the sediment supply is less than the transport capacity, erosion (degradation) will occur in the reach so that the transport capacity at the outlet is satisfied, unless controls exist that limit erosion. Conversely, when the sediment supply is greater than the transport capacity, deposition (aggradation) will occur in the reach.

Controls that limit erosion may either be human induced or natural. Human-induced controls included bank protection works, grade control structures, and stabilized bridge crossings. Natural controls can be geologic, such as outcroppings, or the presence of significant coarse sediment material in the channel. The presence of coarse material can result in the formation of a surface armor layer of larger sediments that are not transported by average flow conditions.

### 2.4.3 Factors Initiating Bed Elevation Changes

Human-induced Changes. Human activities are the major cause of streambed elevation changes. Very few bed elevation changes are due to natural causes, although some may be the result of both natural and human-induced causes. The most common activities which result in bed elevation changes caused by human activity are channel alterations, streambed mining, dams and reservoirs, and land-use changes. Highway construction, including the construction of bridges and channel alterations of limited extent, usually affect stream vertical stability only locally.

Channel Alterations. Dredging, channelization, straightening, the construction of cutoffs to shorten the flow path of a stream, and clearing and snagging to increase channel capacity are the major causes of streambed elevation changes. An increase in slope resulting from a shorter flow path or an increase in flow capacity results in increased velocities and a corresponding increase in sediment transport capacity. If the stream was previously in equilibrium (supply equal to transport capacity) the channel may adjust, either by increasing its length or by reducing its slope by degradation, in order to reestablish equilibrium. The most frequent response is a degrading streambed followed by bank erosion and a new meander pattern.

Constrictions in a stream channel, as in river control projects to maintain a navigation channel or highway crossings, also increase velocities and the sediment transport capacity in the constricted reach. The resulting degradation can be considered local, but it may extend through a considerable reach of stream, depending on the extent of the river control project. Constrictions may also cause local aggradation problems downstream.

The response to an increased sediment load in a stream that was near equilibrium conditions (i.e., supply now greater than transport capacity) is normally deposition in the channel downstream of the alteration. The result is an increase in flood stages and overbank flooding in downstream reaches. In time, the aggradation will progress both upstream and downstream of the end of the altered channel, and the stream reach may become locally braided as it seeks a new balance between sediment supply and sediment transport capacity.

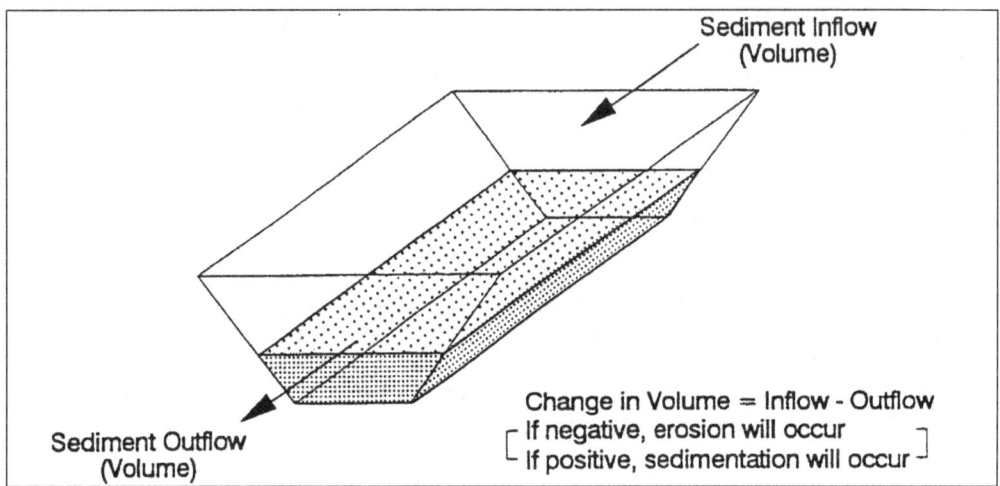

Figure 2.12. Definition sketch of sediment continuity concept applied to a given channel reach over a given time period.

Streambed Mining. Streambed mining for sand or gravel can be beneficial or detrimental, depending on the balance between sediment supply and transport capacity. Where the sediment supply exceeds the stream's transport capacity because of man's activities in the watershed or from natural causes, controlled removal of gravel bars and limited mining may enhance both lateral and vertical stability of the stream.

The usual result of streambed mining is an imbalance between sediment supply and transport capacity. Upstream of the operation, the water surface slope may be increased and bank erosion and headcutting or a nick point may result. The extent of the damage that can result is a function of the volume and depth of the sand and gravel pit relative to the size of the stream, bed material size, flood hydrographs, upstream sediment transport, and the location of the pit. If the size of the borrow pit is sufficiently large, a substantial quantity of the sediment inflow will be trapped in the pit and degradation will occur downstream. If bank erosion and headcutting upstream of the pit produce a sediment supply greater than the trap capacity of the pit and the transport capacity downstream, aggradation could occur. However, this circumstance is unlikely and streambed mining generally causes degradation upstream and downstream of the pit.

Dams and Reservoirs. Storage and flood control reservoirs produce a stream response both upstream and downstream of the reservoir. A stream flowing into a reservoir forms a delta as the sediment load is deposited in the ponded water. This deposition reduces the stream gradient upstream of the reservoir and causes aggradation in the channel. Aggradation can extend many kilometers upstream.

Downstream of reservoirs, stream channel stability is affected because of the changed flow characteristics and because flow releases are relatively sediment-free. Clear-water releases pick up a new sediment load and degradation can result. The stream channel and stream gradient that existed prior to the construction of the dam was the cumulative result of past floods of various sizes and subject to change with each flood. Post-construction flows are usually of lesser magnitude and longer duration, and the stream will establish a new balance in time consistent with the new flow characteristics.

It is possible for aggradation to occur downstream of a reservoir if flow releases are insufficient to transport the size or volume of sediment brought in by tributary streams. Streamflow regulation, which is an objective in dam construction and reservoir operation, is sometimes overlooked in assessing stream system response to this activity. The reduction in flood magnitude and stage downstream of dams as a result of reservoir operation can result in greatly increased hydraulic gradients and degradation in tributaries downstream of the dam. A notorious bridge failure on the Big Sioux River was, in part, attributable to such a condition.

Land Use Changes. Agricultural activities, urbanization, commercial development, and construction activities also contribute to bed elevation problems in streams. Clear cutting of forests and destruction of grasslands by overgrazing, burning and cultivation can accelerate erosion, causing streams draining these areas to become overloaded with sediment (i.e., excess sediment supply). As the overload persists, the stream system aggrades and increases its slope to increase its sediment transport capacity.

Construction and developing urban and commercial areas can affect stream gradient stability. Fully developed urban areas are low sediment producers because of impervious areas and lawns, but tend to increase the magnitude of runoff events and reduce their duration. The response of a small stream system to these changes is degradation, changes in planform (e.g., increased sinuosity), and channel widening downstream of the urbanized area. However, if the urbanized area is small relative to the basin of the stream in which it is located, the net effect will probably be small.

Natural Changes. Natural causes of stream gradient instability are primarily natural channel alterations, earthquakes, tectonic and volcanic activities, climatic change, fire, and channel bed and bank material erodibility.

Cutoffs and chute channel development (as a channel straightens) are the most common natural channel alterations. This results in a shorter flow path, a steeper channel gradient, and an increase in sediment transport capacity. Significant bank erosion and degradation progressing to an upstream control can result. Downstream of the cutoff, aggradation will occur.

Severe landslides, mud flows, uplifts and lateral shifts in terrain, and liquefaction of otherwise semi-stable materials are associated with earthquakes and tectonic activities. The response to these activities include channel changes, scour or deposition locally or system-wide, headcutting and bank instability.

Alluvial fans, discussed under Valley Setting, are among the most common naturally occurring cases of channel aggradation (Schumm and Lagasse 1998).

## 2.5 CHANNEL STABILITY CONCEPTS FOR COHESIVE BOUNDARY CHANNELS

Stream channels flow through and over a variety of boundary materials from bedrock to alluvium, all of which have highly variable erodibilities. Although the designs for bridges should account for the long-term erodibility of the bed and bank materials, both laterally and vertically, most analyses typically focus primarily on pier and abutment scour in non-cohesive materials, for which there are well-established analytical and empirical methodologies. However, more recent research has focused on bed erosion and scour in cohesive materials (e.g., NCHRP 2004c) and erodible bedrock (e.g., NCHRP 2011).

There are many bridges throughout the U.S. that have shallow foundations bearing on bedrock or cohesive materials that may have appeared to be resistant to erosion and scour at the time they were constructed. However, in many cases, long-term channel degradation, bank retreat, and/or channel shifting in these cohesive materials have created problems for bridges. Thus, the responses and types of channel instability at a bridge are dependent on the cohesion of the local bed and bank materials and the duration and magnitude of erosive flows over time.

### 2.5.1 Cohesive Streambeds

Cohesive bed material can include caliche or hardpan (well cemented alluvial or colluvial sediments), loess (slightly cemented wind-blown silt), highly compact and dense clays (e.g., clay plug associated with relict oxbow lakes/abandoned channel segments), and in the broader sense, erodible bedrock (Figure 2.13). The erodibility of these cohesive materials can be highly variable laterally and vertically as well as on a local and regional scale. In general, cohesive materials exposed in a streambed can often impart stability to the streambed if the cohesive materials have been exposed in the streambed for a long period of time and the channel has had time to adjust to them.

Figure 2.13. Stream channel with erodible rock bed and semi-cohesive banks.

Recent or imminent exposure of cohesive streambed materials may produce a response by the river or stream that is highly variable in nature, extent, and intensity, depending on the stability of the downstream channel and the hydrologic characteristics of the watershed. The response of a channel with cohesive material exposed in the bed can include:

- Potential rapid lateral migration (or widening) whereby the erosive energy of the channel is expended on the softer alluvial bank materials, causing "skating" along the bedrock bottom
- Episodic streambed degradation due to local variability of bed materials - plucking of small grains to block-sized material creates variable erosion both laterally and vertically
- Downcutting from abrasion of bed material due to heavy sediment loads
- Episodic channel degradation due to the upstream progression of a headcut or knickzone in cohesive bed materials (which often overlie non-cohesive or less cohesive sediments)

Where cohesive materials underlie non-cohesive channel bed sediments, long-term degradation may result in the exposure of the cohesive materials over time and the future response of the channel may be the same as described above. However, depending on the depth of cohesive materials underlying the alluvial bed sediments, scour depths can be inhibited by the cohesive materials.

In some cases, changes in upstream land use may create long-term aggradation within stable downstream channels segments that contain cohesive bed materials. In these cases, the response of the channel may include:

- Complete burial of the cohesive bed material and loss of channel capacity
- Channel may widen or rate of channel migration may increase
- Same instability problems as those identified for channels with non-cohesive bed materials (see Appendix B)

### 2.5.2 Cohesive Streambanks

Cohesive materials in stream banks include materials such as clays or silty clays, caliche or hardpan, and erodible bedrock (Figure 2.13). In addition, the strength of alluvial streambank sediments can be significantly increased through the development of dense vegetation root systems within the streambank. If the cohesive materials do not occupy all of the bank or at least the lower bank position, the channel will likely respond as if only non-cohesive or composite sediments exist in the streambank.

For those cases where the streambanks consist of cohesive materials:

- If banks consist of cohesive alluvial sediments, modes and rate of failure will be dependent on hydraulic conditions as described in Appendix B
- Potential exists for the channel to entrench or degrade due to confinement of flow and the inability of the streambanks to adjust laterally
- Long-term channel aggradation may result in increased flooding if the channel cannot adjust through widening

Where the streambanks consist of both cohesive and non-cohesive (composite) soils, the modes and rate of failure are dependent on hydraulic conditions as well as on location of non-cohesive units in bank (see Appendix B). Where one bank consists of cohesive materials and the other bank does not:

- There is an increased risk of lateral migration into the non-cohesive bank
- Large, infrequent failures in a cohesive bank, such as occurs with mass wasting, may deflect flow into the non-cohesive bank causing increased rate of bank erosion/retreat

Where clay plugs and fine-grained channel fill deposits (e.g., oxbows, cutoffs, abandoned channels) are present, they can cause temporally and spatially irregular meander bend migration as well as meander bend distortion, compression, and/or deflection during migration.

### 2.6 GRAVEL-BED RIVERS

Gravel-bed rivers contain a larger class of noncohesive bed material than sand-bed rivers (see Table 2.1). Both sand-bed rivers and gravel-bed rivers are considered alluvial; however, in coarser bed channels physical processes and stability considerations can differ considerably from those described earlier in this chapter, which generally apply to sand-bed alluvial channels. Chapter 8 highlights the physical processes related to channel stability in gravel-bed rivers in contrast to the processes common to sand-bed rivers.

(page intentionally left blank)

# CHAPTER 3

# HYDRAULIC FACTORS AND PRINCIPLES

## 3.1 INTRODUCTION

Design of highway stream crossings and countermeasures to prevent damage from streamflow requires assessment of factors that characterize streamflow and channel conditions at each bridge site. The importance of hydraulic or flow factors in the crossing design process is influenced by the importance of the bridge and by land use on the floodplain, among other things. Geometry and location of the highway stream crossing are important considerations in evaluating the interaction of the structure and the flow at the crossing in terms of potential stream instability issues. In addition, hydraulic factors have a significant influence on the design of bridge substructure components when scour and stream stability are considered.

## 3.2 HYDRAULIC DESIGN FOR SAFE BRIDGES

### 3.2.1 Overview

The FHWA Manual "Hydraulics of Bridge Waterways" (HDS 1) was published in 1978 (FHWA 1978c) and, among other supporting topics, contains graphical and hand calculation methods for computing bridge backwater. Bridge hydraulic analysis has advanced significantly over the past three decades to include a prevalence of steady and unsteady state one-dimensional computer models. The use of two-dimensional computer modeling for many complex flow conditions has also become commonplace. The new Hydraulic Design Series document, HDS 7, entitled "Hydraulic Design of Safe Bridges," was developed to serve as a major technical update to HDS 1 to include guidance on one- and two-dimensional hydraulic modeling of bridges in steady and unsteady flow conditions. Additional information is included related to design considerations, regulatory requirements, hydrology, deck drainage, hydraulic forces on bridge elements, and advanced modeling techniques. Important concepts related to scour, stream stability, sediment transport are also included (FHWA 2012a).

HDS 7 defines the state-of-the-practice for bridge hydraulics and is a comprehensive bridge hydraulic design manual. The manual provides necessary background information on open channel flow hydraulics but focuses on providing detailed technical guidance on bridge hydraulic modeling. This guidance includes model selection (one-dimensional versus two-dimensional) steady versus unsteady boundary conditions, bridge method selection (various choices for computing bridge hydraulic losses in one- and two-dimensional models), and best practices for model development (cross-section and mesh requirements, determining boundary roughness coefficients, establishing boundary conditions, and model calibration, validation and verification). Because bridges over waterways can impact aquatic and riparian habitats, floodplains and floodways, and navigable waterways, HDS 7 also provides discussion of the range of design considerations, environmental considerations, and regulatory requirements that can be encountered during bridge design and construction.

As the title of HDS 7 implies, this manual is not limited to the technical aspects of bridge hydraulic modeling, but serves as a guide through the hydraulic design process to improve bridge safety. The document improves bridge safety by promoting better methods for computing hydraulic variables through more frequent use of two-dimensional models, and by making engineers more aware of important design considerations ranging from stream instability and scour to deck drainage and loads caused by hydraulic forces.

### 3.2.2 Hydraulic Modeling Criteria and Selection

Any hydraulic model, whether it is numerical or physical, has assumptions and requirements. It is important for the hydraulic engineer to be aware of and understand the assumptions because they form the limitations of that approach. It is the goal of any hydraulic model study to accurately simulate the actual flow condition. Violating the assumptions and ignoring the limitations will result in a poor representation of the actual hydraulic condition. Treating the model as a black box will often produce inaccurate results. This is not acceptable given the cost of bridges and the potential consequences. Therefore, the approach should be selected based primarily on its advantages and limitations, though also considering the importance of the structure, potential project impacts, cost, and schedule.

One-dimensional modeling requires that variables (velocity, depth, etc.) change predominantly in one defined direction, x, along the channel. Because channels are rarely straight, the computational direction is along the channel centerline. Two-dimensional models compute the horizontal velocity components ($V_x$ and $V_y$) or, alternatively, velocity vector magnitude and direction throughout the model domain. Therefore, two-dimensional models avoid many of assumptions required by one-dimensional models, especially for the natural, compound channels (low-flow channel with floodplains) that make up the vast majority of bridge crossings over water.

As emphasized in HDS 7, the advantages of two-dimensional modeling include a significant improvement in calculating hydraulic variables at bridges. Therefore FHWA has a strong preference for the use of two-dimensional models over one-dimensional models for bridge hydraulic analyses. One-dimensional models are best suited for in-channel flows and when floodplain flows are minor. They are also frequently applicable to small streams. For extreme flood conditions, one-dimensional models generally provide accurate results for narrow to moderate floodplain widths. They can also be used for wide floodplains when the degree of bridge constriction is small and the floodplain vegetation is not highly variable. In general, where lateral velocities are small one-dimensional models provide reasonable results. Avoiding significant lateral velocities is the reason why cross section placement and orientation are so important for one-dimensional modeling.

Two-dimensional models generally provide more accurate representations of:

- Flow distribution
- Velocity distribution
- Water Surface Elevation
- Backwater
- Velocity magnitude
- Velocity direction
- Flow depth
- Shear stress

Although this list is general, these variables are essential information for new bridge design, evaluating existing bridges for scour potential, and countermeasure design. Two-dimensional models should be used when flow patterns are complex and one-dimensional model assumptions are significantly violated. If the hydraulic engineer has great difficulty in visualizing the flow patterns and setting up a one-dimensional model that realistically represents the flow field, then two-dimensional modeling should be used.

### 3.3 BASIC HYDRAULIC PRINCIPLES

The basic equations of flow are continuity, energy, and momentum. They are derived from the laws of (1) conservation of mass; (2) conservation of energy; and (3) conservation of linear momentum, respectively. Analyses of flow problems are much simplified if there is no acceleration of the flow or if the acceleration is primarily in one direction (one-dimensional flow), that is, the accelerations in other directions are negligible. However, a very inaccurate analysis may occur if one assumes accelerations are small or zero when in fact they are not. In the simplest cases, or as a first approximation of flow conditions, steady, uniform flow can be assumed; that is hydraulic variables do not change with time at a cross section or with distance along the channel.

Applications of the basic principles of flow are reviewed in detail in FHWAɪs "Introduction to Highway Hydraulics" (HDS 4) (FHWA 1997b). The user is referred to standard fluid mechanics texts or "River Engineering for Highway Encroachments" (HDS 6) for their derivations (FHWA 2001). Additional discussion of basic hydraulic principles as related to hydraulic computer modeling can be found in "Hydraulic Design for Safe Bridges" (HDS 7) (FHWA 2012a). The continuity and energy equations are particularly useful in evaluating potential stream stability problems. In addition, the basic Manning equation for open channel flow introduces the important concept of hydraulic resistance to flow. These are reviewed briefly in the following sections.

### 3.3.1 Continuity Equation

The continuity equation is based on conservation of mass, that is, matter can neither be created or destroyed (except for mass-energy interchange). For steady flow of incompressible fluids it is:

$$V_1 A_1 = V_2 A_2 = Q = VA \qquad (3.1)$$

where:

$V$ = Average velocity in the cross section perpendicular to the area, ft/s (m/s)
$A$ = Area perpendicular to the velocity, $ft^2$ ($m^2$)
$Q$ = Volume flow rate or discharge, $ft^3$/s ($m^3$/s)

Equation 3.1 is applicable when the fluid density is constant, the flow is steady, there is no significant lateral inflow or seepage (or they are accounted for) and the velocity is perpendicular to the flow area (Figure 3.1). Note that the product of area (A) and velocity (V), which is the discharge, is constant from section to section.

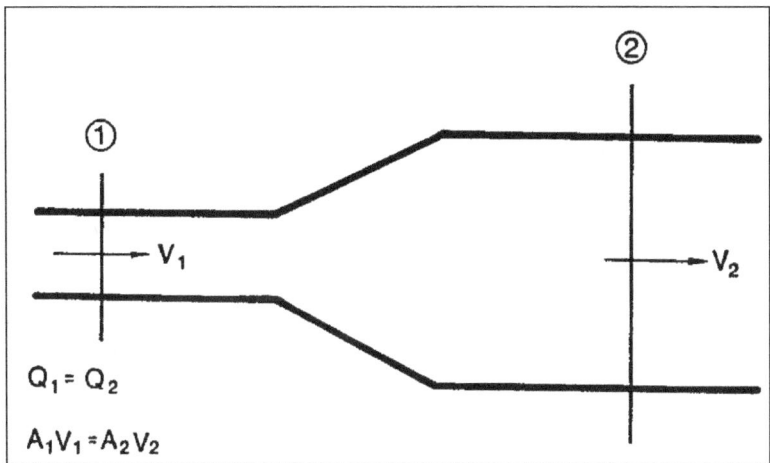

Figure 3.1. Sketch of the continuity concept.

### 3.3.2 Energy Equation

The energy equation is derived from the first law of thermodynamics which states that energy must be conserved at all times. In practical terms, this means that the energy at one section of the flow is equal to the energy at another section plus the losses in between. The energy equation for steady incompressible flow is:

$$Z_1 + y_1 + \frac{V_1^2}{2g} = Z_2 + y_2 + \frac{V_2^2}{2g} + h_L \qquad (3.2)$$

where:

$Z$ = Elevation above some datum, ft (m)
$y$ = Depth of flow at a point, ft (m)
$V$ = Mean velocity, ft/s (m/s)
$h_L$ = Energy head loss, ft (m)

The terms of the energy equation are illustrated in Figure 3.2.

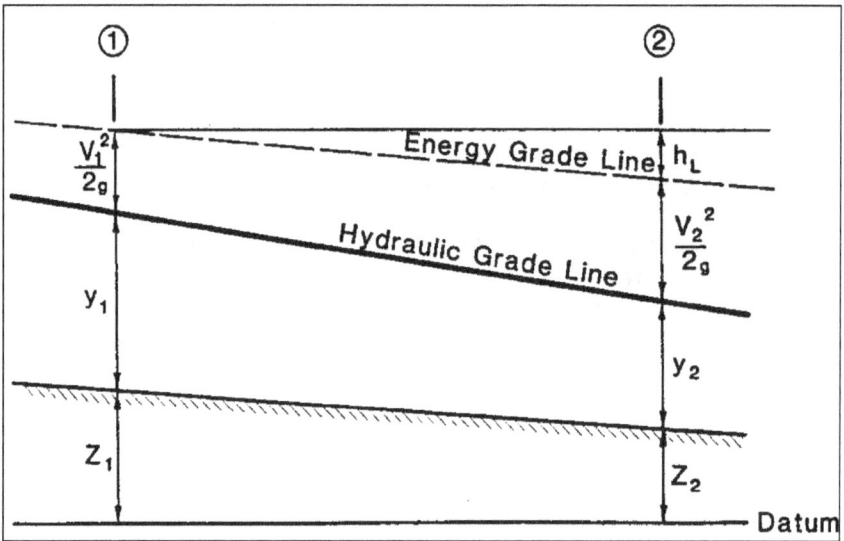

Figure 3.2. Sketch of the energy concept for open channel flow.

### 3.3.3 Manning Equation

Water flows in a sloping channel because of the force of gravity. The flow is resisted by the friction between the water and wetted surface of the channel. For uniform flow, the volume of water flowing (Q), the depth of flow (y), and the velocity of flow (V) depend upon the channel shape, roughness (n), and slope of the channel bed ($S_o$). Various equations have been devised to determine the velocity and discharge under steady, uniform flow conditions in open channels. A useful equation is the one that is named for Robert Manning, an Irish engineer. Manning's equation for the velocity of flow in open channels is:

$$V = \frac{K_u}{n} R^{2/3} S^{1/2} \tag{3.3}$$

where:

- $V$ = Mean velocity, ft/s (m/s)
- $n$ = Manning coefficient of channel roughness
- $R$ = Hydraulic radius, ft (m)
- $S$ = Energy slope, ft/ft (m/m)

For steady, uniform flow $S = S_0$

- $K_u$ = 1.486 English
- $K_u$ = 1.0 SI

Over many decades, a catalog of values of Manning n has been assembled so that an engineer can estimate the appropriate value by knowing the general nature of the channel boundaries (USGS 1984). A pictorial guide for assisting with selection of an appropriate roughness coefficient is given in USGS Water Supply Paper 1849 (USGS 1967). Methods for estimating resistance to flow are discussed in detail in Section 3.4.4

The hydraulic radius, R, is a shape factor that depends only upon the channel dimensions and the depth of flow. It is computed by the equation:

$$R = \frac{A}{P} \tag{3.4}$$

where:

- $A$ = Cross-sectional area of flowing water in ft² (m²) perpendicular to the direction of flow
- $P$ = Wetted perimeter or the length, ft (m), of wetted contact between a stream of water and its containing channel, perpendicular to the direction of flow

By combining the continuity equation for discharge (Equation 3.1) with Equation 3.3, the Manning equation can be used to compute discharge directly

$$Q = \frac{K_u}{n} A R^{2/3} S^{1/2} \tag{3.5}$$

In many computations, it is convenient to group the cross-sectional properties into a term called conveyance, K,

$$K = \frac{K_u}{n} A R^{2/3} \tag{3.6}$$

or

$$Q = K \sqrt{S} \tag{3.7}$$

and

$$K = \frac{Q}{\sqrt{S}} \qquad (3.8)$$

Conveyance can be considered a measure of the carrying capacity of the channel, since it is directly proportional to discharge (Q).

## 3.4 HYDRAULIC FACTORS AFFECTING STREAM STABILITY

### 3.4.1 Overview

Hydraulic, location, and design factors important to the highway engineer are introduced in Figure 3.3. Each of the hydraulic factors has an effect on stream stability at a bridge crossing. Since the geometry and location of the bridge crossing can also affect stream stability, the most significant factors related to bends, confluences, alignment, and highway profile are also summarized. In addition, some general concepts related to the hydraulic design of bridges are discussed in Section 3.6.

### 3.4.2 Magnitude and Frequency of Floods

The hydrologic analysis for a stream crossing consists of establishing peak flow-frequency relationships and such flow-duration hydrographs as may be necessary. Flood-frequency relationships are generally defined on the basis of a regional analysis of flood records, a gaging station analysis, or both. Regional analyses have been completed for all states by the USGS, and the results are generally applicable to watersheds which are unchanged by man. Flood-frequency relationships at gaged sites can be established from station records which are of sufficient length to be representative of the total population of flood events on that particular stream. The Pearson Type III distribution with log transformation of flood data is recommended by the Water Resources Council (1981) for station flood data analysis. Where flood estimates by regional analysis vary from estimates by station analysis, factors such as gaging station record length and the applicability of the regional analysis to that specific site should be considered, as well as high water information, flood data, and information of floods at existing bridges on the stream.

The term "design flood" is purposely avoided in the above discussion because of the implication that a stream crossing can be designed for a unique flood event. In reality, a range of events should be examined to determine which design condition is most advantageous, insofar as costs and risks are concerned (see HEC-18, Chapter 2 (FHWA 2012b)). If a design flood is designated for purposes of stream stability analysis, it probably should be that event which causes the greatest stress to the highway stream crossing system, that is, the flood magnitude and stage which is at incipient overtopping of the highway.

Hydrologic analysis establishes the probability of occurrence of a flood of given magnitude in any one year period. It also is the first step in establishing the probability of occurrence of the flood event which will pass through bridge waterways in the highway-stream crossing system without overtopping the highway. FHWAıs HDS 2 should be referred to for more detailed information and guidelines on hydrologic analysis (FHWA 2002). The second step is the determination of the stage-discharge relationship, flow and velocity distributions, backwater, scour, etc., (i.e., the hydraulics of the crossing system, as discussed in the remainder of this section).

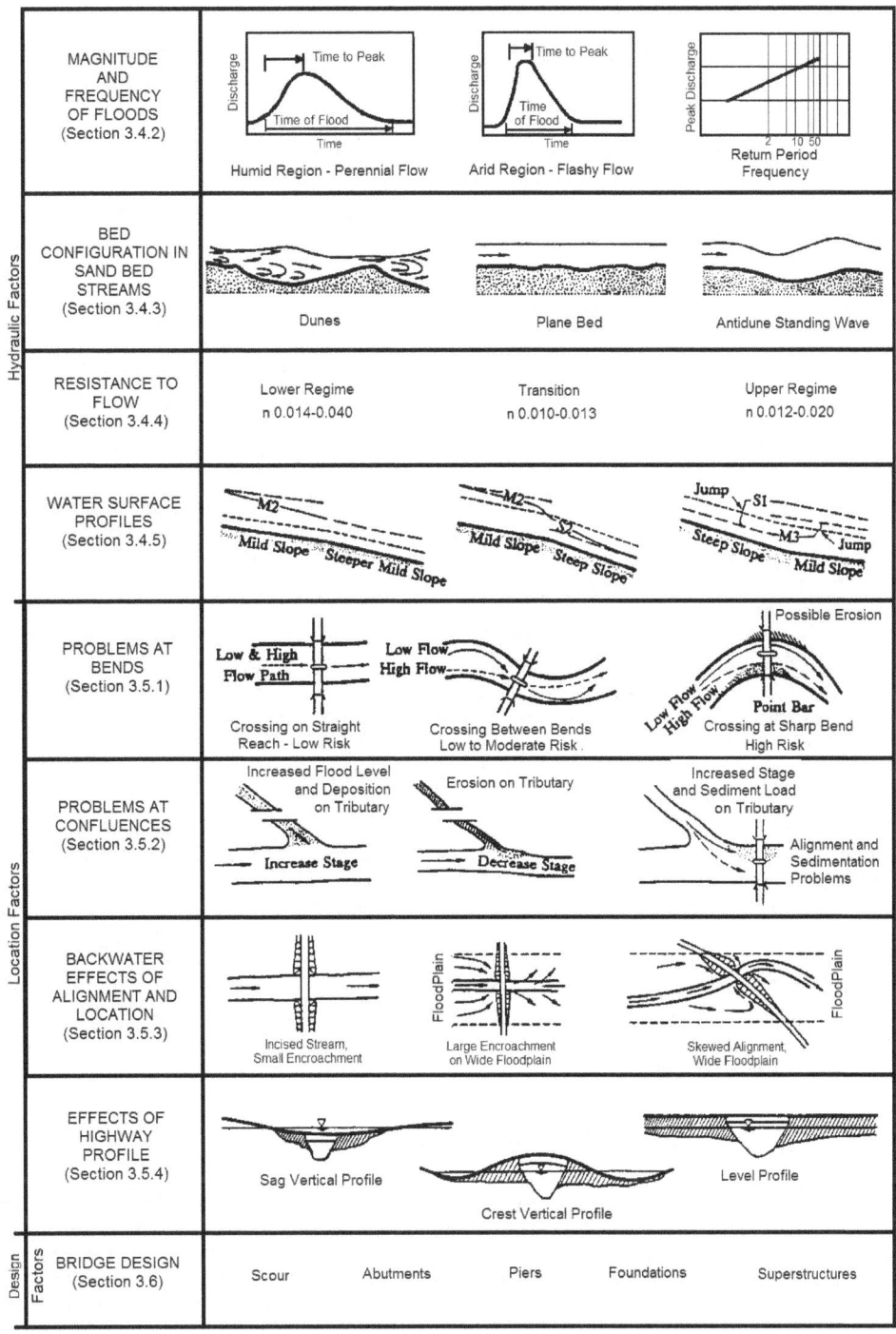

Figure 3.3. Hydraulic, location, and design factors that affect stream stability.

### 3.4.3 Bed Configurations in Sand-Bed Streams

In sand-bed streams, sand material is easily eroded and is continually being moved and shaped by the flow. The interaction between the flow of the water-sediment mixture and the sand-bed creates different bed configurations which change the resistance to flow, velocity, water surface elevation and sediment transport. Consequently, an understanding of the different types of bed forms that may occur and a knowledge of the resistance to flow and sediment transport associated with each bed form can help in analyzing flow in an alluvial channel. More specific to this discussion, it is necessary to understand what bed forms will be present so that the resistance to flow can be estimated and flood stages and water surface profiles can be computed.

Flow Regime. Flow in alluvial sand-bed channels is divided into two regimes separated by a transition zone (FHWA 2001). Forms of bed roughness in sand channels are shown in Figure 3.4a, while Figure 3.4b shows the relationships between water surface and bed configuration. The flow regimes are:

- The lower flow regime, where resistance to flow is large and sediment transport is small. The bed form is either ripples or dunes or some combination of the two. Water-surface undulations are out of phase with the bed surface, and there is a relatively large separation zone downstream from the crest of each ripple or dune. The velocity of the downstream movement of the ripples or dunes depends on their height and the velocity of the grains moving up their backs.

- The transition zone, where the bed configuration may range from that typical of the lower flow regime to that typical of the upper flow regime, depending mainly on antecedent conditions. If the antecedent bed configuration is dunes, the depth or slope can be increased to values more consistent with those of the upper flow regime without changing the bed form; or, conversely, if the antecedent bed is plane, depth and slope can be decreased to values more consistent with those of the lower flow regime without changing the bed form.

- Resistance to flow and sediment transport also have the same variability as the bed configuration in the transition. This phenomenon can be explained by the changes in resistance to flow and, consequently, the changes in depth and slope as the bed form changes.

- The upper flow regime, in which resistance to flow is small and sediment transport is large. The usual bed forms are plane bed or antidunes. The water surface is in phase with the bed surface except when an antidune breaks, and normally the fluid does not separate from the boundary.

- There is no direct relationship between the classification of upper and lower flow regime and Froude Number (supercritical/subcritical flow).

Effects of Bed Forms at Stream Crossings. At high flows, most sand-bed stream channels shift from a dune bed to a transition or a plane bed configuration. The resistance to flow is then decreased to one-half to one-third of that preceding the shift in bed form. The increase in velocity and corresponding decrease in depth may increase scour around bridge piers, abutments, spurs, or guide banks and may increase the required size of riprap. However, maximum scour depth with a plane bed can be less than with dunes because of the absence of dune troughs. On the other hand, the decrease in stage resulting from planing out of the bed will decrease the required elevation of the bridge, the height of embankments across the floodplain, the height of any dikes, and the height of any channel control works that may be needed. The converse is also true.

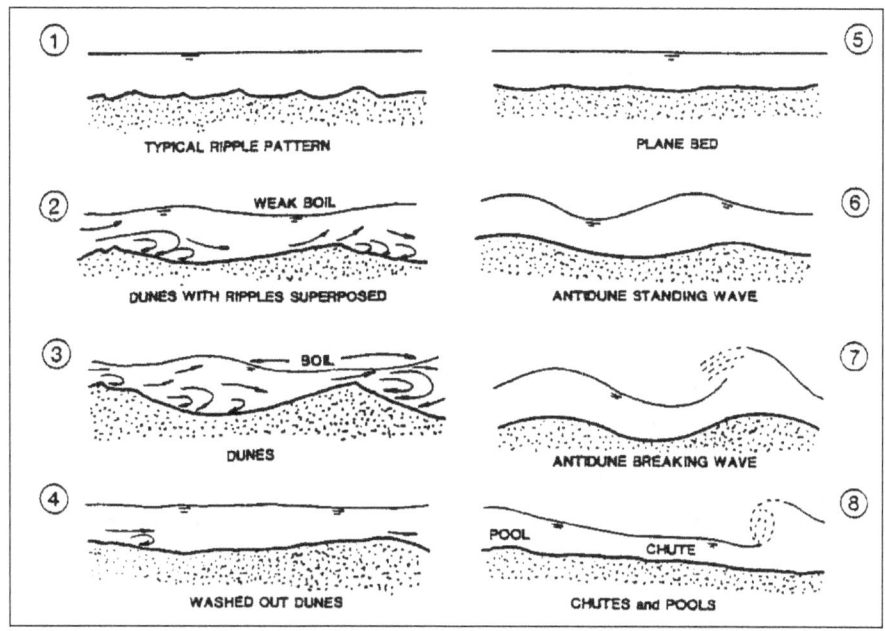

Figure 3.4(a). Forms of bed roughness in sand channels (FHWA 2001).

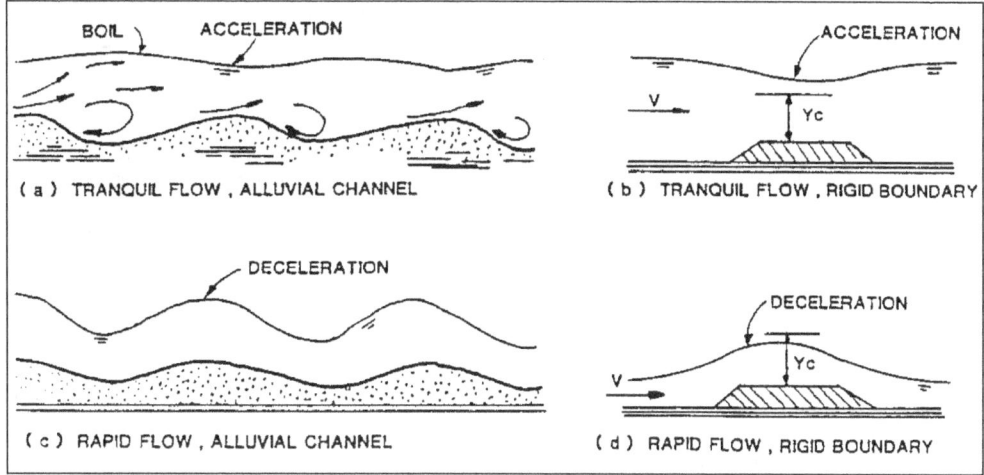

Figure 3.4(b). Relation between water surface and bed configuration (FHWA 2001).

Another effect of bed forms on highway crossings is that with dunes on the bed, there is a fluctuating pattern of scour on the bed and around piers. The average height of dunes is approximately one-third of the average depth of flow, and the maximum height of a dune may approach one-half the average depth of flow. With the passage of a dune through a bridge opening, an increase in local scour would be anticipated when the trough of the dune arrives at the bridge. It has been determined experimentally that local scour increases by 30% or more over equilibrium scour depth with the passage of a large dune trough (FHWA 2012a and b) (see also Section 7.4.3).

A very important effect of bed forms and bars is the change of flow direction in channels. At low flow, the bars can be residual and cause high velocity flow along or at a pier or other structures in the streambed, causing deeper than anticipated scour.

Care must be used in analyzing crossings of sand-bed streams in order to anticipate changes that may occur in bed forms and the impact of these changes on the resistance to flow, sediment transport, and the stability of the reach and highway structures. As described in Section 3.4.4, with a dune bed, the Manning n (see Section 3.3.3) could be as large as 0.040. Whereas, with a plane bed, the n value could be as low as 0.010. A change from a dune bed to a plane bed, or the reverse, can have an appreciable effect on depth and velocity. In the design of a bridge or a stream stability or scour countermeasure, it is good engineering practice to assume a dune bed (large n value) when establishing the water surface elevations, and a plane bed (low n value) for calculations involving velocity.

**3.4.4 Resistance to Flow**

Use of the Manning equation (Section 3.3.3) to compute flow in open channels and floodplains assumes one-dimensional flow. Procedures for summing the results of computations for subsections to obtain results for the total cross section involve use of the following assumptions: (1) mean velocity in each subsection is the same, (2) the total force resisting flow is equal to the sum of forces in the subsections, and (3) total flow in the cross section is equal to the sum of the flows in the subsections. This implies that the slope of the energy grade line is the same for each subsection (Figure 3.2). Assumption (3) is the basis for computing total conveyance for a cross section by adding conveyances of subsections (see Section 3.3.3).

Resistance to Flow in Channels. The general approach for estimating the resistance to flow in a stream channel is to select a base n value for materials in the channel boundaries assuming a straight, uniform channel, and then to make corrections to the base n value to account for channel irregularities, sinuosity, and other factors which affect the resistance to flow (Cowan 1956, FHWA 2001). Equation 3.9 is used to compute the equivalent material roughness coefficient "n" for a channel:

$$n = (n_b + n_1 + n_2 + n_3 + n_4)m \tag{3.9}$$

where:

$n_b$ = Base value for straight, uniform channel
$n_1$ = Value for surface irregularities in the cross section
$n_2$ = Value for variations in shape and size of the channel
$n_3$ = Value for obstructions
$n_4$ = Value for vegetation and flow conditions
$m$ = Correction factor for sinuosity of the channel

Table 3.1 provides base n values for stable channels and sand channels, while Table 3.2 provides adjustment factors for use in Equation 3.9. HDS 6 and Arcement and Schneider provide more detailed descriptions of conditions that affect the selection of appropriate values (FHWA 2001, USGS 1984).

<u>Resistance to Flow in Sand-Bed Channels</u>. The value of n varies greatly in sand-bed channels because of the varying bed forms that occur with lower and upper flow regimes. Figure 3.5 shows the relative resistance to flow in channels in lower regime, transition, and upper regime flow and the bed forms which exist for each regime.

Sand-bed channels with bed materials having a median diameter from 0.14 to 0.4 mm usually plane out during high flows. Manning n values change from as large as 0.040 at low flows to as small as 0.010 at high flow. Table 3.3 provides typical ranges of n values for sand-bed channels.

| Table 3.1. Base Values of Manning n ($n_b$). | | | | |
|---|---|---|---|---|
| Channel or Floodplain Type | Median Size, Bed Material | | Base n Value | |
| | Millimeters (mm) | Inches (in) | Benson and Dalrymple | Chow |
| Sand Channels* | 0.2 | -- -- | 0.012 | -- -- |
| | .3 | -- -- | 0.017 | -- -- |
| | .4 | -- -- | 0.020 | -- -- |
| | .5 | -- -- | 0.022 | -- -- |
| | .6 | -- -- | 0.023 | -- -- |
| | .8 | -- -- | 0.025 | -- -- |
| | 1.0 | -- -- | 0.026 | -- -- |
| **Stable Channels and Floodplains** | | | | |
| Concrete | -- -- | -- -- | 0.012 - 0.018 | 0.011 |
| Rock cut | -- -- | -- -- | -- -- | 0.025 |
| Firm soil | -- -- | -- -- | 0.025 - 0.032 | 0.020 |
| Coarse sand | 1 - 2 | -- -- | 0.026 - 0.035 | -- -- |
| Fine gravel | -- -- | -- -- | -- -- | 0.024 |
| Gravel | 2 - 64 | 0.08 – 2.5 | 0.028 - 0.035 | -- -- |
| Coarse gravel | -- -- | -- -- | -- -- | 0.026 |
| Cobble | 64 - 256 | 2.5 – 10.1 | 0.030 - 0.050 | -- -- |
| Boulder | > 256 | > 10.1 | 0.040 - 0.070 | -- -- |
| * For Sand Channels note only for upper regime flow where grain roughness is predominant | | | | |

## Table 3.2. Adjustment Factors for the Determination of n Values for Channels.

| n factor | Conditions | n Value | Remarks |
|---|---|---|---|
| $n_1$ | Smooth | 0 | Smoothest channel |
| | Minor | 0.001-0.005 | Slightly eroded side slopes |
| | Moderate | 0.006-0.010 | Moderately rough bed and banks |
| | Severe | 0.011-0.020 | Badly sloughed and scalloped banks |
| $n_2$ | Gradual | 0 | Gradual Changes |
| | Alternating Occasionally | 0.001-0.005 | Occasional shifts from large to small sections |
| | Alternating Frequently | 0.010-0.015 | Frequent changes in cross-sectional shape |
| $n_3$ | Negligible | 0-0.004 | Obstructions < 5% of cross-section area |
| | Minor | 0.005-0.015 | Obstructions < 15% of cross-section area |
| | Appreciable | 0.020-0.030 | Obstructions 15-50% of cross-section area |
| | Severe | 0.040-0.060 | Obstructions > 50% of cross-section area |
| $n_4$ | Small | 0.002-0.010 | Flow depth > 2 x vegetation height |
| | Medium | 0.010-0.025 | Flow depth > vegetation height |
| | Large | 0.025-0.050 | Flow depth < vegetation height |
| | Very Large | 0.050-0.100 | Flow depth < 0.5 vegetation height |
| m | Minor | 1.00 | Sinuosity < 1.2 |
| | Appreciable | 1.15 | 1.2 < Sinuosity < 1.5 |
| | Severe | 1.30 | Sinuosity > 1.5 |

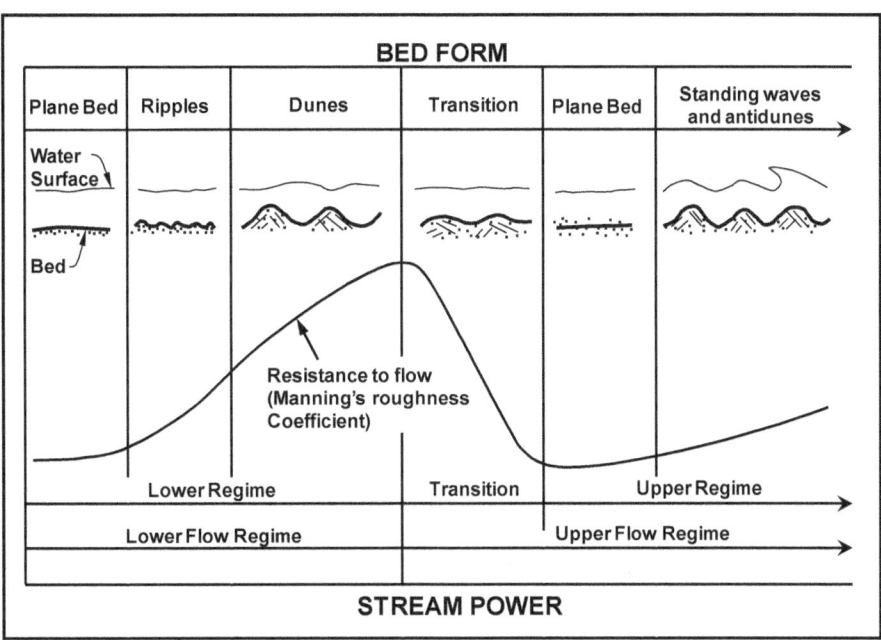

Figure 3.5. Relative resistance to flow in sand-bed channels (after USGS 1984).

| Table 3.3. Manning n ($n_b$) Roughness Coefficients for Alluvial Sand-bed Channels (no vegetation[1]). | | |
|---|---|---|
| | Bed Form | Manning n |
| **Lower Flow Regime** | Plane bed | 0.014 - 0.020 |
| | Ripples | 0.018 - 0.030 |
| | Dunes | 0.020 - 0.040 |
| **Transition** | Washed out dunes | 0.014 - 0.025 |
| **Upper Flow Regime** | Plane bed | 0.010 - 0.013 |
| | Standing Waves | 0.012 - 0.015 |
| | Antidunes | 0.012 - 0.020 |
| [1]Data are limited to sand channels with $D_{50} < 1.0$ mm. | | |

Resistance to Flow in Coarse Material Channels. A coarse material channel may range from a gravel bed channel up to the cobble-boulder channels typical of mountainous regions. The latter type channels may have bed material that is only partly submerged making it difficult to determine the channel roughness. However, for gravel and small cobble and boulder-bed channels analysis of data from many rivers, canals and flumes shows that channel roughness can be predicted by the following equation (NCHRP 1970):

$$n = K_u \, D_{50}^{1/6} \tag{3.10}$$

where:

$D_{50}$ is measured in ft (m)

$K_u$ = 0.0395 English
$K_u$ = 0.0482 SI

Alternately, Limerinos developed Equation 3.11 from samples on streams having bed materials ranging in size from small gravel to medium size boulders (USGS 1970).

$$n = \frac{K_u R^{1/6}}{1.16 + 2.08 \log\left(\frac{R}{D_{84}}\right)} \tag{3.11}$$

where:

$R$ = Hydraulic radius, ft (m)
$D_{84}$ = 84th percentile (percent finer) size of bed material, ft (m)
$K_u$ = 0.0926 English
$K_u$ = 0.113 SI

Flow depth, Y, may be substituted for the hydraulic radius, R in wide channels (W/Y > 10). Note that Equation 3.11 also applies to sand-bed channels in upper regime flow (USGS 1984).

The alternative to use of Equations 3.10 or 3.11 for gravel-bed streams is to select a value of n from Table 3.1. Because of the range of values in the table, it would be advisable to verify the selected value by use of one of the above equations if flow depth or velocities will significantly affect a design. HDS 6 (FHWA 2001) also gives equations for this case and Chapter 8 provides additional discussion of fluvial processes in gravel-bed rivers.

<u>Resistance to Flow on Floodplains</u>. Arcement and Schneider modified Equation 3.9 for channels to make it applicable for the estimation of n values for floodplains (USGS 1984). The correction factor for sinuosity, m, becomes 1.0 for floodplains, and the value for variations in size and shape, $n_2$, is assumed equal to zero. Equation 3.9, adapted for use on floodplains, becomes:

$$n = n_b + n_1 + n_3 + n_4 \tag{3.12}$$

where:

$n_b$ = Base value of n for a bare soil surface
$n_1$ = Value to correct for surface irregularities
$n_3$ = Value for obstructions
$n_4$ = Value for vegetation

Selection of the base n value for floodplains is the same as for channels. The USGS Water Supply Paper 2339 is recommended for a detailed discussion of factors which affect flow resistance in floodplains (USGS 1984).

## 3.4.5 Water Surface Profiles

The water surface profile in a stream or river is a combination of gradually varied flow over long distances, and rapidly varied flow over short distances. Due to various obstructions in the flow, such as bridges, the actual flow depth over longer reaches is either larger or smaller than the normal depth defined by Manning's uniform flow equation. In the immediate vicinity of the obstruction, the flow can be rapidly varied.

Gradually Varied Flow. In gradually varied flow, changes in depth and velocity take place slowly over a large distance, resistance to flow dominates and acceleration forces are neglected. The calculation of a gradually varied flow profile is well defined by analytical procedures (e.g., see HDS 6 and HDS 7), which can be implemented manually or more commonly by computer programs such as the U.S. Army Corps of Engineers (USACE) HEC River Analysis System (RAS) (FHWA 2001, FHWA 2012a, and USACE 2010). A qualitative analysis of the general characteristics of the backwater curve is often useful prior to quantitative evaluation. Such an analysis requires locating control points, determining the type of profile upstream and downstream of the control points, and then sketching the backwater curves. For example, Figure 3.3 illustrates several typical profiles that would result from a control represented by a change in bed slope. HDS 6 and HDS 7 provide detailed discussions of water surface profiles for gradually varied flow (FHWA 2001 and 2012a).

Rapidly Varied Flow. In rapidly varied flow, changes in depth and velocity take place over short distances, acceleration forces dominate and resistance to flow may be neglected. The calculation of certain types of rapidly varied flow are well defined by analytical procedures, such as the analysis of hydraulic jumps, but analysis of other types of rapidly varied flow, such as flow through bridge openings (Figure 3.6) are a combination of analytical and empirical relationships. HDS 1 provides a procedure for manual calculation of the backwater created by certain types of flow conditions at bridge openings (FHWA 1978c). Gradually varied flow computer programs, such as HEC-RAS include analysis of bridge backwater, but do not calculate undular jump conditions or the flow through the bridge when flow accelerations are large, that is, large change in velocity either in magnitude or direction (USACE 2010).

Superelevation of Water Surface at Bends. Because of the change in flow direction which results in centrifugal forces, there is a superelevation of the water surface in bends. The water surface is higher at the concave bank than at the convex bank (Figure 3.7). The total superelevation is measured from waters edge to waters edge. Half this amount is added to the average water surface elevation to obtain the water surface elevation at the concave (outside) bank. The resulting transverse slope can be evaluated quantitatively. By assuming velocity equal to average velocity, the following equation was derived for superelevation for subcritical flow (Woodward 1920): Other equations for superelevation are given in HDS 6 (FHWA 2001).

$$\Delta Z = \frac{V^2}{gr_c}(r_o - r_i) \tag{3.13}$$

where:

| | | |
|---|---|---|
| $g$ | = | Acceleration of gravity, 32.2 ft/s² (9.81 m/s²) |
| $r_o$ | = | Radius of outside bank at bend, ft (m) |
| $r_i$ | = | Radius of inside bank, ft (m) |
| $r_c$ | = | Radius of center of stream, ft (m) |
| $\Delta Z$ | = | Difference in water surface elevation between concave and convex banks, ft (m) |
| $V$ | = | Average velocity, ft/s (m/s) |

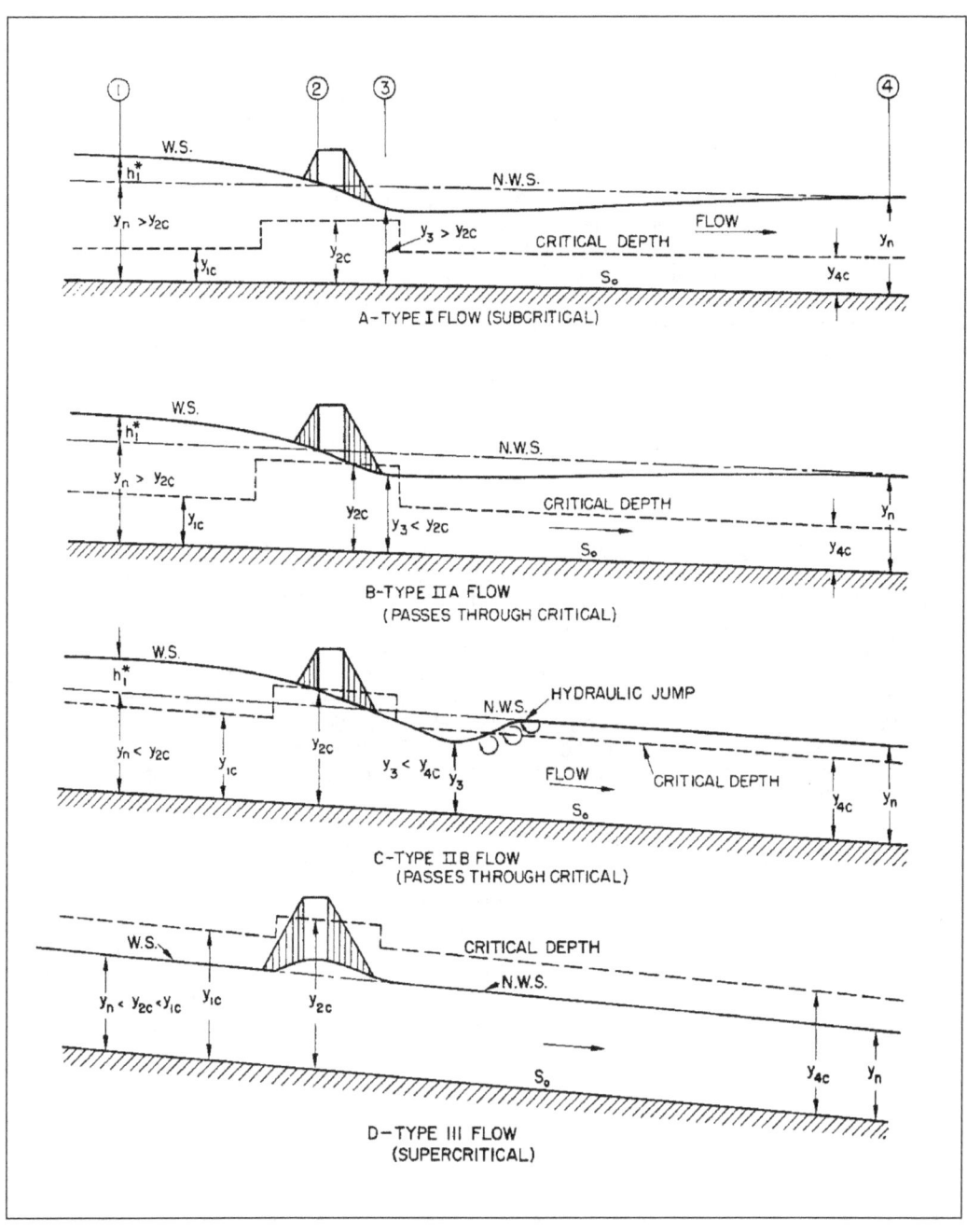

Figure 3.6. Types of water surface profiles through bridge openings (after Bradley (FHWA 1978c)).

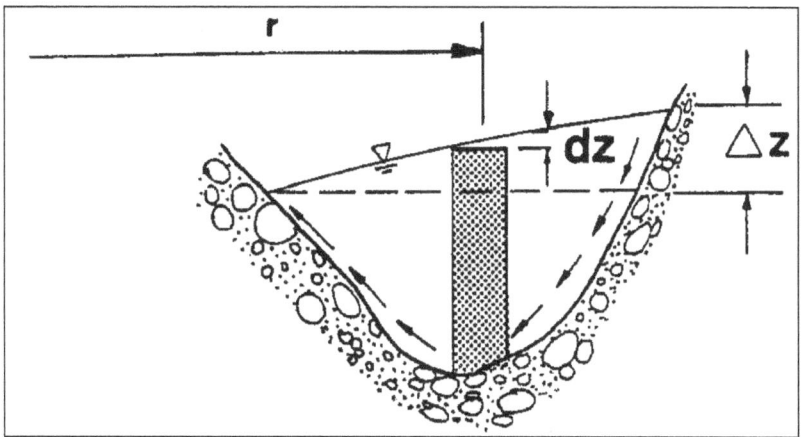

Figure 3.7. Superelevation of water surface in a bend.

## 3.5 GEOMETRY AND LOCATION OF HIGHWAY STREAM CROSSINGS

### 3.5.1 Problems at Bends

The location of a highway stream crossing is important because of the inherent instability of streams at some locations (see Chapter 2) and because the crossing system can contribute to instability. In general, a crossing on a straight reach is preferred because stability problems are usually minor. Low-flow and high-flow paths (thalwegs) are generally similar for a straight reach, reducing the risk of problems related to alignment and orientation of bridge piers and superstructures (Figure 3.3).

For a relatively stable meandering stream, a bridge crossing at the inflection point between bends generally reduces the risk of instability problems. At the inflection point, the low-flow and high-flow paths are comparable (Figure 3.3) and the crossing is in a zone where deposition and erosion are usually moderate. However, countermeasures against meander migration may still be required.

More hydraulic problems occur at alluvial stream crossings at or near bends than at all other locations because bends are naturally unstable. In addition, ice and floating debris tend to create greater problems in bends than in straight reaches. Other problems at bends include the shifting of the thalweg which can result in unanticipated scour at piers because of changes in flow direction and velocities, and nonuniform velocity distribution which could cause scour of the bed and bank at the outside of the bend and deposition in the inside of the bend (Figure 3.3). The high velocities at the outside of the bend or downstream of the bend can contribute substantially to local scour on abutments and piers.

### 3.5.2 Problems at Confluences

Hydraulic problems may also be experienced at crossings near stream confluences. Crossings of tributary streams are affected by the stage of the main stream (See Chapter 2). Aggradation of the channel of the tributary may occur if the stage of the main stream is high during a flood on the tributary, and scour in the tributary may occur if the stage in the main stream is low. Similarly, problems at a crossing of the larger stream can result from varying flow distribution and flow direction at various stages in the stream and its tributary, and from sediment deposited in the stream by the tributary (Figure 3.3). Tributaries entering the main channel upstream of a main channel bridge can also cause varying flow distribution and direction at various stages (flows) in the main channel and the tributary.

### 3.5.3 Backwater Effects of Alignment and Location

As flow passes through a channel constriction, most of the energy losses occur as expansion losses downstream of the contraction. This loss of energy is reflected by a rise in the water surface and the energy line upstream from the constriction. Upstream of bridges, the rise in water level above the normal water surface (that which would exist without the bridge) is referred to as the bridge backwater (Figure 3.6). However, many bridges do not cause backwater even at high flows even though they constrict the flows (FHWA 2001). Hydraulic engineers are concerned with backwater with respect to flooding upstream of the bridge; backwater elevation with respect to the highway profile; and the effects on sediment deposition upstream, scour around embankments, contraction scour due to the constriction, and local scour at piers.

The effects of highway-stream crossing alignments on backwater conditions shown in Figure 3.8 are based on:

- Backwater resulting from a long skewed or curved roadway embankment (Figure 3.8a) may be quite large for wide floodplains. In effect, the bridge opening is located up-valley from one end of the embankment and the water level at the downstream extreme of the approach roadway, as at point A in Figure 3.8a, can be significantly higher than at the bridge.

- Backwater in an incised stream channel without substantial overbank flow (Figure 3.8b) is seldom large, but contraction and local scour may be severe. Backwater results from encroachment in the channel by approach embankments and from piers located in the channel.

- Backwater resulting from a normal crossing of the valley where road approach embankments block overbank flow (Figure 3.8c) may be significant. General and local scour may be severe if a significant quantity of flow is diverted from the floodplain to the bridge waterway.

### 3.5.4 Effects of Highway Profile

A highway stream crossing is a system consisting of the stream and its floodplain, the bridge(s) and the approach roadways on the floodplain. All floods which occur during the life of the crossing system will pass either through the bridge waterways provided or through the waterways and over the highway. The highway profile and alignment control the quantity of flow which must pass through waterway openings. Flood frequency should be considered in the design of bridge components and may influence highway profile and alignment.

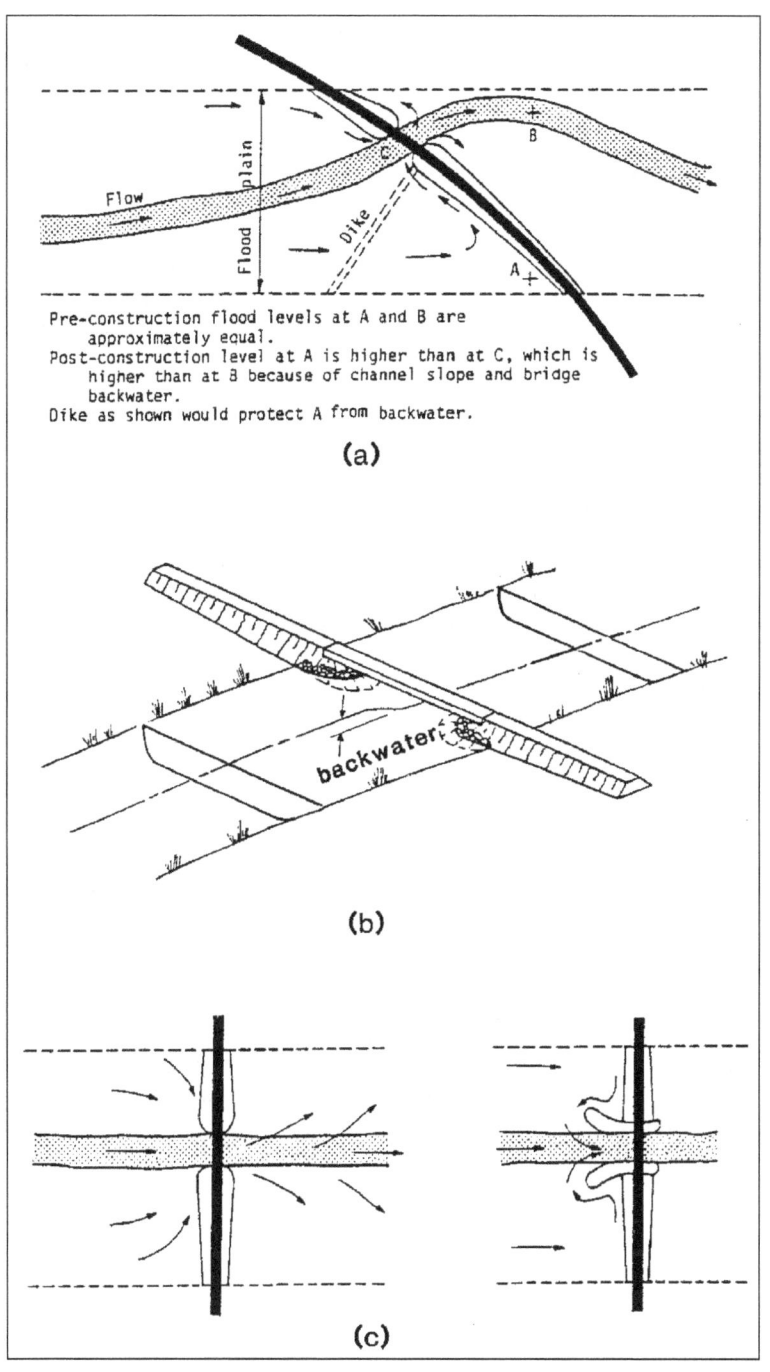

Figure 3.8. Backwater effect associated with three types of stream crossings: (a) a skewed alignment across a floodplain, (b) constriction of channel flow, and (c) constriction of overbank flow (after Neill 1972).

The stage-discharge relationship for the stream and backwater associated with a crossing design are the hydraulic considerations for establishing the highway profile. Profile alternatives are dependent on site topography and other site constraints, such as land use, traffic requirements, and flood damage potential. Figures 3.9a, b, and c illustrate profile alternatives, namely, a sag vertical curve, a crest vertical curve on the bridge or a rolling profile, and a level profile. A distinctive aspect of the sag vertical curve, as depicted in Figure 3.9a, is the certainty that the bridge structure will be submerged before overflow of the roadway will occur. Therefore, the magnitude and probability of occurrence of such a flood event should be considered in the design of the waterway opening and bridge components. A variation of the sag vertical curve where the low point of the curve is located on a floodplain rather than on the bridge affords relief to the bridge waterway. Bridges on level profiles and sag vertical curves are susceptible to debris accumulation on the superstructure, impact forces, buoyant forces, and accentuated contraction and local scour.

The crest vertical profile illustrated in Figure 3.9b provides protection to the bridge in that flood events exceeding the stage of the low point in the sag vertical curve will, in part, flow over the roadway. This relieves the bridge and the bridge waterway of stresses to which bridges on sag vertical curves and level profiles are subjected.

Regardless of the profile, when the superstructure is submerged (pressure flow through the bridge), pier scour is increased. In some cases the local scour with pressure flow will be two to three times deeper than for free flow (FHWA 2012b).

## 3.6 BRIDGE DESIGN

The design of bridge components must consider the effects on the local stability of a stream because of scour caused by the bridge encroachment on the stream (Figure 3.3). It is prudent to utilize designs which minimize undesirable stream response, to the extent practicable. This applies to component design as well as to the design of the total crossing system, including countermeasures against stream instability. The term countermeasure, as used here, is not necessarily an appurtenance to the highway stream crossing, but may be an integral part of the highway or bridge (for further discussion see HEC-23 (FHWA 2009)).

The location and size of bridge openings influence stream stability. Encroachment in the stream channel by abutments and piers reduces the channel section and may cause significant contraction scour. Severe constriction of floodplain flow may cause approach embankment failures and serious contraction scour in the bridge waterway. Auxiliary (relief) openings should be carefully designed to avoid excessive diversion of floodplain flow to main channel bridge openings on wide floodplains and at skewed crossings of floodplains.

### 3.6.1 Scour at Bridges

Scour at bridges consists of three components: (1) long-term aggradation or degradation of the stream channel (natural or human-induced), (2) contraction scour due to constriction or the location of the bridge, and (3) local scour. In general, the three components are additive (for further discussion see HEC-18 (FHWA 2012b).

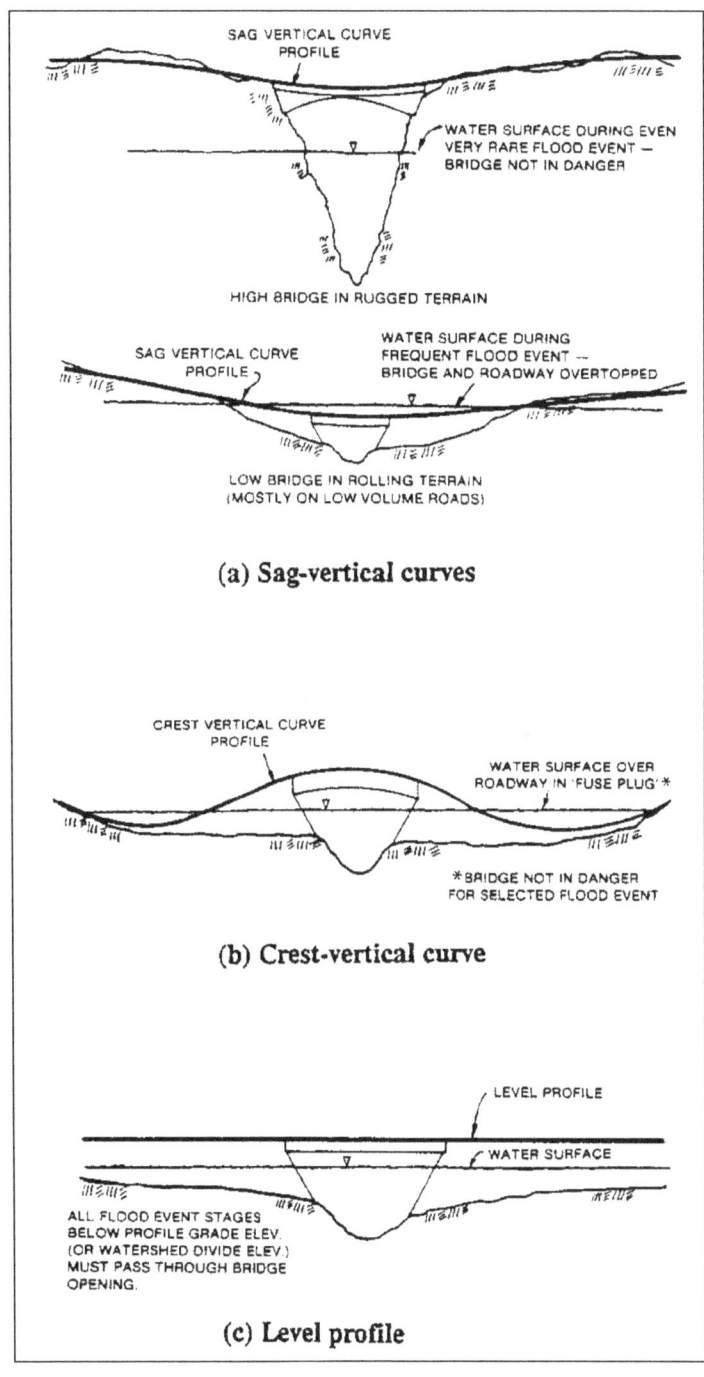

Figure 3.9. Various highway profiles: (a) sag-vertical curves, (b) crest-vertical curve, and (c) level profile (AASHTO 1982).

Scour can be related to the following factors: (1) channel slope and alignment, (2) channel shifting, (3) bed sediment size distribution, (4) antecedent floods and surging phenomena, (5) accumulation of debris, logs, or ice, (6) flow contraction, flow alignment, and flow depth, (7) pier and abutment geometry and location, (8) type of foundation, (9) natural or man-induced modification of the stream, and (10) failure of a nearby structure.

The rate of scour depends on the erosive forces exerted on the channel boundary and the resistance of the material to erosion. Resistance to erosion in fine cohesive material results from molecular forces. Resistance in noncohesive material depends primarily on bed sediment size distribution and density.

Under steady flow conditions, scour processes gradually approach an equilibrium condition; however, equilibrium scour conditions are not necessarily attained during a single flow event. Bridge crossings are generally subjected to unsteady flow conditions, and a series of events are often required to reach equilibrium or maximum scour depth. Deposition often occurs during the recession of the hydrograph, and the maximum scour depth measured after the flood is generally less than the maximum depth of scour reached during the flood event.

Gravel mining in the streambed can cause severe stream instability. Therefore, it is essential to monitor sand and gravel mining so that countermeasures can be installed to stabilize the stream in the vicinity of a highway facility. Where possible, mining should be managed so that instabilities in the stream system will be minimized (see additional discussion in Section 2.4.3).

Methods and equations for determining scour at piers and abutments are given in HEC-18 (FHWA 2012b). Countermeasures for stream instability, pier scour, and abutment scour are discussed in HEC-23 (FHWA 2009).

### 3.6.2 Abutments

Bridge abutments are classified as spill-through, vertical wall, or vertical wall with wingwalls. Abutments are susceptible to damage by scour depending on flow distribution, foundation materials, velocities and other factors. However, scour at spill-through abutments is about 50 percent smaller than at vertical wall abutments subjected to the same scouring actions.

In addition to the effects of abutment shape, scour at abutments is affected by the skew of approach flow at the abutment, soils materials, encroachment on the floodplain and in the channel, and the amount of overbank flow diverted to the bridge waterway by approach fills to the bridge. Equations and methods for computing abutment scour are presented in HEC-18 (FHWA 2012b).

### 3.6.3 Piers

The number of piers in any stream channel should be limited to a practical minimum and, if possible, piers should not be located in the channel of small streams. Piers properly oriented with the flow do not contribute significantly to bridge backwater, but they can contribute to contraction scour. Piers should be aligned with flow direction at flood stage in order to minimize the opportunity for debris collection, to reduce the contraction effect of piers in the waterway, to minimize ice forces and the possibility of ice dams forming at the bridge, and to minimize backwater and local scour.

Pier orientation is difficult where flow direction changes with stage or time (see Figure 3.3 - Problems at Bends). Cylindrical piers or some variation thereof, are probably the best alternative if orientation is critical. A solid pier will not collect as much debris as a pile bent or a multiple-column bent. Rounding or streamlining the leading edges of piers helps to decrease the accumulation of debris and reduces local scour. Recent studies have provided additional data on the effects of footings and the behavior of pile groups (FHWA 1989a). Guidance pertaining to pier foundations is presented in HEC-18 (FHWA 2012b).

Piers located on a bank or in the stream channel near the bank are likely to cause lateral erosion of the bank. Piers located near the streambank on the floodplain are vulnerable to undermining by bank scour and meander migration. Piers which must be placed in such locations should be founded at elevations safe from undermining (FHWA 2012b).

### 3.6.4 Bridge Foundations

The types of foundations used for bridges include piles, piles with pile caps, spread footings, footings on piles or drilled shafts, drilled shafts, and caissons. Spread footings are used where sound rock is relatively shallow, but failures have occurred where spread footings were set in erodible rock.

Piling usually are dependent on the surrounding material for skin friction and lateral stability. In some locations, they can be carried to bedrock or other dense materials for bearing capacity. Tip elevation for piling should be based on estimates of potential scour depths as well as bearing in order to avoid losing lateral support and load carrying capacity after scour. Pile bearing capacity derived from driving records has little validity if the material through which the piles were driven is scoured away during a flood.

Caissons are used in large rivers and are usually sunk to dense material by excavation inside the caisson. Founding depths are such that scour is not usually a problem after construction is completed; however, severe contraction scour has developed at some bridges, because of contraction of flow from the large piers.

Attention should be given to potential scour resulting from channel shifts in designing foundations on floodplains. Also, the thalweg in channels should not be considered to be in a fixed location. Consideration should be given, therefore, to duplicating the foundation elevations of the main channel piers on adjacent floodplain piers. The history of stream channel activity can be very useful in establishing foundation elevations (see Chapter 2).

### 3.6.5 Superstructures

Hydraulic forces that should be considered in the design of a bridge superstructure include buoyancy, drag, and impact from ice and floating debris (for discussion, see HEC-18, Chapter 2 (FHWA 2012b)). The configuration of the superstructure should be influenced by the highway profile, the probability of submergence, expected problems with ice and debris, and flow velocities, as well as the usual economic, structural and geometric considerations. Superstructures over waterways should provide structural redundancy, such as continuous spans (rather than simple spans). The catastrophic bridge failures on Schoharie Creek and the Hatchie River due to scour and stream instability involved non-redundant bridges (NTSB 1988 and 1990).

(page intentionally left blank)

# CHAPTER 4

# ANALYSIS PROCEDURES FOR STREAM INSTABILITY

## 4.1 INTRODUCTION

A stable stream does not change in size, form, or position with time; however, all alluvial channels change to some extent and are somewhat unstable. For highway engineering purposes, a stream channel can be considered unstable if the rate or magnitude of change is great enough that the planning, location, design, or maintenance considerations for a highway encroachment are significantly affected. The kinds of changes that are of concern are:

- Lateral bank erosion, including the erosion that occurs from meander migration
- Aggradation or degradation of the streambed that progresses with time
- Short-term fluctuations in streambed elevation that are usually associated with the passage of a flood (scour and fill)

These changes are associated with instability in a stream system or in an extensive reach of stream.

Local instability caused by the construction of a highway crossing or encroachment on a stream is also of concern. This includes scour caused by contraction of the flow and local scour due to the disturbance of streamlines at an object in the flow, such as at a pier or an abutment, or the passage of bed forms (ripples and dunes). The purpose of this chapter is to outline the analysis procedures that may be utilized to evaluate stream instability. These analysis procedures provide details on many of the general analysis steps of the comprehensive analysis flow chart of Figure 1.1.

## 4.2 GENERAL SOLUTION PROCEDURE

The analysis of any complex problem should begin with an overview or general evaluation, including a qualitative assessment of the problem and its solution. This fundamental initial step should be directed towards providing insight and understanding of significant physical processes, without being too concerned with the specifics of any given component of the problem. The understanding generated from such analyses assures that subsequent detailed analyses are properly designed.

The progression to more detailed analyses should begin with application of basic principles, followed as required, with more complex solution techniques. This solution approach, beginning with qualitative analysis, proceeding through basic quantitative principles and then utilizing, as required, more complex or state-of-the-art solution procedures assures that accurate and reasonable results are obtained while minimizing the expenditure of time and effort.

The inherent complexities of a stream stability analysis, further complicated by highway stream crossings, require such a solution procedure. The evaluation and design of a highway stream crossing or encroachment should begin with a qualitative assessment of stream stability. This involves application of geomorphic concepts to identify potential problems and alternative solutions. This analysis should be followed with quantitative analyses using basic hydrologic, hydraulic and sediment transport engineering concepts.

Such analyses could include evaluation of flood history, channel hydraulic conditions (up to and including, for example, water surface profile analysis) and basic sediment transport analyses such as evaluation of watershed sediment yield, incipient motion analysis and scour calculations. This analysis can be considered adequate for many locations if the problems are resolved and the relationships among different factors affecting stability are adequately explained. If not, a more complex quantitative analysis based on detailed mathematical modeling and/or physical hydraulic models should be considered.

In summary, the general solution procedure for analyzing stream stability could involve the following three levels of analysis:

Level 1: Application of Simple Geomorphic Concepts and other Qualitative Analyses
Level 2: Application of Basic Hydrologic, Hydraulic and Sediment Transport Engineering Concepts
Level 3: Application of Mathematical or Physical Modeling Studies

## 4.3 DATA NEEDS

The types and detail of data required to analyze a highway crossing or encroachment on a stream channel are highly dependent on the relative instability of the stream and the depth of study required to obtain adequate resolution of potential problems. More detailed data are needed where quantitative analyses are necessary, and data from an extensive reach of stream may be required to resolve problems in complex and high risk situations.

### 4.3.1 Data Needs for Level 1 Qualitative and Other Geomorphic Analyses

The data required for preliminary stability analyses include maps, aerial photographs, notes and photographs from field inspections, historic channel profile data, information on man's activities, and changes in stream hydrology and hydraulics over time.

The National Bridge Inspection Standards (NBIS) Program requires inspections on a 2-year cycle of the approximately 600,000 bridges on the National Bridge Inventory. The FHWA publication the "Recording and Coding Guide for the Structure Inventory and Appraisal of the Nation's Bridges" specifies the bridge and channel hydraulics and scour data that are evaluated and reported within the NBIS (FHWA 1995). Item 60, substructure, Item 61, Channel and Channel Protection, Item 71, Waterway Adequacy, and Item 113, Scour Critical Bridges, are among the items reported in the NBIS. These items can be used to aid the highway engineer in generating data needed for analysis.

Typically, a cross section of the bridge waterway at the time of each inspection will provide a chronological picture of changes in the bridge waterway. Area, vicinity, site, geologic, soils, and land use maps each provide essential information. Unstable stream systems upstream or downstream of the encroachment site can cause instability at the bridge site. Area maps are needed to locate unstable reaches of streams relative to the bridge site. Vicinity maps help to identify more localized problems. They should include a sufficient reach of stream to permit identification of stream characteristics, and to locate bars, braids, and channel controls. Site maps are needed to determine factors that influence local stability and flow alignment, such as bars and tributaries. Geologic maps provide information on deposits and rock formations and outcrops that control stream stability. Soils and land use maps provide information on soil types, vegetative cover, and land use which affect the character and availability of sediment supply.

Aerial photographs record much more ground detail than maps and are generally available at frequent intervals. This permits measurement of the rate of progress of bend migration and other stream changes that cannot be measured from maps made less frequently. A highway agency should periodically obtain aerial photographs of actively unstable streams that threaten highway facilities, including immediately after major floods. However, aerial photographs taken after the passage of an ice jam or immediately after a major flood must be interpreted with care since they may provide misleading information regarding the rate of change.

Notes and photographs from field inspections are important to gain an understanding of stream stability problems, particularly local stability. Field inspections should be made during high- and low-flow periods to record the location of bank cutting or sloughing and deposition in the channel. Flow directions should be sketched, signs of aggradation or degradation noted, properties of bed and bank materials estimated or measured, and the locations and implications of impacting activities recorded.

If historic stream profile data are available, it will provide information on channel stability. Stage trends at stream gaging stations and comparisons of streambed elevations with elevations before construction at structures will provide information on changes in stream profile. As-built bridge data and cross sections are frequently useful. Structure-induced scour should be taken into consideration where such comparisons are made.

Human activities in a watershed are frequently the cause of stream instability. Information on urbanization, land clearing, snagging in stream channels, channelization, bend cutoffs, streambed mining, dam construction, reservoir operations, navigation projects, and other activities, either existing or planned, are necessary to evaluate the impact on stream stability.

Data on changes in morphology are important because change in a stream rarely occurs at a constant rate. Stream instability can often be associated with an event, such as an extreme flood or a particular activity in the watershed or stream channel. If association is possible, the rate of change can be more accurately assessed. Similarly, information on changes in hydrology or hydraulics can sometimes be associated with activities that caused the change. Where changes in stream hydraulics are associated with an activity, changes in stream morphology are also likely to have occurred.

### 4.3.2 Data Needs for Level 2 Basic Engineering Analyses

Data requirements for basic hydrologic, hydraulic and sediment transport engineering analyses are dependent on the types of analyses that must be completed. Hydrologic data needs include dominant discharge (or bankfull flow), flow duration curves, and flow frequency curves. Discussion of hydrologic methods is beyond the scope of this manual; however, information can be obtained from the FHWA publication HDS 2 (FHWA 2002) and Department of Transportation manuals. Hydraulic data needs include cross sections, channel and bank roughness estimates, channel alignment, and other data for computing channel hydraulics, up to and including water surface profile calculations. Analysis of basic sediment transport conditions requires information on land use, soils, and geologic conditions, sediment sizes in the watershed and channel, and available measured sediment transport rates (e.g., from USGS gaging stations).

More detailed quantitative analyses require data on the properties of bed and bank materials and, at times, field data on bed-load and suspended-load transport rates. Properties of bed and bank materials that are important to a study of sediment transport include size, shape, fall velocity, cohesion, density, and angle of repose.

Chapter 5 outlines stream reconnaissance techniques and provides checklists that can assist in obtaining and organizing much of the data needed for Level 1 and Level 2 analyses. Chapter 5 also provides reference to rapid assessment procedures that will support a preliminary evaluation of potential scour and channel stability problems using limited site data. Chapter 6 contains additional quantitative techniques that will assist in determining the extent of lateral and vertical instability problems and in-channel stability analysis.

### 4.3.3 Data Needs for Level 3 Mathematical and Physical Model Studies

Application of mathematical and physical model studies requires the same basic data as a Level 2 analysis, but typically in much greater detail. For example, water and sediment routing by mathematical models (e.g., HEC-RAS), and construction of a physical model, would both require detailed channel cross-sectional data (USACE 2010). The more extensive data requirements for either mathematical or physical model studies, combined with the additional level of effort needed to complete such studies, results in a relatively large scope of work.

### 4.4 DATA SOURCES

Preliminary stability data may be available from government agencies such as the USACE, Natural Resources Conservation Service (formerly Soil Conservation Service, SCS), USGS, local river basin commissions, and local watershed districts. These agencies may have information on historic streambed profiles, stage-discharge relationships, and sediment load characteristics. They may also have information on past and planned activities that affect stream stability. Table 4.1 provides a list of sources for the various types of data that may be useful for assessing stream stability at a site. Table 4.2 provides a supplementary list of Internet data sources.

### 4.5 LEVEL 1: QUALITATIVE GEOMORPHIC ANALYSES

A flow chart of the typical steps in qualitative geomorphic analyses is provided in Figure 3.1. The six steps are generally applicable to most stream stability problems. As shown on Figure 4.1, the qualitative evaluation leads to a conclusion regarding the need for more detailed (Level 2) analysis or a decision to complete a screening or evaluation based on the Level 1 analysis. A Level 1 qualitative analysis is a prerequisite for a Level 2 engineering analysis for bridge design, evaluation, or rehabilitation (see also Chapter 1, Figure 1.1).

### 4.5.1 Step 1. Define Stream Characteristics

The first step in stability analysis is to identify stream characteristics according to the factors discussed in Chapter 2, Geomorphic Factors and Principles. Defining the various geomorphic characteristics of the stream provides insight into stream behavior and response (see Chapter 5 for additional stream channel reconnaissance and classification techniques).

Table 4.1. List of Data Sources (after FHWA 2001).

**Topographic Maps:**

(1) Quadrangle maps - U.S. Department of the Interior, Geological Survey, Topographic Division; and U.S. Department of the Army, Army Map Service.

(2) River plans and profiles - U.S. Department of the Interior, Geological Survey, Conservation Division.

(3) National parks and monuments - U.S. Department of the Interior, National Park Service.

(4) Federal reclamation project maps - U.S. Department of the Interior, Bureau of Reclamation.

(5) Local areas - commercial aerial mapping firms.

(6) American Society of Photogrammetry.

**Planimetric Maps:**

(1) Plats of public land surveys - U.S. Department of the Interior, Bureau of Land Management

(2) National forest maps - U.S. Department of Agriculture, Forest Service.

(3) County maps - State DOTs.

(4) City plats - city or county recorder.

(5) Federal reclamation project maps - U.S. Department of the Interior, Bureau of Reclamation.

(6) American Society of Photogrammetry.

(7) ASCE Journal - Surveying and Mapping Division.

Table continues

| Table 4.1. List of Data Sources (after FHWA 2001) continued. |
|---|
| **Aerial Photographs:** |
| (1)     The following agencies have aerial photographs of portions of the United States: U.S. Department of the Interior, Geological Survey, Topographic Division; U.S. Department of Agriculture, Commodity Stabilization Service, Soil Conservation Service and Forest Service; U.S. Air Force; various state agencies; commercial aerial survey; National Oceanic and Atmospheric Administration; and mapping firms. <br><br> (2)     American Society of Photogrammetry. <br><br> (3)     Photogrammetric Engineering. <br><br> (4)     Earth Resources Observation System (EROS) - Photographs from Gemini, Apollo, Earth Resources Technology Satellite (ERTS) and Skylab. |
| **Transportation Maps:** |
| (1)     State DOTs |
| **Triangulation and Benchmarks:** |
| (1)     State Engineer. <br><br> (2)     State DOTs |
| **Geologic Maps:** |
| (1)     U.S. Department of the Interior, Geological Survey, Geologic Division; and state geological surveys or departments. (Note - some regular quadrangle maps show geological data also). |
| **Soils Data:** |
| (1)     County soil survey reports - U.S. Department of Agriculture, Soil Conservation Service. <br><br> (2)     Land use capability surveys - U.S. Department of Agriculture, Soil Conservation Service. <br><br> (3)     Land classification reports - U.S. Department of the Interior, Bureau of Reclamation. <br><br> (4)     Hydraulic laboratory reports - U.S. Department of the Interior, Bureau of Reclamation. |
| Table continues |

Table 4.1. List of Data Sources (after FHWA 2001) continued.

**Climatological Data:**

(1) National Weather Service Data Center.

(2) Hydrologic bulletin - U.S. Department of Commerce, National Oceanic and Atmospheric Administration.

(3) Technical papers - U.S. Department of Commerce, National Oceanic and Atmospheric Administration.

(4) Hydro-meteorological reports - U.S. Department of Commerce, National Oceanic and Atmospheric Administration; and U.S. Department of the Army, Corps of Engineers.

(5) Cooperative study reports - U.S. Department of Commerce, National Oceanic and Atmospheric Administration; and U.S. Department of the Interior, Bureau of Reclamation.

**Streamflow Data:**

(1) Water supply papers - U.S. Department of the Interior; Geological Survey, Water Resources Division.

(2) Reports of state engineers.

(3) Annual reports - International Boundary and Water Commission, United States and Mexico.

(4) Annual reports - various interstate compact commissions.

(5) Hydraulic laboratory reports - U.S. Department of the Interior, Bureau of Reclamation.

(6) Bureau of Reclamation.

(7) U.S. Army Corps of Engineers, Flood control studies.

**Sedimentation Data:**

(1) Water supply papers - U.S. Department of the Interior, Geological Survey, Quality of Water Branch.

(2) Reports - U.S. Department of the Interior, Bureau of Reclamation; and U.S. Department of Agriculture, Soil Conservation Service.

(3) Geological Survey Circulars - U.S. Department of the Interior, Geological Survey.

Table continues

Table 4.1. List of Data Sources (after FHWA 2001) continued.

**Quality of Water Reports:**

(1) Water supply papers - U.S. Department of the Interior, Geological Survey, Quality of Water Branch.

(2) Reports - U.S. Department of Health, Education, and Welfare, Public Health Service.

(3) Reports - state public health departments

(4) Water resources publications - U.S. Department of the Interior, Bureau of Reclamation.

(5) Environmental Protection Agency, regional offices.

(6) State water quality agency.

**Irrigation and Drainage Data:**

(1) Agriculture census reports - U.S. Department of Commerce, Bureau of the Census.

(2) Agricultural statistics - U.S. Department of Agriculture, Agricultural Marketing Service.

(3) Federal reclamation projects - U.S. Department of the Interior, Bureau of Reclamation.

(4) Reports and progress reports - U.S. Department of the Interior, Bureau of Reclamation.

**Basin and Project Reports and Special Reports**

(1) U.S. Army Corps of Engineers.

(2) U.S. Department of the Interior, Bureau of Land Management, Bureau of Mines, Bureau of Reclamation, Fish and Wildlife Service, and National Park Service.

Table 4.2 List of Internet Data Sources.

| Source | Aerial Imagery | Historical Aerial Imagery | Elevation Data | Topographic Maps | Land Use/Cover | Planimetrics | Soils Data | Geologic Maps | Stream Flow Data | Climatological Data | Drainage Data | Water Quality | Benchmarks |
|---|---|---|---|---|---|---|---|---|---|---|---|---|---|
| **Free Data Sources** ||||||||||||||
| USGS Seamless Server | ■ |  | ■ |  | ■ | ■ |  |  |  |  |  |  |  |
| USGS National Hydrography Dataset (NHD) |  |  |  |  |  | ■ |  |  | ■ |  | ■ |  |  |
| USGS WaterData |  |  |  |  |  |  |  |  | ■ | ■ |  | ■ |  |
| USGS National Geologic Map Data Base |  |  |  | ■ |  |  |  | ■ |  |  |  |  |  |
| USDA Data Gateway | ■ |  | ■ | ■ | ■ |  | ■ |  |  |  |  |  |  |
| Aerial Photo Field Office (APFO) | ■ | ■ |  |  |  |  |  |  |  |  |  |  |  |
| USDA Web Soil Survey |  |  |  |  |  |  | ■ |  |  |  |  |  |  |
| National Geodetic Survey (NGS) |  |  |  |  |  |  |  |  |  |  |  |  | ■ |
| Geospatial One Stop | ■ |  | ■ |  |  |  |  |  |  |  |  |  |  |
| National Atlas | ■ |  | ■ |  |  |  |  |  |  |  |  |  |  |
| USGS EROS | ■ |  | ■ |  |  |  |  |  |  |  |  |  |  |
| **Data Viewers** ||||||||||||||
| Google Earth* | ■ |  |  | ** | ** | ** | ** | ** | ** | ** | ** | ** | ** |
| Google Maps | ■ |  |  |  |  |  |  |  |  |  |  |  |  |
| Bing Maps | ■ |  |  |  |  |  |  |  |  |  |  |  |  |
| ArcGIS Online | ■ |  | ■ | ■ |  |  |  |  |  |  |  |  |  |
| ArcGIS Explorer | ■ |  | ■ | ■ |  |  |  |  |  |  |  |  |  |

\* Google Earth is a desktop software application that is available, but requires a download and install.

\** This data is not in Google Earth by default, but is available from several third parties such as the USGS.

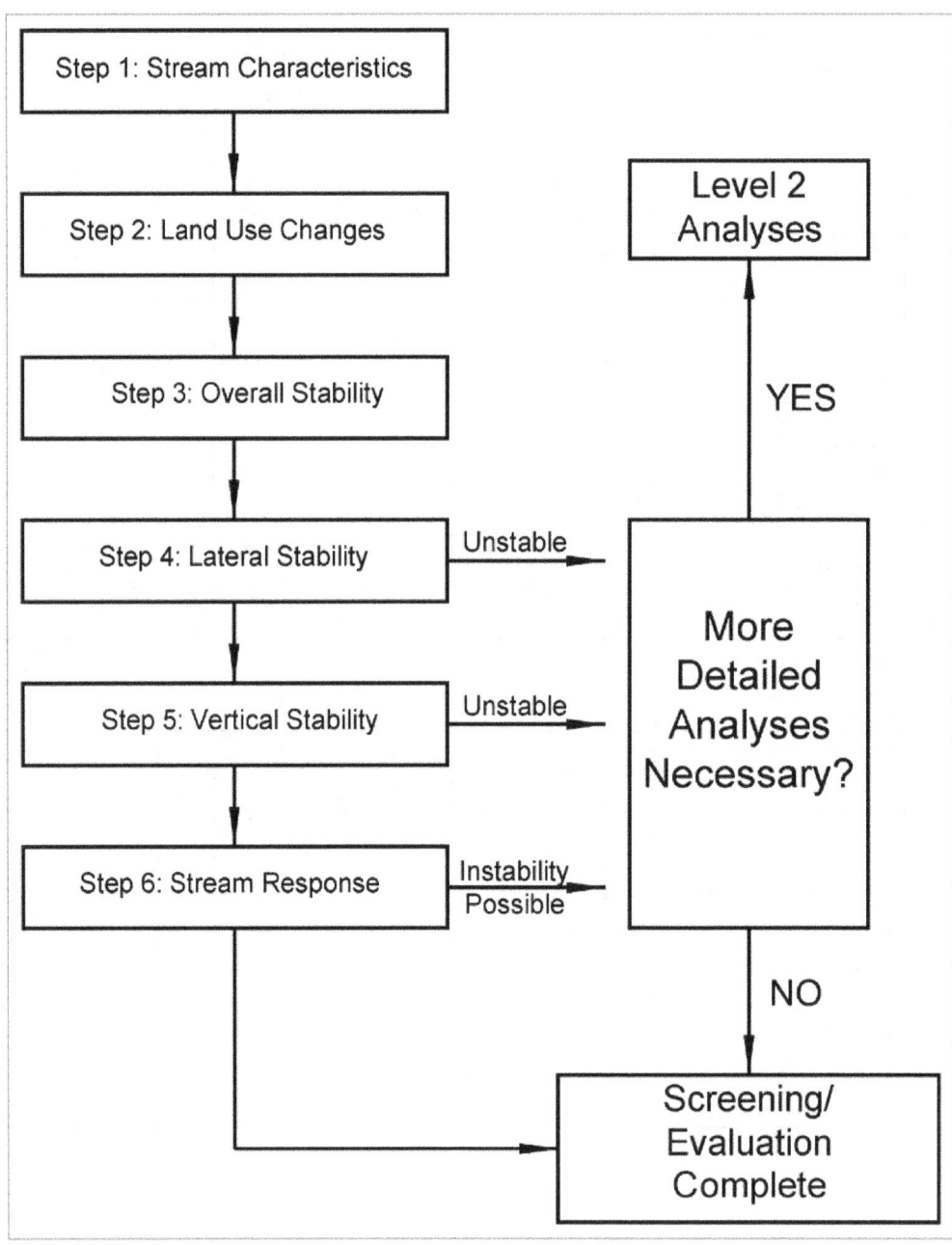

Figure 4.1. Flow chart for Level 1: Qualitative Geomorphic Analyses.

### 4.5.2 Step 2. Evaluate Land Use Changes

Water and sediment yield from a watershed is a function of land-use practices. Thus, knowledge of the land use and historical changes in land use is essential to understanding conditions of stream stability and potential stream response to natural and human-induced changes.

The presence or absence of vegetative growth can have a significant influence on the runoff and erosional response of a fluvial system. Large scale changes in vegetation resulting from fire, logging, land conversion and urbanization can either increase or decrease the total water and sediment yield from a watershed. For example, fire and logging tend to increase water and sediment yield, while urbanization promotes increased water yield and peak flows, but decreased sediment yield from the watershed. Urbanization may increase sediment yield from the channel.

Information on land use history and trends can be found in Federal, State and Local government documents and reports (i.e., census information, zoning maps, future development plans, etc.). Additionally, analysis of historical aerial photographs can provide significant insight on land use changes. Land use change due to urbanization can be classified based on estimated changes in pervious and impervious cover. Changes in vegetative cover can be classified as simply as no change, vegetation increasing, vegetation damaged and vegetation destroyed. The relationship or correlation between changes in channel stability and land use changes can contribute to a qualitative understanding of system response mechanisms.

### 4.5.3 Step 3. Assess Overall Stream Stability

Table 4.3 summarizes possible channel stability interpretations according to stream characteristics discussed in Chapter 2 (Figure 2.6), as well as additional factors that commonly influence stream stability. Figure 4.2 is also useful in making a qualitative assessment of stream stability based on stream characteristics. It shows that straight channels are relatively stable only where flow velocities and sediment load are low. As these variables increase, flow meanders in the channel causing the formation of alternate bars and the initiation of a meandering channel pattern. Similarly, meandering channels are progressively less stable with increasing velocity and bed load. At high values of these variables, the channel becomes braided. The presence and size of point bars and middle bars are indications of the relative lateral stability of a stream channel.

Bed material transport is directly related to stream power, and relative stability decreases as stream power increases as shown by Figure 4.2. Stream power is the product of shear stress at the bed and the average velocity in the channel section. Shear stress can be determined from the average shear stress equation ($\gamma RS$). See Section 6.4.2 or HDS 6 (FHWA 2001) for further discussion.

Table 4.3. Interpretation of Observed Data (after FHWA 1980b).

|  | Observed Condition | Channel Response | | | |
|---|---|---|---|---|---|
|  |  | Stable | Unstable | Degrading | Aggrading |
| **Alluvial Fan[1]** | Upstream |  | X |  | X |
|  | Downstream |  | X | X |  |
| **Dam and Reservoir** | Upstream |  | X |  | X |
|  | Downstream |  | X | X |  |
| **River Form** | Meandering | X | X | Unknown | Unknown |
|  | Straight |  | X | Unknown | Unknown |
|  | Braided |  | X | Unknown | Unknown |
|  | Bank Erosion |  | X | Unknown | Unknown |
|  | Vegetated Banks | X |  | Unknown | Unknown |
|  | Head Cuts |  | X | X |  |
| **Diversion** | Clear water diversion |  | X |  | X |
|  | Overloaded w/sediment |  | X | X |  |
|  | Channel Straightened |  | X | X |  |
|  | Deforest Watershed |  | X |  | X |
|  | Drought Period | X |  |  | X |
|  | Wet Period |  | X | X |  |
| **Bed Material Size** | Increase |  | X |  | X |
|  | Decrease |  | X | Unknown | X |

[1]The observed condition refers to location of the bridge on the alluvial fan, i.e., on the upstream or downstream portion of the fan.

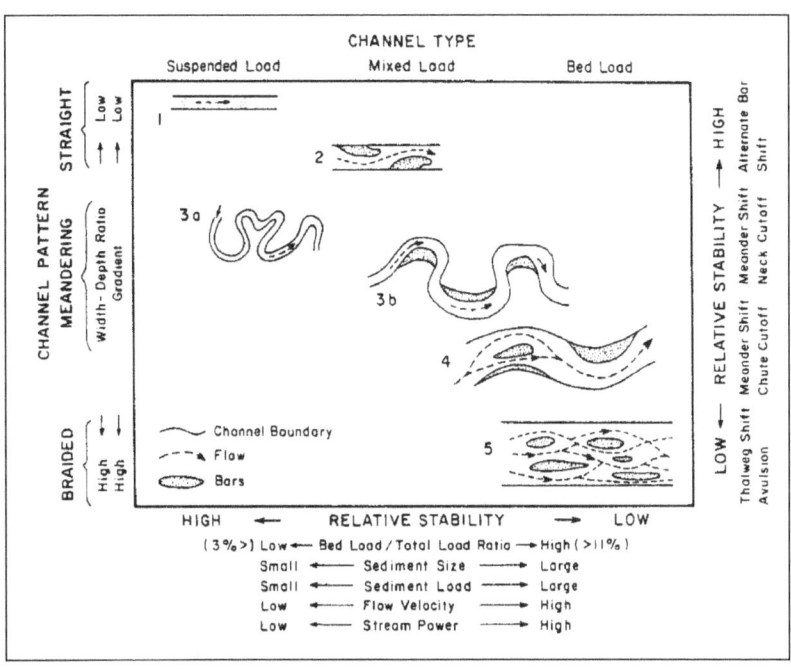

Figure 4.2. Channel classification and relative stability as hydraulic factors are varied (after FHWA 1981).

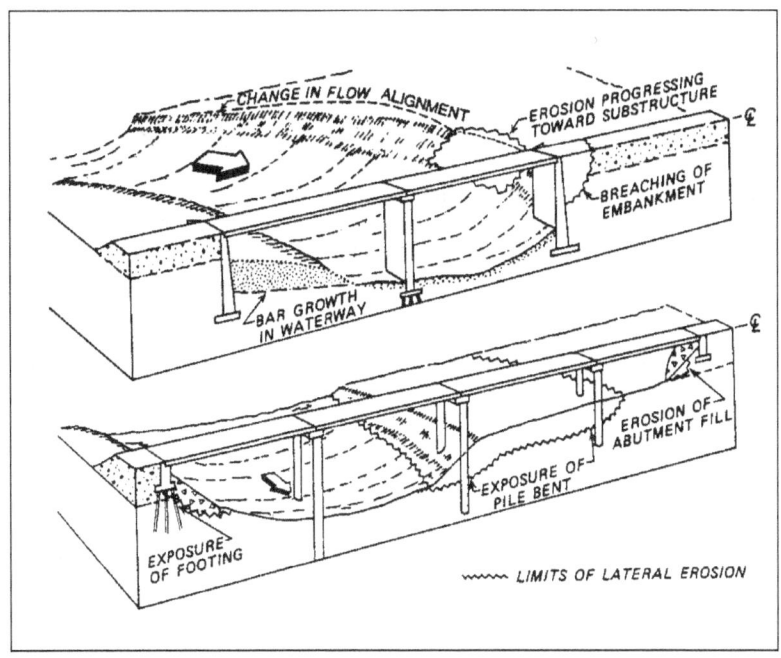

Figure 4.3. Hydraulic problems at bridges attributed to erosion at a bend or to lateral migration of the channel (after FHWA 1978a and b).

### 4.5.4 Step 4. Evaluate Lateral Stability

The effects of lateral instability of a stream at a bridge are dependent on the extent of the bank erosion and the design of the bridge. Bank erosion can undermine piers and abutments located outside the channel and erode abutment spill slopes or breach approach fills. Where bank failure is by a rotational slip, lateral pressures on piers located within the slip zone may cause cracks in piers or piling or displacement of pier foundations. Migration of a bend through a bridge opening changes the direction of flow through the opening so that a pier designed and constructed with a round-nose acts as a blunt-nosed, enlarged obstruction in the flow, thus accentuating local and contraction scour. Also, the development of a point bar on the inside of the migrating bend can increase contraction at the bridge if the outside bank is constrained from eroding. Figure 4.3 illustrates some of the problems of lateral erosion at bridges.

A field inspection is a critical component of a qualitative assessment of lateral stability. A comparison of observed field conditions with the descriptions of stable and unstable channel banks presented in Section 2.3.9 helps qualify bank stability. Similarly, field observations of bank material, composition and existing failure modes can provide insight on bank stability, based on the descriptions of cohesive, noncohesive and composite banks given in Section 2.3.9 (see also Appendix B). An evaluation of lateral stability in conjunction with the design of a bridge should take the performance of existing nearby bridges into account. The experience of such structures which have been subjected to the impacts of the stream can provide insight into response at a nearby structure.

Lateral stability assessment can also be completed from records of the position of a bend at two or more different times; aerial photographs or maps are usually the only records available. Surveyed cross sections are extremely useful when available. Some progress is being made on the numerical prediction of loop deformation and bend migration (Level 3 type analyses). At present, however, the best available estimates are based on past rates of lateral migration at a particular reach (see Section 6.3 Predicting Meander Migration). In using the estimates, it should be recognized that erosion rates may fluctuate substantially from one period of years to the next.

Measurements of bank erosion on two time-sequential aerial photographs (or maps) require the identification of reference points which are common to both. Useful reference points include roads, buildings, irrigation canals, bridges and fence corners. This analysis of lateral stability is greatly facilitated by a drawing of changes in bankline position with time. To prepare such a drawing, aerial photographs are matched in scale and the photographs are superimposed holding the reference points fixed. For additional discussion of comparative techniques, see Section 6.3.

A site of potential avulsion (channel shifting to new flow path) in the vicinity of a highway stream crossing should be identified so that steps can be taken to mitigate the effects of avulsion when it occurs. A careful study of aerial photographs will show where overbank flooding has been taking place consistently and where a channel exists that can capture the flow in the existing channel. In addition, topographic maps and special surveys may show that the channel is indeed perched above the surrounding alluvial surface, with the inevitability of avulsion. Generally, avulsion, as the term is used here, will only be a hazard on alluvial fans, alluvial plains, deltas, and wide alluvial valleys. In a progressively aggrading situation, as on an alluvial fan, the stream will build itself out of its channel and be very susceptible to avulsion. In other words, in a cross profile on an alluvial fan or plain, it may be found that the river is flowing between natural levees at a level somewhat higher than the surrounding area. In this case, avulsion is inevitable.

### 4.5.5 Step 5. Evaluate Vertical Stability

The typical effects associated with bed elevation changes at highway bridges are erosion at abutments and the exposure and undermining of foundations from degradation, or a reduction in flow area from aggradation under bridges resulting in more frequent flow over the highway. Bank caving associated with degradation poses the same problems at bridges as lateral erosion from bend migration, but the problems may be more severe because of the lower elevation of the streambed. Aggrading stream channels also tend to become wider as aggradation progresses, eroding floodplain areas and highway embankments on the floodplain. The location of the bridge crossing upstream, downstream, or on tributaries may cause bed elevation problems.

Brown et al. reported that their study indicated that there are serious problems at about three degradation sites for every aggradation site (FHWA 1980a). This is a reflection of the fact that degradation is more common than aggradation, and also the fact that aggradation does not endanger the bridge foundation. It is not an indication that aggradation is not a serious problem in some areas of the United States.

Problems other than those most commonly associated with degrading channels include the undermining of cutoff walls, other flow-control structures, and bank protection. Bank sloughing because of degradation often greatly increases the amount of debris carried by the stream and increases the potential for blocked waterway openings and increased scour at bridges. The hazard of local scour becomes greater in a degrading stream because of the lower streambed elevation.

Aggradation in a stream channel increases the frequency of backwater that can cause damage. Bridge decks and approach roadways become inundated more frequently, disrupting traffic, subjecting the superstructure of the bridge to hydraulic forces that can cause failure, and subjecting approach roadways to overflow that can erode and cause failure of the embankment. Where lateral erosion or increased flood stages accompanying aggradation increase the debris load in a stream, the hazards of clogged bridge waterways and hydraulic forces on bridge superstructures are increased.

Data records for at least several years are usually needed to detect bed elevation problems. This is due to the fact that the channel bottom often is not visible and changes in flow depth may indicate changes in the rate of flow rather than bed elevation changes. Bed elevation changes develop over long periods of time even though rapid change can occur during an extreme flood event. The data needed to assess bed elevation changes include historic streambed profiles, and long-term trends in stage-discharge relationships. Occasionally, information on bed elevation changes can be gained from a series of maps prepared at different times. Bed elevations at railroad, highway and pipeline crossings monitored over time may also be useful. On many large streams, the long-term trends have been analyzed and documented by agencies such as the USGS and the USACE.

### 4.5.6 Step 6. Evaluate Channel Response to Change

The knowledge and insight developed from evaluation of present and historical channel and watershed conditions, as developed above through Steps 1 through 5, provide an understanding of potential channel response to previous impacts and/or proposed changes, such as construction of a bridge. Additionally, the application of simple, predictive geomorphic relationships, such as the Lane relationship (see Section 5.5.2) can assist in evaluating channel response mechanisms. Sections 5.5.2 and 5.5.3 illustrates the evaluation of stream response based on geomorphic and other qualitative considerations. Additional applications of Level 1 analysis techniques to bridge related stream stability problems can be found in Chapters 5 and 9 of HDS 6 (FHWA 2001).

## 4.6 LEVEL 2: BASIC ENGINEERING ANALYSES

A flow chart of the typical steps in basic engineering analyses is provided in Figure 4.4. The flow chart illustrates the typical steps to be followed if a Level 1 qualitative analysis resulted in a decision that Level 2 analysis is required (Figure 4.1). The eight basic engineering steps are generally applicable to most stream stability problems. The basic engineering analysis steps lead to a conclusion regarding the need for more detailed (Level 3) analysis or a decision to proceed to bridge design, selection and design of countermeasures, or channel restoration design without more complex studies. Selection and design of countermeasures are discussed in HEC-23 (FHWA 2009).

### 4.6.1 Step 1. Evaluate Flood History and Rainfall-Runoff Relations

Detailed discussion of hydrologic analysis techniques, in particular the analysis of flood magnitude and frequency, is presented in HDS 2 (FHWA 2002). However, several hydrologic concepts of particular significance to evaluation of stream stability are summarized in the following paragraphs.

Consideration of flood history is an integral step in attempting to characterize watershed response and morphologic evolution. Analysis of flood history is of particular importance to understanding arid region stream characteristics. Many dryland streams flow only during the spring and immediately after major storms. For example, Leopold et al. found that arroyos near Santa Fe, New Mexico, flow only about three times a year (USGS 1966b). As a consequence, dryland stream response can be considered to be more hydrologically dependent than streams located in a humid environment. Whereas the simple passage of time may be sufficient to cause change in a stream located in a humid environment, time alone, at least in the short term, may not necessarily cause change in a dryland system due to the infrequency of hydrologically significant events. Thus, the absence of significant morphological changes in a dryland stream or river, even over a period of years, should not necessarily be construed as an indication of system stability.

Although the occurrence of single large storms can often be directly related to system change in any region of the country, this is not always the case. In particular, the succession of morphologic change may be linked to the concept of geomorphic thresholds. Under this concept, although a single major storm may trigger an erosional event in a system, the occurrence of such an event may be the result of a cumulative process leading to an unstable geomorphic condition.

Where available, the study of flood records and corresponding system responses, as indicated by time-sequenced aerial photography or other physical information, may help determine the relationship between morphological change and flood magnitude and frequency. Evaluation of wet-dry cycles can also be beneficial to an understanding of historical system response. Observable historic change may be found to be better correlated with the occurrence of a sequence of events during a period of above average rainfall and runoff than with a single large event. The study of historical wet-dry trends may explain certain complex aspects of system response. For example, a large storm preceded by a period of above-average precipitation may result in less erosion, due to better vegetative cover, than a comparable storm occurring under dry antecedent conditions; however, runoff volumes might be greater due to saturated soil conditions.

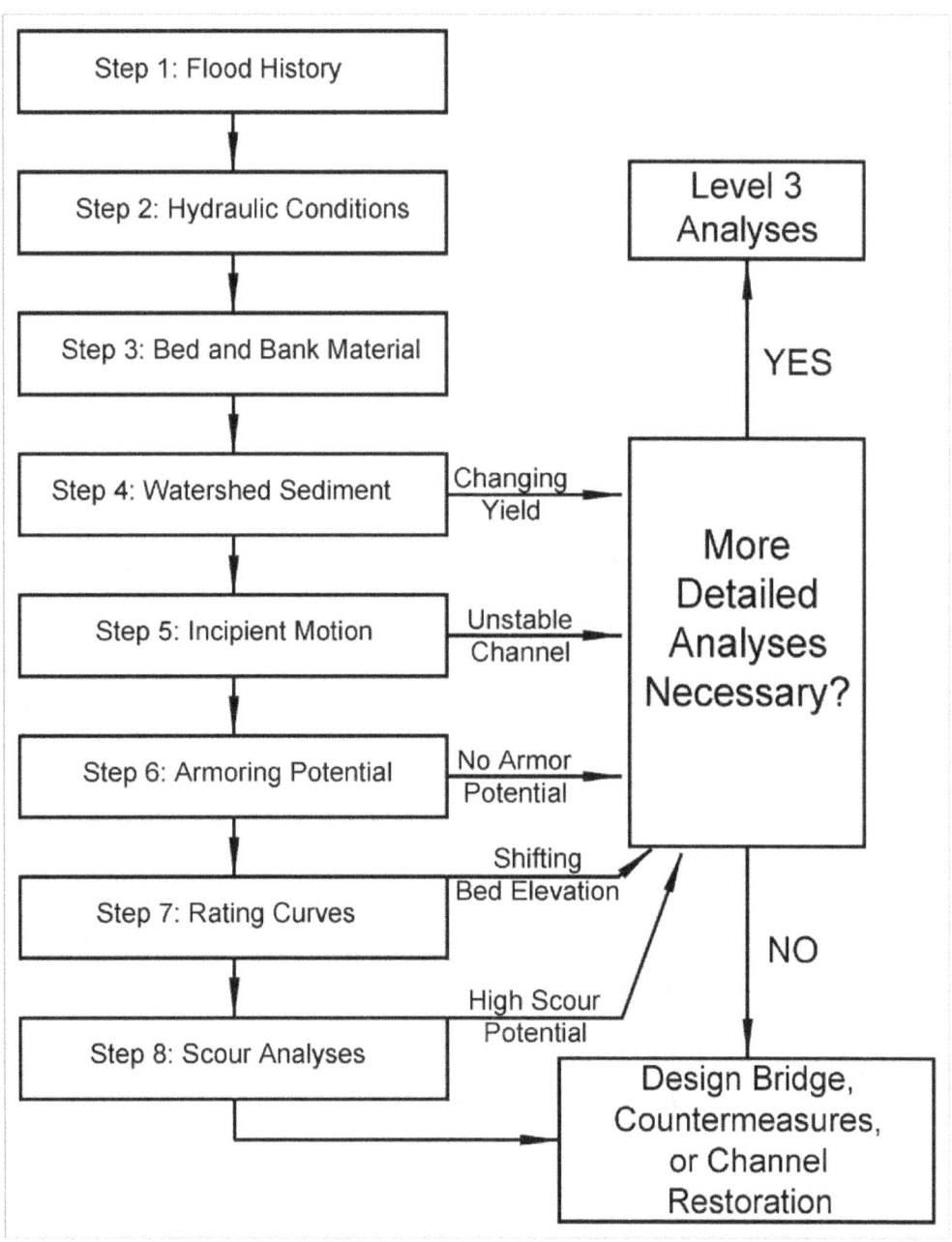

Figure 4.4. Flow chart for Level 2: Basic Engineering Analyses.

A good method to evaluate wet-dry cycles is to plot annual rainfall amounts, runoff volumes and maximum annual mean daily discharge for the period of record. A comparison of these graphs will provide insight into wet-dry cycles and flood occurrences. Additionally, a plot of the ratio of rainfall to runoff is a good indicator of watershed characteristics and historical changes in watershed condition.

### 4.6.2 Step 2. Evaluate Hydraulic Conditions

Knowledge of basic hydraulic conditions, such as velocity, flow depth and top width, etc., for given flood events is essential for completion of Level 2 stream stability analyses. Incipient motion analysis, scour analysis, assessment of sediment transport capacity, etc. all require basic hydraulic information. Hydraulic information is sometimes required for both the main channel and overbank areas, such as in the analysis of contraction scour.

Evaluation of hydraulic conditions is based on the factors and principles reviewed in Chapter 3. For many river systems, particularly near urban areas, hydraulic information may be readily available from previous studies, such as flood insurance studies, channel improvement projects, etc., and complete re-analysis may not be necessary. However, in other areas, hydraulic analysis based on appropriate analytical techniques will be required prior to completing other quantitative analyses in a Level 2 stream stability assessment. The most common computer model for analysis of water surface profiles and hydraulic conditions is the Corps of Engineers HEC-RAS (River Analysis System) (USACE 2010).

### 4.6.3 Step 3. Bed and Bank Material Analysis

Bed material is the sediment mixture of which the streambed is composed. Bed material ranges in size from huge boulders to fine clay particles. The erodibility or stability of a channel largely depends on the size of the particles in the bed. Additionally, knowledge of bed sediment is necessary for most sediment transport analyses, including evaluation of incipient motion, armoring potential, sediment transport capacity and scour calculations. Many of these analyses require knowledge of particle size gradation, and not just the median ($D_{50}$) sediment size.

Bank material usually consists of particles the same size as, or smaller than, bed particles. Thus, banks are often more easily eroded than the bed, unless protected by vegetation, cohesion, or some type of protection, such as revetment.

Of the various sediment properties, size has the greatest significance to the hydraulic engineer, not only because size is the most readily measured property, but also because other properties, such as shape and fall velocity, tend to vary with particle size. A comprehensive discussion of sediment characteristics, including sediment size and its measurement, is provided in HDS 6 (FHWA 2001). The following information briefly discusses sediment sampling considerations.

Important factors to consider in determining where and how many bed and bank material samples to collect include:

- Size and complexity of the study area
- Number, lengths and drainage areas of tributaries
- Evidence of or potential for armoring
- Structural features that can impact or be significantly impacted by sediment transport
- Bank failure areas

- High bank areas
- Areas exhibiting significant sediment movement or deposition (i.e., bars in channel).

Tributary sediment characteristics can be very important to channel stability, since a single major tributary or tributary source area could be the predominant supplier of sediment to a system.

The depth of bed material sampling depends on the homogeneity of surface and subsurface materials. Where possible it is desirable to dig down some distance to establish bed-material characteristics. For example, in sand/gravel bed systems the potential existence of a thin surface layer of coarser sediments (armor layer) on top of relatively undisturbed subsurface material must be considered in any sediment sampling. Samples containing material from both layers would contain materials from two populations in unknown proportions, and thus it is typically more appropriate to sample each layer separately. If the purpose of the sampling is to evaluate hydraulic friction or initiation of bed movement, then the surface sample will be of most interest. Conversely, if bed-material transport during a large flood (i.e., large enough to disturb the surface layer) is important, then the underlying layer may be more significant. Methods of analysis are given in HDS 6 (FHWA 2001).

### 4.6.4 Step 4. Evaluate Watershed Sediment Yield

Evaluation of watershed sediment yield, and in particular, the relative increase in yield as a result of some disturbance, can be an important factor in stream stability assessment. Sediment eroded from the land surface can cause silting problems in stream channels resulting in increased flood stage and damage. Conversely, a reduction in sediment supply can also cause adverse impacts to river systems by reducing the supply of incoming sediment, thus promoting channel degradation and headcutting. A radical change in sediment yield as a result of some disturbance, such as a recent fire or long-term land use changes, would suggest that stream instability conditions either already exist or might readily develop.

Assessment of watershed sediment yield requires understanding the sediment sources in the watershed and the types of erosion that are most prevalent. The physical processes causing erosion can be classified as sheet erosion, rilling, gullying and stream channel erosion. Other types of erosional processes are classified under the category of mass movement, e.g., soil creep, mudflows, landslides, etc. Data from publications and maps produced by the NRCS and the USGS can be used along with field observations to evaluate the area of interest.

Quantification of sediment yield is at best an imprecise science. The most useful information is typically obtained not from analysis of absolute magnitude of sediment yield, but rather the relative changes in yield as a result of a given disturbance. One useful approach to evaluating sediment yield from a watershed was developed by the Pacific Southwest Interagency Committee (PSIAC 1968). This method, which was designed as an aid for broad planning purposes only, consists of a numerical rating of factors affecting sediment production in a watershed, which then defines ranges of annual sediment yield. The factors are surficial geology, soil climate, runoff, topography, ground cover, land use, upland erosion, and channel erosion and transport.

Other approaches to quantifying sediment yield are based on regression equations, as typified by the Universal Soil Loss Equation (USLE). The USLE is an empirical formula for predicting annual soil loss due to sheet and rill erosion and is perhaps the most widely recognized method for predicting soil erosion. The USDA Agricultural Handbook 537 provides detailed descriptions of this equation and its terms (USDA 1978).

### 4.6.5 Step 5. Incipient Motion Analysis

An evaluation of relative channel stability can be made by evaluating incipient motion parameters. The definition of incipient motion is based on the critical or threshold conditions where hydrodynamic forces acting on one grain of sediment have reached a value that, if increased even slightly, will move the grain. Under critical conditions, or at the point of incipient motion, the hydrodynamic forces acting on the grain are just balanced by the resisting forces of the particle.

Evaluation of the incipient motion size for various discharge conditions provides insight on channel stability and the magnitude of the flood that might potentially disrupt channel stability. The results of such an analysis are generally more useful for analysis of gravel or cobble-bed systems. When applied to a sand-bed channel, incipient motion results usually indicate that all particles in the bed material are capable of being moved for even very small discharges, a physically realistic result. An equation and techniques for incipient motion analyses are provided in Chapter 6.

### 4.6.6 Step 6. Evaluate Armoring Potential

The armoring process begins as the non-moving coarser particles segregate from the finer material in transport. The coarser particles are gradually worked down into the bed, where they accumulate in a sublayer. Fine bed material is leached up through this coarse sublayer to augment the material in transport. As sediment movement continues and degradation progresses, an increasing number of non-moving particles accumulate in the sublayer. Eventually, enough coarse particles can accumulate to shield or "armor" the entire bed surface.

An armor layer sufficient to protect the bed against moderate discharges can be disrupted during high flow, but may be restored as flows diminish. Therefore, as in any hydraulic design, the analysis must be based on a certain design event. If the armor layer is stable for that design event, it is reasonable to conclude that no degradation will occur under design conditions. However, flows exceeding the design event may disrupt the armor layer, resulting in further degradation. While armoring of the bed by the coarser material size fraction can temporarily reduce the rate of degradation and stabilize the stream system, armoring cannot be counted on as a long-term solution.

Potential for development of an armor layer can be assessed using incipient motion analysis and a representative bed-material composition. In this case the representative bed-material composition is that which is typical of the depth of anticipated degradation. For given hydraulic conditions the incipient motion particle size can be computed as referenced above in Step 5. If no sediment of the computed size or larger is present in significant quantities in the bed, armoring will not occur.

The $D_{90}$ or $D_{95}$ size of the representative bed material is frequently found to be the size armoring the channel when degradation is arrested. Armoring is probable when the computed incipient motion size is equal to or smaller than the $D_{95}$ size in the bed material. A simple equation for determining armoring potential is given in Chapter 6.

### 4.6.7 Step 7. Evaluation of Rating Curve Shifts

When stream gage data are available, such as that collected by the USGS, an analysis of the stage-discharge rating curve over time can provide insight on stream stability. For example, a rating curve that was very stable for many years, but suddenly shifts might indicate a change in watershed conditions causing increased channel erosion or sedimentation, or some other change related to channel stability. Similarly, a rating curve that shifts continually would be a good indicator that channel instability exists. However, it is important to note that not all rating curve shifts are the result of channel instability. Other factors promoting a shift in a rating curve include changes in channel vegetation, ice conditions, or beaver activity.

The most common cause of rating curve shifts in natural channel control sections is generally scour and fill (USGS 1982). A positive shift in the rating curve results from scour, and the depth and, hence, the discharge are increased for a given stage. Conversely, a negative shift results from fill, and the depth and discharge will be less for a given stage.

Shifts may also be the result of changes in channel width. Channel width may increase due to bank-cutting or decrease due to undercutting of steep stream banks. In meandering streams, changes in channel width can occur as point bars are created or destroyed.

Analysis of rating curve shifts is typically available from the agency responsible for the stream gage. If such information is not available, field inspection combined with the methods described by Rantz can be utilized to analyze observed rating curve shifts (USGS 1982). If the shifts can be traced to scour, fill, or channel width changes, such information will be a reliable indicator of potential channel instability.

Gaging stations at which continuous sediment data are collected may also provide clues to the existence of gradation problems. Any changes in the long-term sediment load may indicate lateral movement of the channel, gradation changes, or a change in sediment supply from the watershed.

Where an extended historical record is available, one approach to using gaging station records to determine long-term bed elevation change is to plot the change in stage through time for a selected discharge. This approach is often referred to as establishing a "specific gage" record.

Figure 4.5 shows a plot of specific gage data for a discharge of 500 cfs (14 $m^3$/sec) from about 1910 to 1980 for Cache Creek in California. Cache Creek has experienced significant gravel mining with records of gravel extraction quantities available since about 1940. When the historical record of cumulative gravel mining is compared to the specific gage plot, the potential impacts are apparent. The specific gage record shows more than 10 ft (3 m) of long-term degradation in a 70-year period.

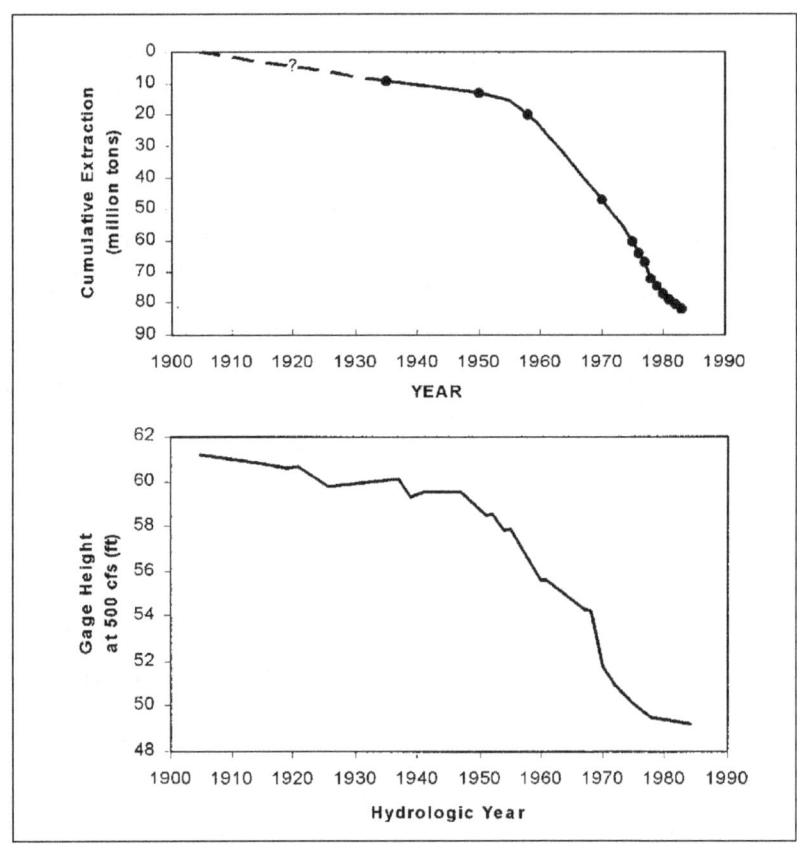

Figure 4.5. Specific gage data for Cache Creek, California.

### 4.6.8 Step 8. Evaluate Scour Conditions

Section 3.6.1 provides an overview of scour at bridge crossings and HEC-18 provides detailed computational procedures (FHWA 2012b). Figure 4.6 illustrates common scour related problems at bridges. These problems are attributable to the effects of obstructions to the flow (local scour) and contraction of the flow or channel deepening at the outside of a bend. Calculation of the three components of scour, i.e., local scour, contraction scour and aggradation/ degradation, quantifies the potential instability at a bridge crossing. As shown in the comprehensive analysis flow chart (Figure 1.1), HEC-18 (FHWA 2012b) is the primary source for guidance on these issues.

## 4.7 LEVEL 3: MATHEMATICAL AND PHYSICAL MODEL STUDIES

Detailed evaluation and assessment of stream stability can be accomplished using either mathematical or physical model studies. A mathematical model is simply a quantitative expression of the relevant physical processes involved in stream channel stability. Various types of mathematical models are available for evaluation of sediment transport, depending on the application (watershed or channel analysis) and the level of analysis required. The use of such models can provide detailed information on erosion and sedimentation throughout a study reach and allows evaluation of a variety of "what-if" questions. HDS 7 (FHWA 2012a) provides a survey of 1- and 2-dimensional mathematical models available for alluvial river analyses and HEC-18 (FHWA 2012b) summarizes the capabilities of 1- and 2-dimensional mathematical models for unsteady flow tidal hydraulic analyses.

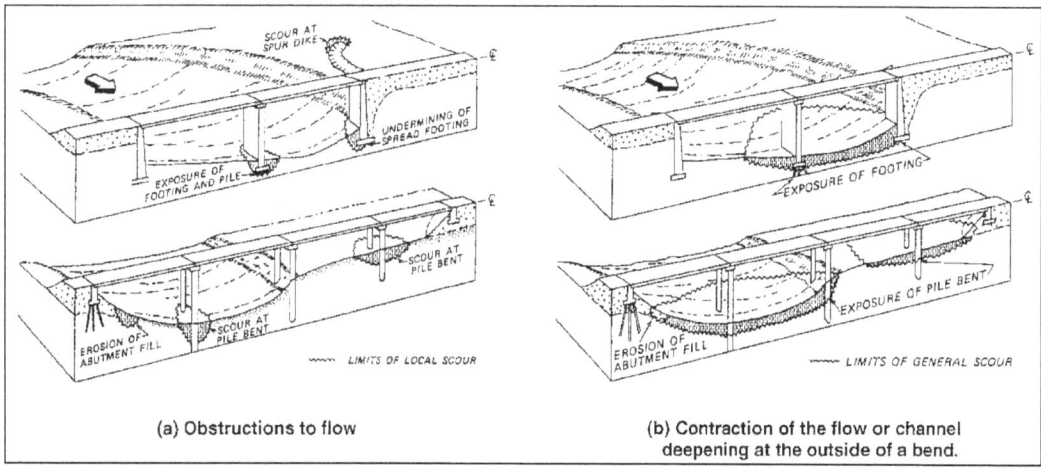

Figure 4.6. Local scour and contraction scour related hydraulic problems at bridges related to (a) obstructions to the flow or (b) contraction of the flow or channel deepening at the outside of a bend (FHWA 1978a and b).

Similarly, physical model studies completed in a hydraulics laboratory can provide detailed information on flow conditions and, to some extent, sediment transport conditions at a bridge crossing. The hydraulic laws and principles involved in scaling physical models are well defined and understood, allowing accurate extrapolation of model results to prototype conditions. Physical model studies can sometimes provide better information on complex flow conditions than mathematical models, due to the complexity of the process and the limitations of 2- and 3-dimensional mathematical models. Often the use of both physical and mathematical models can provide complementary information (see HDS 6) (FHWA 2001).

The need for detailed information and accuracy available from either mathematical or physical model studies must be balanced by the time and resources available. As the analysis becomes more complicated, accounting for more factors, the level of effort necessary becomes proportionally larger. The decision to proceed with a Level 3 type analysis has historically been made only for high risk locations, extraordinarily complex problems, and for forensic analysis where losses and liability costs are high; however, the importance of stream stability to the safety and integrity of all bridges suggests that Level 3 type analyses should be considered more routinely. The widespread use of personal computers and the continued development of more sophisticated software have greatly facilitated completion of Level 3 type investigations and have reduced the level of effort and cost required.

## 4.8 ILLUSTRATIVE EXAMPLES

The FHWA manual, "River Engineering for Highway Encroachments," provides a discussion of design considerations for highway encroachment and river crossings in Chapter 9 (FHWA 2001). This discussion includes principal factors for design, procedures for evaluation and design, and conceptual examples. The procedures for evaluation and design of river crossings and encroachments parallel the three-level approach of this chapter. A series of short conceptual discussions in Chapter 9 of HDS 6 illustrate the application of qualitative (Level 1) techniques, and a series of short case studies provide various applications. Finally, Chapter 10 of HDS 6 presents two overview examples which illustrate various steps in the three-level approach (FHWA 2001).

(page intentionally left blank)

# CHAPTER 5

# RECONNAISSANCE, CLASSIFICATION, ASSESSMENT, AND RESPONSE

## Application of Level 1 Analysis Procedures

## 5.1 INTRODUCTION

The design and protection of a structure at a stream crossing requires identification of the cause and extent of channel instability problems. The problems may result from a wide variety of geomorphic processes operating at various scales within the watershed. Some of these processes may be operating locally, others may be active within a given reach, and still others may be associated with the response of the entire fluvial system to changes in rainfall-runoff and sediment yield within the entire basin. Therefore, it is important to understand the relationship of any project site to the stream system and the basin geomorphology, and to see the channel within the project reach as part of an interlinked system with complex feedback mechanisms.

Identification of the geomorphic factors that can affect stream stability in the bridge reach provides a useful first step in detecting existing or potential channel instability. Consideration of fundamental geomorphic principles can lead to a qualitative prediction, in terms of trends, of the most likely direction of channel response to natural and human-induced change in the watershed and river system. However, more general methods of river classification can also provide insight on potential instability problems common to a given stream type.

A necessary first step in any channel classification or stability analysis is a field site visit. Geomorphologists have developed stream reconnaissance guidelines and specific techniques, including checklists, which will be useful to the highway engineer during a site visit. In addition, a rapid assessment methodology which uses both geomorphic and hydraulic factors can help identify the most likely sources of stability problems in a stream reach. This chapter extends the geomorphic concepts introduced in Chapter 2 to include guidance and checklists for geomorphic reconnaissance, a consideration of stream channel classification concepts, a rapid assessment method to evaluate channel stability, and qualitative techniques for evaluating channel response. The geomorphic reconnaissance, in particular, provides a systematic approach to gathering the data necessary to apply the quantitative analysis techniques of Chapter 6.

## 5.2 STREAM RECONNAISSANCE

As indicated in Chapter 2, there are numerous geomorphic factors that influence stream stability and, potentially, bridge stability at highway stream crossings. It is important to document these factors and existing conditions not only at the proposed project site, but also within a reasonable distance upstream and downstream of the site as well. Identifying the linkages within a fluvial system, as outlined in the Level 1 and 2 analysis procedures of Chapter 4, involves observation and interpretation of data obtained during a site visit, i.e., through the use of stream reconnaissance. Appendix C presents reconnaissance techniques and checklists to support a field site visit. In addition, a stream reconnaissance offers the best opportunity to evaluate the potential for a watershed and stream system to produce potentially damaging quantities of drift (vegetative debris), which can have a catastrophic impact on bridge stability. Techniques for recognizing the potential for drift accumulations are also presented in this section.

Systematic data collection is an integral part of conducting a reconnaissance along a stream for assessing channel stability. The amount of data required depends on the level of detail desired. There are a wide range of types of data that are useful in assessing stream channel conditions, including topographic maps, aerial photos, bridge inspection reports, hydrologic and hydraulic reports, stream gage data, and other geomorphic reports. Aerial photos, satellite imagery, and topographic maps are enormously useful in providing an overall view of the bridge, the stream below it, and the watershed conditions (see Chapter 4 for data sources and Chapter 6 for a comparative methodology). These tools help to visualize the location of the bridge relative to the location of meanders as well as the bridge alignment. Given the relative ease of checking aerial and satellite imagery, this should be done as a standard part of any survey. In addition to photos and maps, examination of prior reports on assessments conducted at or near the bridge are useful to determine trends. Given that bridge inspections are conducted at least every two years, typically with a rough cross section or two measured, these are good reports to compare for changes over a longer period of time. Geomorphic assessments that have been conducted along the stream, although they may not be concerned with the bridge, are also excellent sources of information.

### 5.2.1 Stream Reconnaissance Techniques

The most comprehensive method of documenting stream and watershed conditions is through the use of a detailed geomorphological stream reconnaissance. Although there have been many methods of stream reconnaissance proposed, they have often been unstructured, primarily qualitative in nature, and have been tailored to the specific needs of the project for which they were being conducted. Thorne (1998) has developed a comprehensive handbook that can be used to document stream channel and watershed conditions (see Appendix C).

The purpose of a stream reconnaissance is summarized as follows (Thorne 1998):

- Supply a methodological basis for field studies of channel form and process.
- Present a format for the collection of qualitative information and quantitative data on the stream system.
- Provide a basis for progressive morphological studies that start with a broadly focused watershed baseline study, continue through a fluvial audit of the channel system, and culminate with a detailed investigation of the geomorphological forms and processes in critical reaches.
- Supply the data and input information to support techniques of geomorphological classification, analysis, and prediction necessary to support sustainable river engineering, conservation, and management.

Using Thorne's (1998) approach, data collection begins with geological and watershed level observations, then continues to focus in on the stream corridor and hill slopes and finally the actual bed and banks of the channel or water body. The data set developed through this reconnaissance provides complete documentation of current conditions. In addition, photographs are taken to assist in documentation of the current conditions. According to Thorne (1998), a reconnaissance could range from a very detailed study over 5 to 10 river widths that would include one pool-riffle couplet, individual meander, or primary bifurcation-bar-confluence unit in braided channel, to a low-level detail study over a much longer reach in which channel form and processes do not change significantly.

Although individual items on the data sheets of Appendix C may not be directly indicative of channel stability or instability, all of the data are collectively important in assessing long-term stability. The relationships between data collected as part of Thorne's reconnaissance and long-term indications of stability are provided in Appendix D (Johnson 2005). These relationships supported the development of simplified reconnaissance sheets in terms of using the stability assessment method described in Section 5.4.

Field forms have been developed to assist the observer in making observations on specific stability-related aspects of a stream, to provide consistently in those observations, and to systematically record the observations. Johnson (FHWA 2006) modified the field sheets developed by Thorne (Appendix C) so that they were relevant for highway related purposes, reduced the highly detailed and time-consuming level of data collection, and simplified the observations required. Interpretive observations, while critical to communicating between observers, are neglected in the revised sheets. In addition, items on the sheet that cannot be assessed in a very brief site visit are excluded from the revised sheets. The simplified and revised data collection sheets, based strongly on Thorne's reconnaissance sheets, are provided in Appendix E.

In addition to the stability assessment, keeping a record of channel dimensions upstream and downstream of the bridge will provide a record of changes in width and depth. A simple measurement of station and elevation upstream of the bridge taken annually will provide adequate cross-sectional information to assess longer-term changes that are taking place. Without this information, gradual but continual changes in the channel may be overlooked.

## 5.2.2 Specific Applications

Geomorphological stream reconnaissance sheets (Appendices C and E) serve different purposes and have a wide variety of applications. Some of those applications include:

Field Identification of Channel Instability Near Structures – With regard to bridges and other highway related structures in the stream environment, the stream reconnaissance sheets can be used to ensure that rapid and accurate assessments of stream channel stability are conducted by the engineers most concerned with inspection and maintenance of these structures (see Section 5.4).

Stream Classification – The qualitative and quantitative information gathered on the stream reconnaissance sheets can be applied to almost any existing stream classification system (see Section 5.3) including those of Brice (USARO 1975), Schumm (1977), Rosgen (1994, 1996), and Figure 2.6 in Chapter 2.

Engineering-Geomorphic Analysis of Streams – The broader spatial scale and scope of the stream reconnaissance provides a basis for subsequent quantitative work, thereby increasing the efficiency and utility of future hydraulic and sediment transport studies (see applications in Chapter 6).

Supplying Input to Stable Channel Design Techniques – Compilation of selective qualitative and semi-quantitative data through the use of the sheets is necessary to characterizing existing channels, identifying flow and sediment processes, and estimating the severity of any flow or sediment-related problems. These are important steps in pre-feasibility and feasibility studies prior to the design of bridges, channel stabilization, and other engineering works (see Chapters 6, 7, and 9).

Assessment, Modeling, and Control of Bank Retreat – Because conventional geotechnical engineering analyses of bank stability are site-specific and require a detailed site investigation, the data gathered on the stream reconnaissance sheets can be used in engineering-geomorphic bank erosion and stability analyses developed for reach-scale and, possibly, system-wide assessments (see Appendix B).

As a Training Aid – The sheets can be used to train staff inexperienced in field methods and techniques.

To Establish a Permanent Record of Stream Condition – The record sheets provide a medium that permanently documents the results of a stream reconnaissance trip and can provide the input for an expert system or GIS database.

### 5.2.3 Assessment of Drift Accumulation Potential

Accumulations of large woody debris, primarily floating debris (drift), can cause increased backwater, increased local scour at bridge piers and abutments, increased lateral forces on bridges, and promote bed and bank scour (Figure 5.1). The results of detailed studies by Diehl and Bryan (1993) and Diehl (FHWA 1997a) throughout the United States revealed several conclusions about drift accumulation. The potential for drift accumulation depends on channel, bridge, and basin characteristics. Drift that accumulates at bridges comes primarily from trees undermined by bank erosion. Rivers with unstable channels have the most bank erosion and the most drift. In addition, abundant drift in the channel can aggravate channel instability. Most drift floats along the thread of the stream where flow is deepest and fastest. Logs longer than the channel width accumulate in jams or are broken into shorter pieces. Drift piles up against obstacles such as bridge piers that divide the flow at the water surface. Drift is trapped most effectively by groups of obstacles separated by narrow gaps that are narrower than the longest logs within the drift accumulation. Accumulations of drift begin at the water surface, but may grow downward toward the streambed through accretion.

Figure 5.1. Increased scour at bridge piers as a result of debris.

As an extension of the original work by Diehl (FHWA 1997a), guidelines and flowcharts were developed for estimating (1) the potential for debris production and delivery from the contributing watershed of a selected bridge and (2) the potential for accumulation on individual bridge elements under NCHRP Project 24-26 (NCHRP 2010). The objectives of this study were to develop guidelines for predicting size and geometry of debris accumulations at bridge piers and methods for quantifying scour at bridge piers resulting from debris accumulations. The methods for quantifying effects of scour at bridge piers resulting from debris are presented in HEC-18 (FHWA 2012b). This section summarizes guidelines for predicting the size and geometry of debris accumulations at bridges as a reconnaissance activity.

As originally suggested by Diehl (FHWA 1997a) and modified in NCHRP Report 653 (NCHRP 2010), there are three major phases for assessing the potential for debris production and accumulation at a bridge site: (1) estimate the potential for debris production and delivery, (2) estimate the potential for debris accumulation on individual bridge elements, and (3) delineate bridge segments or zones that have the same accumulation potential ratings. These major phases and the associated tasks and subtasks that are required to assess the potential for debris production and accumulation at a bridge site provide the basis for the following guidelines and are described in Table 5.1. There may be direct or indirect evidence for the degree of debris production and delivery potential at any given bridge site, however, direct evidence should be evaluated first and given greater weight than indirect evidence.

## Potential for Debris Production

Observations of the channel upstream of a bridge site as well as observations and knowledge of the physical conditions of the watershed upstream of the site and nearby watersheds can assist in determining the potential for debris production and delivery at a given site. A lack of debris at a bridge site does not indicate that there is a low potential for the production and delivery of debris at a site. Even if debris is relatively sparse at a particular site, infrequent or catastrophic events may produce significant debris available for transport to the site. Figure 5.2 provides a flow chart for use in evaluating the potential for debris production upstream of a bridge site.

Table 5.1. Major Phases, Tasks, and Subtasks for Assessing the Potential for Debris Production and Accumulation at a Bridge (NCHRP 2010).

| Phase | Task | Subtask |
|---|---|---|
| 1. Evaluate potential for debris production and delivery. | a. Evaluate potential for debris production upstream of the bridge site. | • Direct evidence<br>• Indirect evidence |
| | b. Evaluate the potential for debris delivery to the site. | • Direct evidence<br>• Indirect evidence |
| | c. Estimate the size of the largest debris that could be delivered to the site. | • Channel width and depth<br>• Maximum design log length |
| | d. Assign location categories to all parts of the bridge crossing. | • Sheltered<br>• Bank/Floodplain<br>• In the channel (bend or straight reach)<br>• In the path (bend or straight reach) |
| 2. Evaluate the potential for debris accumulation on individual elements. | a. Assign bridge structure characteristics to all submerged parts of the bridge. | • Horizontal and vertical gaps between fixed bridge elements<br>• Pier and substructure in flow field |
| | b. Determine accumulation potential for each part of the bridge. | • Assume maximum size of potential accumulations |
| 3. Delineate bridge segments that have the same accumulation potential ratings. | a. Identify elements with accumulation potential ratings of low, medium, high, and chronic. | • Delineate zones of low, medium, high, and chronic potential where adjacent elements have the same rating |

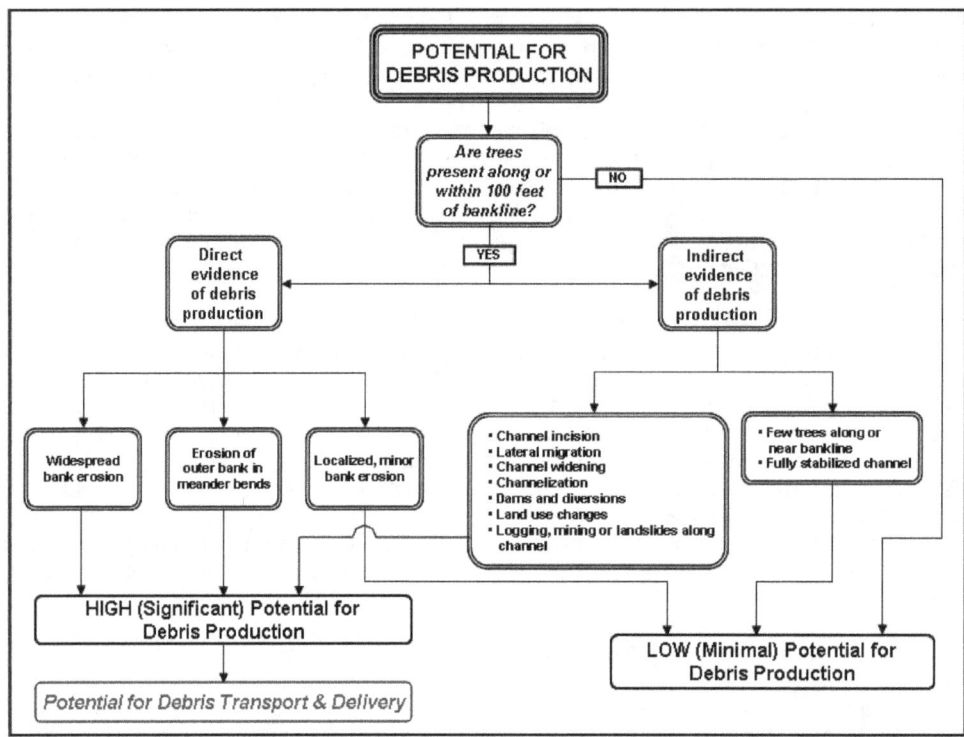

Figure 5.2. Flow chart for evaluating debris production potential (NCHRP 2010).

<u>Direct Evidence</u>. The primary method of debris production is through bank erosion that results in woody vegetation being introduced into the channel. Therefore, existing bank erosion along forested streams provides direct evidence of the potential for debris production (Figure 5.2). Bank erosion may be extensive and severe or localized and minor. Extensive and severe bank erosion may be evident along straight and meandering reaches of streams that are currently unstable and undergoing incision or downcutting and/or widening. Moderate bank erosion may occur along the outer banks of actively migrating meander bends and in reaches where bars are well developed such that the bars, and any trapped debris, cause flow to impinge directly on the bank. Localized, minor bank erosion may occur at any location and is generally insufficient in magnitude to contribute a significant amount of debris.

The most direct evidence for a high potential for debris production is the presence or absence and density of a riparian forest or corridor along a stream channel upstream of a bridge site. In order for a river or stream corridor to produce debris that may be available for transport, the channel must contain a forested area or trees along its banks. Streams with cleared banks or sparse or intermittent riparian zones will provide little debris to a channel unless the channel has become unstable because of land use practices.

Of course, the upstream channel must have trees in close proximity to the channel banks (generally within 100 feet (30 m)) for their introduction through bank erosion to occur. Trees that are leaning over the water at the bankline, incorporated in failed bank sediments at the bank toe, or lying in the water are direct indicators of ongoing bank failure and potentially high debris production.

Debris may not be present at a bridge site or in view from the site, but debris may be stored in significant quantities at sites in the upstream channel. These storage areas may not have had sufficient flows to mobilize the stored debris. Sites of debris storage include the heads of flow splits, on bars and islands, and along the banks, especially along the outer banks of actively migrating meander bends.

Rare, large magnitude or catastrophic events such as ice storms, hurricanes, tornados, microbursts or wind shear, and extreme floods can produce considerable amounts of deadfall available for introduction into a channel. If overbank flooding occurs regularly, deadfall in close proximity to the channel may be introduced to the channel as large rafts if the path of movement from the floodplain to the channel is relatively unobstructed.

Indirect Evidence. Indirect evidence for a high potential for debris production includes historic or ongoing channel changes that affect channel stability and ultimately bank erosion, which is the primary method of debris introduction to a channel. These include:

- Downcutting or incision
- Lateral migration
- Channel widening
- Channelization
- Dams and diversions
- Widespread timber harvesting in the basin
- Existing or potential changes in land use practices

Indirect evidence for a low potential for debris production includes:

- Absence or scarcity of woody vegetation growing along the channel and on the slopes leading down to the stream
- Channel may be fully stabilized (both vertically and laterally) and is unlikely to undergo significant change

## Potential for Debris Transport and Delivery

The potential for debris transport and delivery is dependent on the availability of debris and the channel geometry. If debris is readily available, the potential for transport is dependent on the size of the debris relative to the channel width, depth, and planform. There is a high potential for transport if the channel width and depth exceed the maximum design log length and diameter. If there is a possibility that the width and depth of the upstream channel could increase in the future, the potential dimensions should be accounted for as well. Yet, even if the channel can accommodate the transport of debris, the channel planform may restrict delivery of the debris to the bridge site. Figure 5.3 provides a flow chart for determining the potential for debris transport and delivery to a bridge site.

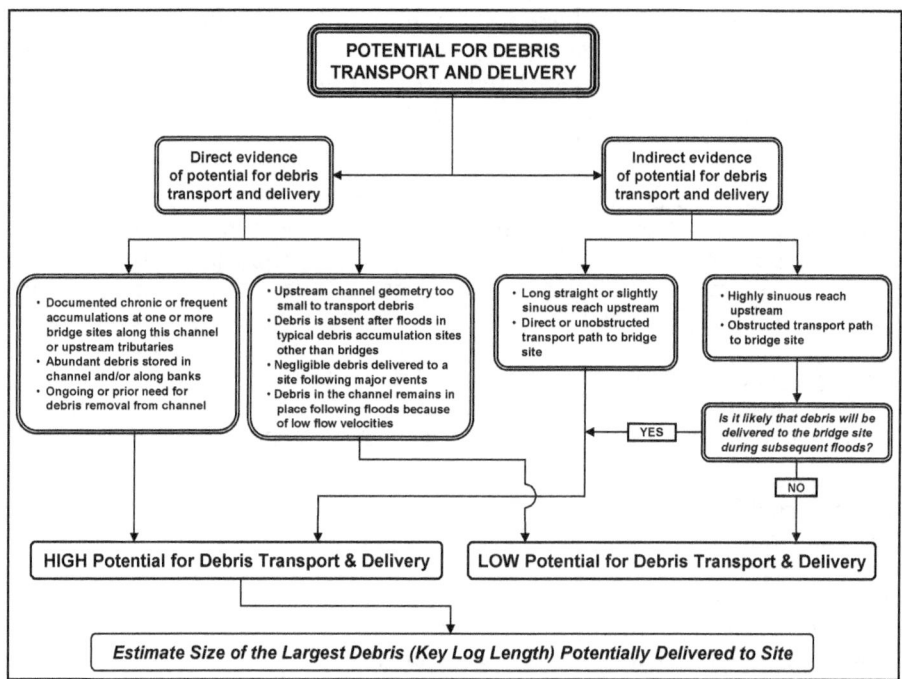

Figure 5.3. Flow chart for determining the potential for debris transport and delivery. (NCHRP 2010)

Direct Evidence. Observations of existing debris in the channel and at a site provide the most direct evidence for assessing the potential for debris transport and delivery to a site. Direct evidence for a high potential for debris delivery to a bridge site may include the following observations:

- Documented chronic or frequent debris accumulations at one or more bridges in the area
- Abundant debris stored in the channel and along the banks
- Ongoing or prior need for debris removal from the channel

Direct evidence of low debris delivery potential includes:

- The upstream channel is narrower and/or shallower during flood flows than most of the debris produced
- Debris is absent after floods in typical debris accumulation sites other than bridges (e.g., on bars and along the outer bank of meander bends)
- Negligible debris delivered to a site following large floods or other catastrophic events
- Debris in the channel remains in place following floods because of low flow velocities

Indirect Evidence. Long straight reaches upstream of a bridge site will provide the greatest potential for debris delivery. The thalweg of a straight channel is generally near the centerline and debris transported in the channel will generally follow that path. Some debris may become lodged along the banks or on bars as it moves downstream. Streams or rivers with low sinuosity or long, high radius bends can also provide a high potential for debris delivery to a bridge site since the flood path and, consequently, the debris path, will generally follow a relatively straight down valley path (Figure 5.4).

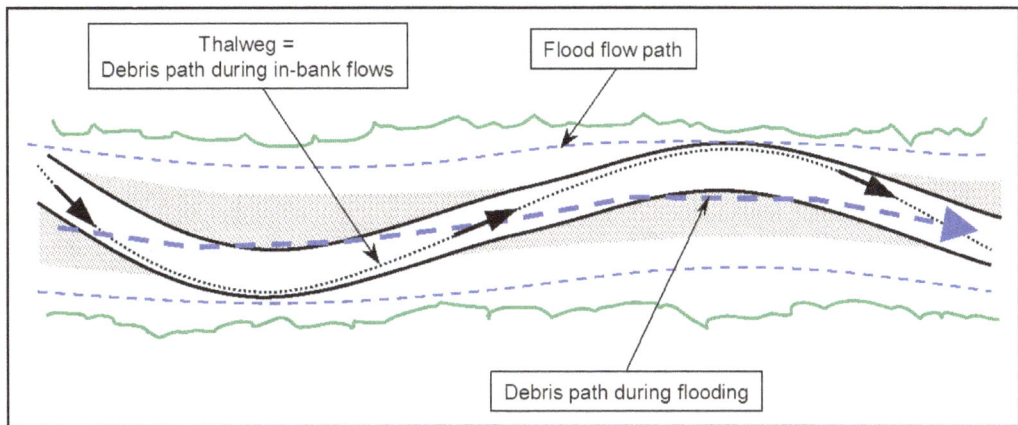

Figure 5.4. Hypothetical debris and flood flow paths during in-bank and out-of-bank flood flows for a low sinuosity channel (NCHRP 2010).

On forest-lined streams or rivers, as channel sinuosity increases and bend radius decreases, the potential for debris transport and delivery during any given flood decreases since debris is generally transported along the thalweg. Consequently, debris in highly sinuous, well-forested rivers or streams will not be transported far from its source area during any given event and is often deposited along or on top of the outer bank or on the point bar of the next downstream bend. However, some of this debris may eventually be moved downstream to a bridge site during subsequent flows.

Streams with actively migrating meander bends may be fairly sinuous, but can produce substantially more debris than less sinuous or relatively straight streams because of the bank erosion associated with lateral migration processes. Yet because of the planform geometry of forested meandering streams, delivery of debris on highly sinuous streams may take longer to reach a bridge in comparison to a low sinuosity or straight channel. Therefore, fairly sinuous streams with evident bend migration or bank erosion and significant debris production can be considered to have a moderate to high delivery potential, depending on existing conditions and sound engineering judgment.

### Size of Debris Potentially Delivered to Site

The potential for a channel to deliver debris to a site will be controlled by the ability of the stream to transport it. Existing or future channel dimensions, particularly width, upstream of a site determine the size of debris that can be transported and influences the potential size of accumulations.

The maximum design (or key) log length (Figure 5.5) is estimated either by examining the largest existing logs in the channel or on the basis of the channel width upstream of the site as measured at inflection points between bends (see Figure 2.10). Diehl (FHWA 1997a) states that the maximum log length on wide channels for much of the United States is about 80 feet (24 m) and that channels less than 40 feet (12 m) wide transport logs with lengths equal to or less than the upstream channel width. In the eastern United States, channels that are 40 to 200 feet (12 to 60 m) wide transport logs with an estimated design (or key) log length of 30 feet (9 m) plus one quarter of the channel width. Depth is important as well; the depth sufficient to float logs (Figure 5.5) is the diameter of the butt of the tree plus the distance the root mass extends out from the butt or approximately 3 to 5% of the estimated log length (FHWA 1997a). Diehl also indicates that the length of transported logs with attached rootwads rarely exceeds about 30 times the maximum flow depth.

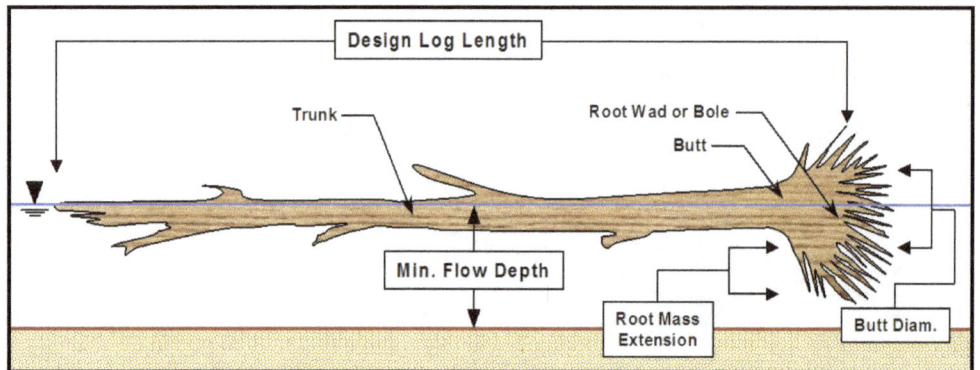

Figure 5.5. Schematic of design (or key) log length, butt diameter, and root mass extension.

## Drift Delivery Locations at a Site

Drift delivery at a highway crossing is localized and the location of accumulations can vary among piers and spans. Therefore, the potential for drift delivery should be evaluated at each pier and span. In general, floating material is transported along a relatively narrow drift path defined by secondary circulation currents converging at the surface within the channel. Piers located in this position are the most common sites of accumulation. The drift path typically coincides with the thread (thalweg) or center of the channel in straight reaches. The middle of the drift path generally lies between the thalweg and the outer bank in curved reaches. The drift path should be evaluated relative to flooding since it may not remain within the confines of the channel of meandering streams during out of bank flooding.

The potential for drift accumulation can also be strongly influenced by bridge characteristics. The width of horizontal opening and elevation of vertical openings between fixed elements of a bridge opening affects potential drift accumulation. Pieces of drift in the longest size fraction delivered to a site may come in contact with a bridge element, rotate downstream, and become lodged against another element, thereby trapping other debris. Skew also reduces the effective width of horizontal openings and increases the potential for debris trapping. Since most drift is transported at the surface, drift can become trapped between the bridge superstructure and the streambed when the water level is at or above the bottom of the superstructure. Drift can accumulate along or below the superstructure or may become lodged between the superstructure and streambed if they rotate vertically.

Narrow openings of structural elements of the bridge at the water surface also determine whether drift is deflected or trapped. Multiple closely spaced pier or pile groups, closely spaced rows of piers, or exposed pier footing piles are examples of narrow flow-carrying openings that can trap and accumulate drift. Where a pile bent or a pier composed of a single row of columns is skewed to the flow, either during normal or flood flows, debris can become trapped and accumulate within the narrow intervening apertures. Existing drift accumulations will also trap additional drift. The bridge superstructure or other bridge elements with flow-carrying openings or protrusions at or below the water surface can also accumulate drift if flood stages are sufficient. Freeboard (the distance between the water surface and elevation of the lowest element of the superstructure) should be large enough to pass the largest expected tree root ball.

There is considerable direct and indirect evidence of drift generation that can be collected and used to evaluate the potential for drift accumulation at a site. Most of this information can be collected as part of a stream reconnaissance. A comparable qualitative estimate can be made during a field reconnaissance or by bridge maintenance personnel with an intimate knowledge of the site. For detailed guidance, flow charts and an illustrative case study on debris accumulation potential on specific bridge sub-structure elements, reference to NCHRP Report 653 (NCHRP 2010) is suggested. This report also provides field reconnaissance data sheets designed specifically for assessing woody debris accumulation potential at a bridge site.

## 5.3 STREAM CHANNEL CLASSIFICATION

### 5.3.1 Overview

A stream classification scheme is a method of classifying a stream according to a set of observations. Streams are usually classified for the purpose of communication. A description of a stream by a classification gives the reader or audience an immediate "picture" of the appearance and condition of that stream channel and possibly its relationship to the surrounding floodplain and other streams in the system. More recently, classification schemes are also being used as a basis for channel restoration designs. There are a variety of classification schemes; the choice is problem dependent. For example, Niezgoda and Johnson (2005) provide a listing of 27 classification schemes devised over time, beginning in 1899, and a brief description of each.

Thus, channel classification systems provide engineers with useful information on typical characteristics associated with a given river type and establish a common language as a basis for communication. Classification requires identifying a range of geomorphological channel types that minimizes variability within them and maximizes variability between them (Thorne 1997). Given the complexity of natural systems, inevitably some information is sacrificed in the attempt to simplify a continuum of channel geomorphic characteristics into discrete intervals for classification. However, enough useful information can result from stream channel classification to make the effort worthwhile.

Although classifications are initially useful for clarity of communication and as an index of the numerous types of channels that exist, it is the characteristics of an individual channel that are important in defining channel processes and response. From a practical perspective, measurements of sinuosity, width-depth ratio, gradient, dimensions (width, depth), and sediment type (bed and bank) when combined with measurements or calculation of discharge, flow velocity and stream power will provide the information necessary for the understanding of a river and the knowledge required to evaluate stability and predict future change. When such quantitative information about a river is available, classifications are only the first step in evaluating channel stability and predicting channel change (Figures 1.1 and 4.1).

The most basic form of river classification uses channel pattern or planform to define three river types: straight, meandering, or braided. The discussion of geomorphic factors that affect stream stability in Chapter 2 uses this simple classification (Figure 2.6), and the "typical" planform characteristics of each stream type are discussed in Sections 2.3.10 and 2.3.11. Section 5.5.2 illustrates how the definitions of a simple classification can be extended with empirical data to provide reasonably definitive conclusions regarding stream type. Other approaches to channel classification use independent variables, such as discharge and sediment load to determine stream type.

### 5.3.2 Channel Classification Concepts

One of the most commonly used and useful classification schemes is stream order. The order of a stream basically describes the relationship of the stream to all other streams in the watershed. Streams that have no tributaries flowing into them are ranked as number one, or first-order streams. A second-order stream is one that is formed by the junction of two first-order streams or by the junction of a first- and a second-order stream. This ranking scheme is continued for all channels within the drainage basin. Stream order increases in the downstream direction and only one stream channel can have the highest ranking. This process of ranking is a rather simple matter for a small drainage basin, but can become cumbersome for a large, complex basin. Various stream characteristics have been related to stream order. For example, channel slope and channel length can be related to stream order in a given basin. First-order streams are typically steeper and shorter than second-order or higher streams. Fourth-order streams are typically relatively large, wide, low-gradient streams. This can be useful information in determining various characteristics about drainage basins, particularly large basins where extensive data gathering is impractical.

Rivers are often categorized as either straight, meandering, or braided. These categories identify the three major alluvial river types. An alluvial river is one that is flowing in a channel that has bed and banks composed of sediment transported by the river. That is, the channel is not confined by bedrock or terraces, but it is flanked by a floodplain. In addition to these three basic river "types," there are also anabranching alluvial rivers and rivers that are termed wandering. Brice (USARO 1975) illustrates the range of channel types for meandering, braided, and anabranching channels (see Figure 5.6, which expands the simpler classification of Figure 2.6).

Figure 5.6 shows the difference between low sinuosity, straight channels and meandering channels, as well as the difference between bar-braided and island-braided channels. It also demonstrates that the braided river occupies one channel whereas the anabranching channel has multiple channels separated by a vegetated floodplain.

The majority of the work on rivers has been concentrated on alluvial rivers. In order to develop a broader understanding of these rivers it is necessary to relate them to the independent variables that control channel size and morphology (shape, pattern, gradient). Channel size is clearly related to the volume of water conveyed by the channel. On average, channel width increases downstream as the square root of discharge (USGS 1953), and gradient decreases downstream as discharge increases. Assuming a graded stream, one that is neither progressively aggrading or degrading, the type of sediment transported by the river has a major influence on channel shape, pattern, and gradient. Table 5.2 summarizes a classification of alluvial channels based on the relative proportions of sand and silt-clay transported by a stream. Based on studies of rivers on the great plains of the U.S.A. and the riverine plain of Australia, it was determined that suspended-load streams that transported very little bedload were narrow, deep, gentle, and sinuous whereas bed-load streams were wide, shallow, steep, and relatively straight. This classification related channel characteristics to type of sediment load. During experimental studies it was further determined that valley gradient exerted a major influence on channel patterns.

Figure 4.2 in Chapter 4 suggests that the range of channels from straight through braided forms a continuum, but experimental work and field studies have indicated that within the continuum, river-pattern thresholds can be identified where the pattern changes between straight, meandering, and braided. The pattern changes take place at critical values of stream power, gradient, and sediment load (Schumm and Kahn 1972).

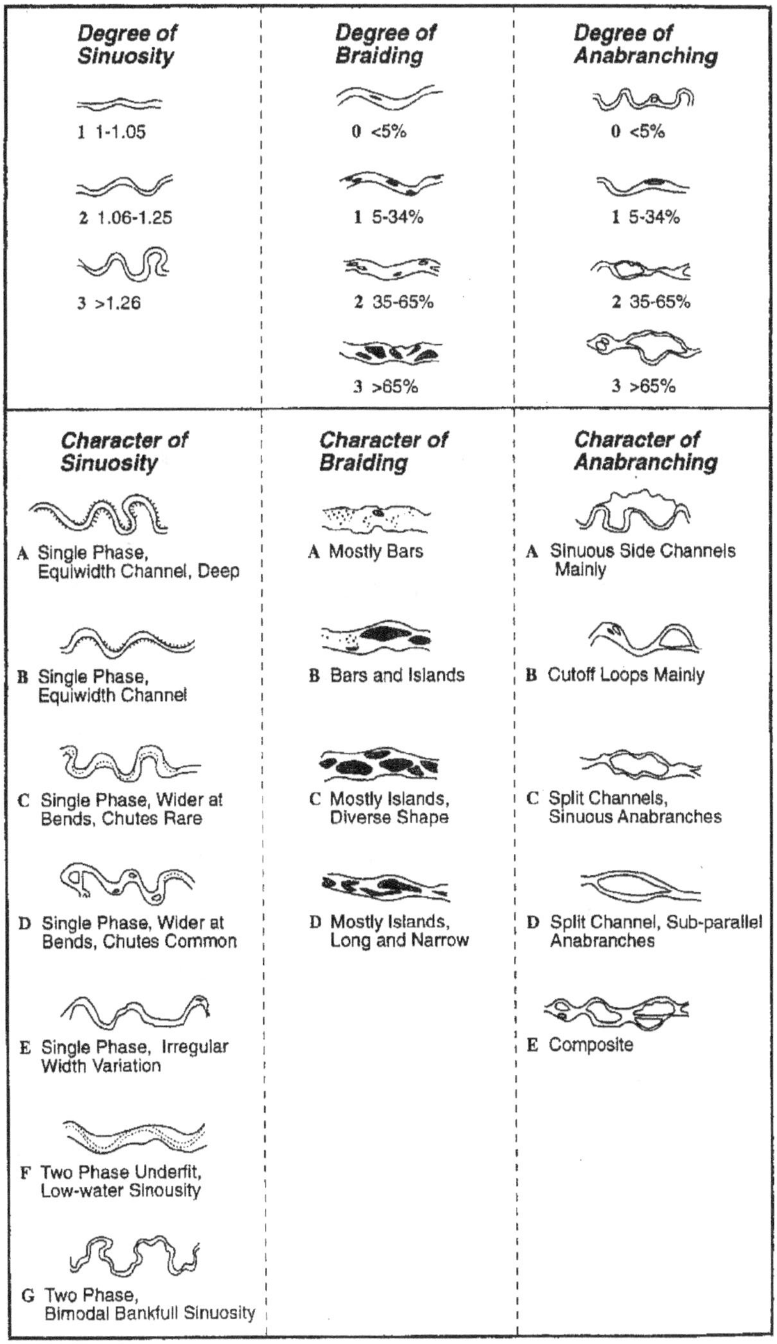

Figure 5.6. Alluvial channel pattern classification devised by Brice.
(after Brice (USARO 1975)).

| Table 5.2. Classification of Alluvial Channels (from Schumm 1977). | | | | |
|---|---|---|---|---|
| Mode of Sediment Transport and Type of Channel | Bedload (percentage of total load) | Channel Stability, Stable (graded stream) | Channel Stability, Depositing (excess load) | Channel Stability, Eroding (deficiency of load) |
| Suspended Load | <3 | Stable suspended-load channel. Width/depth ratio <10; sinuosity usually >2.0; gradient, relatively gentle | Depositing suspended load channel. Major deposition on banks cause narrowing of channel; initial streambed deposition minor. | Eroding suspended-load channel. Streambed erosion predominant; initial channel widening minor. |
| Mixed Load | 3-11 | Stable mixed-load channel. Width/depth ratio >10 <40; sinuosity usually <2.0 >1.3; gradient, moderate | Depositing mixed-load channel. Initial major deposition on banks followed by streambed deposition. | Eroding mixed-load channel. Initial streambed erosion followed by channel widening. |
| Bed Load | >11 | Stable bed-load channel. Width/depth ratio >40; sinuosity usually <1.3; gradient, relatively steep | Depositing bed-load channel. Streambed deposition and island formation. | Eroding bed-load channel. Little streambed erosion; channel widening predominant. |

In addition to the channel patterns shown in Figure 5.6, there are five basic bed-load channel patterns (Figure 5.7A) that have been recognized during experimental studies of channel patterns. These five basic bed-load channel patterns can be extended to mixed-load and suspended-load channels to produce 13 patterns (Figure 5.7). As indicated above, patterns 1-5 are bed-load channel patterns (Figure 5.7A), patterns 6-10 are mixed-load channel patterns (Figure 5.7B), and patterns 11-13 are suspended-load channel patterns (Figure 5.7C). For each channel type, pattern changes can be related to increasing valley slope, stream power, and sediment load.

The different bed-load channel patterns (Figure 5.7A) can be described as follows: Pattern 1: straight, essentially equal-width channel, with migrating sand waves; Pattern 2: alternate-bar channel with migrating side or alternate bars and a slightly sinuous thalweg; Pattern 3: low-sinuosity meandering channel with large alternate bars that develop chutes, and Pattern 4: transitional meandering-thalweg braided channel. The large alternate bars or point bars have been dissected by chutes, but a meandering thalweg can be identified. Pattern 5 is a bar-braided channel.

As compared to the bed-load channel pattern, the five-mixed load patterns (Figure 5.7B) are relatively narrower and deeper, and there is greater bank stability. The higher degree of bank stability permits the formation of narrow, deep straight channels (Pattern 6), and alternate bars stabilize because of the finer sediments, to form slightly sinuous channels (Pattern 7). Pattern 8 is a truly meandering channel, wide on the bends, relatively narrow at the crossings, and subject to chute cutoffs. Pattern 9 maintains the sinuosity of a meandering channel, but due to the greater sediment transport the presence of bars gives it a composite sinuous-braided appearance. Pattern 10 is an island-braided channel that is relatively more stable than that of bedload channel 5.

Suspended-load channels (Figure 5.7C) are narrow and deep. Suspended-load Pattern 11 is a straight, narrow, deep channel. With only small quantities of bed load, this type of channel will have the highest sinuosity of all (Patterns 12 and 13).

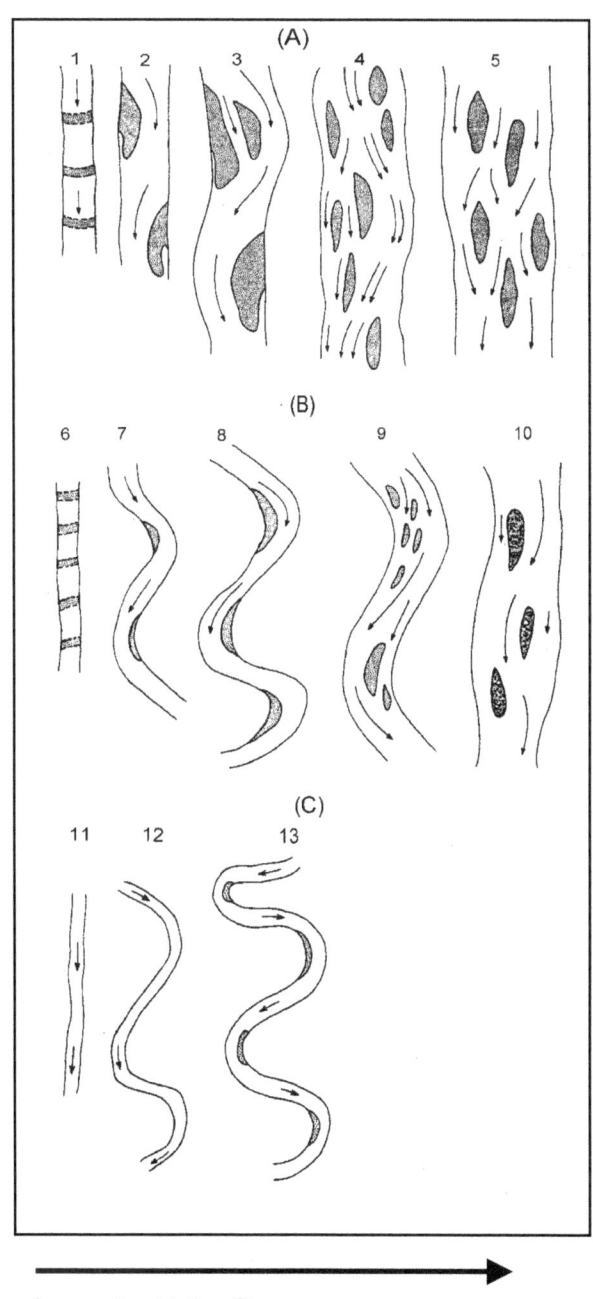

Figure 5.7. The range of alluvial channel patterns: (A) Bed-load channel patterns, (B) Mixed-load channel patterns, (C) Suspended-load channel patterns (from Schumm 1981).

It must be stressed that the preceding classification applies to adjustable alluvial rivers, with sediment loads primarily of sand, silt and clay, which would be considered regime channels by Montgomery and Buffington (1997) who have considered the full range of channels from high mountain bedrock channels to those described previously (Figures 5.6 and 5.7). This classification starts at the drainage divide (Figure 5.8) and moves down through bedrock and colluvial depressions or chutes to the point where one can recognize fluvial channels. Five distinct reach morphologies are identified: cascade, step-pool, plane-bed, pool-riffle, and dune-ripple (regime). Most of these reaches will be confined by valley walls and terraces in contrast to the alluvial regime channels. Thus, the stream type is based on the location within the watershed, the response to the sediment load, and several physical attributes. One of the primary advantages of using this method is that it is relatively simple and provides information on processes according to channel type. The channel types refer to natural, unmodified channels. Table 5.3 summarizes the important characteristics of each channel type.

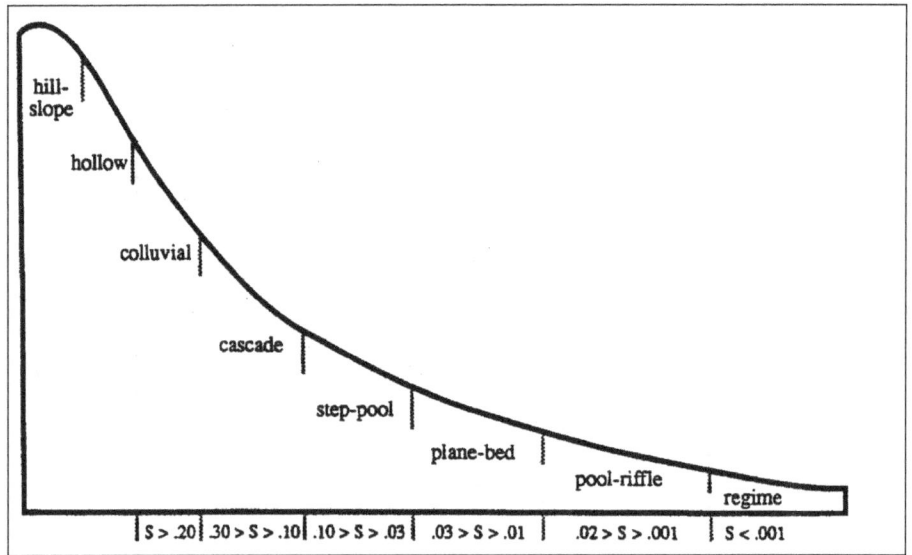

Figure 5.8. Idealized long profile from hillslopes and unchanneled hollows downslope through the channel network showing the general distribution of alluvial channel types (from Montgomery and Buffington 1997).

Rosgen (1994, 1996) developed a comprehensive system for classifying natural rivers. This system divides streams into seven major types on the basis of degree of entrenchment, gradient, width/depth ratio, and sinuosity. Within each major category there are six subcategories depending on the dominant type of bed/bank materials. For example, a C5 stream is a low gradient, meandering stream with a high sinuosity and a sand bed. The classification system shows a distinct bias toward streams that are relatively small and steep. For example, of the stream types categorized based on dominant bed material, seven are braided, 30 are entrenched, in the sense that overbank floods are confined by valley walls or terraces, and four are narrow, sinuous mountain meander-type channels. The basic framework of Rosgen's method is set out in Figures 5.9 and 5.10. This classification is comprehensive in its scope, but requires "a strong geomorphological insight and understanding to apply consistently and usefully" (Thorne 1997). For a discussion of Rosgen's approach to channel restoration, see Chapter 9.

Table 5.3. Classification of Channel-Reach Morphology in Mountain Drainage Basins of the Pacific Northwest. (from Montgomery and Buffington 1997)

| | Dune Ripple | Pool Riffle | Plane Bed | Step Pool | Cascade | Bedrock | Colluvial |
|---|---|---|---|---|---|---|---|
| Typical Bed Material | Sand | Gravel | Gravel-cobble | Cobble-boulder | Boulder | Rock | Variable |
| Bedform Pattern | Multilayered | Laterally oscillatory | Featureless | Vertically oscillatory | Random | Irregular | Variable |
| Dominant Roughness Elements | Sinuosity, bedforms (dunes, ripples, bars) grains, banks | Bedforms (bars, pools), grains, sinuosity, banks | Grains, banks | Bedforms (steps, pools), grains, banks | Grains, banks | Boundaries (bed and banks) | Grains |
| Dominant Sediment Sources | Fluvial, bank failure | Fluvial, bank failure | Fluvial, bank failure, debris flows | Fluvial, hillslope, debris flows | Fluvial, hillslope, debris flows | Fluvial, hillslope, debris flows | Hillslope, debris flows |
| Sediment Storage Elements | Overbank, bedforms | Overbank, bedforms | Overbank | Bedforms | Lee and stoss sides of flow obstructions | Pockets | Bed |
| Typical Confinement | Unconfined | Unconfined | Variable | Confined | Confined | Confined | Confined |
| Typical Pool Spacing (channel widths) | 5 to 7 | 5 to 7 | None | 1 to 4 | <1 | Variable | Confined |

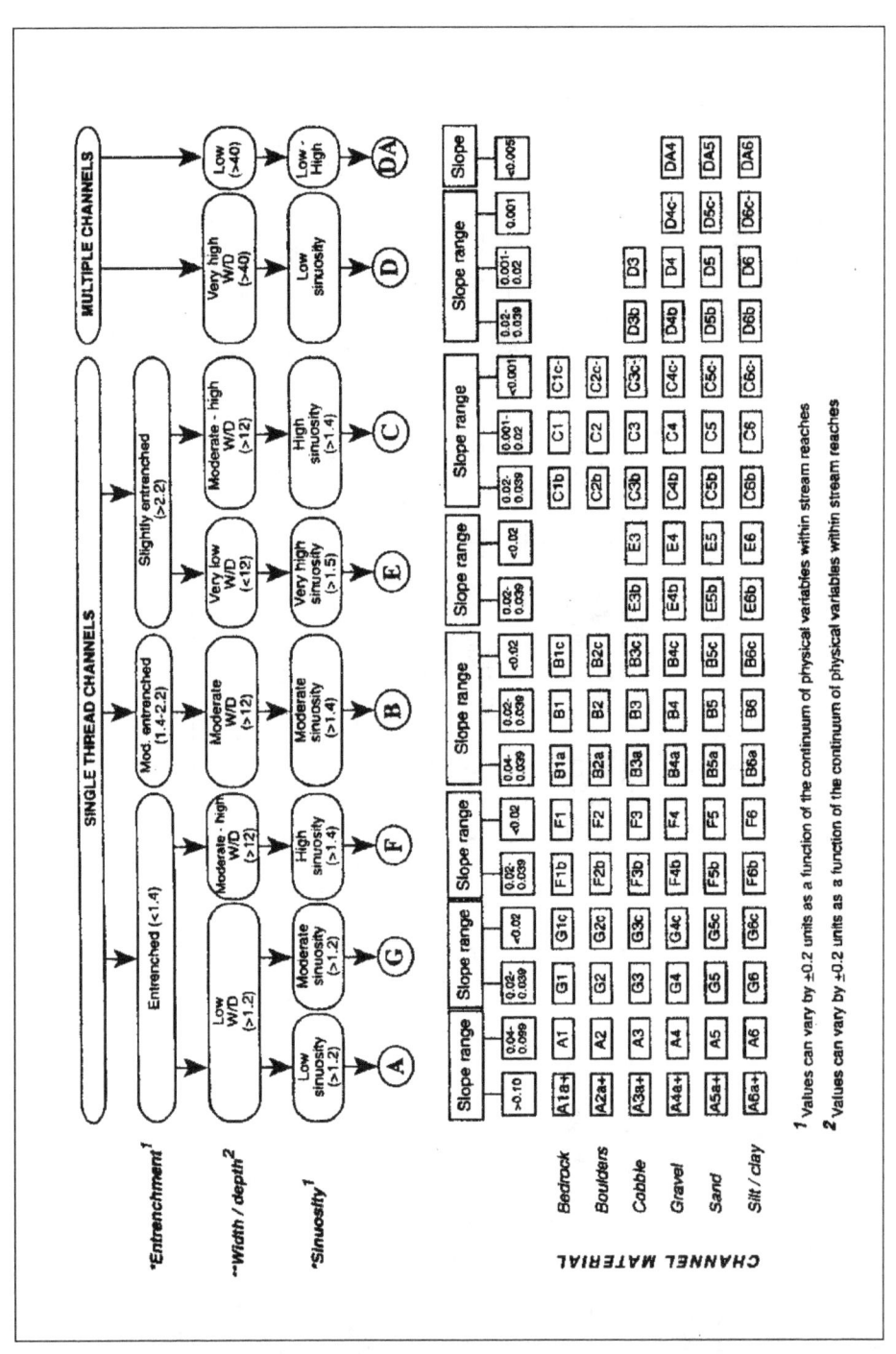

Figure 5.9. Key to classification of rivers in Rosgen's method (modified from Rosgen 1994 by Thorne (1997)).

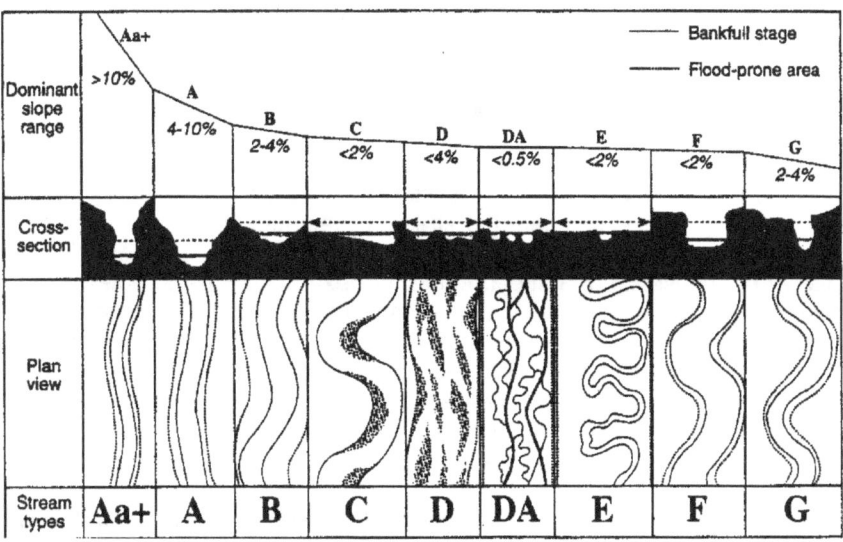

Figure 5.10. Longitudinal, cross-sectional, and planform views of major stream types in Rosgen's method (modified from Rosgen 1994 by Thorne (1997)).

### 5.3.3 Channel Classification and Stream Stability

There are a number of methods for classifying stream channels based on channel stability. Some require the expertise of an experienced geomorphologist while others require only a brief period of training. All of these methods are, at least in part, based on observations of a variety of parameters that describe the characteristics and conditions of the channel and surrounding floodplain. The purpose of each of these methods is to assess the current condition of the channel and possibly identify the processes that are acting to change the condition over at least a reach level or over the entire watershed system. The goal of the assessments is to better understand the processes so that stream restoration, bank stabilization, or a host of other river applications can be designed successfully. These methods are discussed briefly below.

Pfankuch (USDA 1978) developed a method to rate stream stability for mountain streams in the northwestern US. The methodology was developed for the purpose of planning various stream projects on second- to fourth-order streams. The user evaluates the condition of the stream by assessing and rating 15 stability indicators. The total score is then related to a subjective description for the overall stability of the stream as excellent, good, fair, or poor, such that the higher the number, the more unstable the stream. Several of the parameters in this method, such as channel width-depth ratio, may not be useful as stability indicators in channels outside of the area used to develop the method. Myers and Swanson (1992; 1996) applied the method developed by Pfankuch (USDA 1978) to assess and monitor stream channel stability for streams in northern Nevada. They correlated the stream stability ranking to the stream type according to the Rosgen (1994) classification scheme. They found that several of the stability indicators proposed by Pfankuch were not useful in the evaluation. Based on these findings they deleted rock angularity from the rating procedure and separated the combined scour and deposition indicator into two individual indicators. They also made slight adjustments to the scoring procedure. In addition, they found that if the rating was combined with a stream classification the underlying morphological processes could be inferred from the classification which could then be used to indicate the type of engineering response that would be appropriate to mitigate further stream instability.

Based on prior work by Simon (1989), Simon and Downs (1995) developed a method for assessing stability of channels that have been straightened. In this method the data are summarized on a ranking sheet, a weight is assigned, and a total rating is derived by summing the weighted data in each category. The higher the rating, the more unstable the channel. Simon and Downs found that for streams in western Tennessee, a rating of 20 or more indicated an unstable channel that could threaten bridges and land adjacent to the channel. The rating system provides a systematic method for evaluating stability; however, the final ratings cannot be compared to streams evaluated in other geomorphic, geologic, or physiographic regions. In addition, some of the parameters are very difficult to assess, particularly in the absence of a stream gage. For example, considerable weight is placed on the identification of the stage of channel evolution. To properly assess this stage, it is necessary to determine whether the channel is in the process of widening, degrading, or aggrading. Simon and Hupp (USGS 1992) provide a good description of determining bed degradation based on gage data. However, the determination of aggradation or degradation based on a gage analysis typically requires at least several years of stream gage data and cannot usually be determined during a brief site visit (Johnson et al. 1999).

Thorne et al. (1996) expanded upon the method developed by Simon and Downs (1995) by adding a quantitative segment based predominantly on hydraulic geometry analysis. The ranking based on the Simon and Downs method provides a qualitative assessment while the comparison of measured hydraulic geometry to that calculated from equations developed for stable channels provides a quantitative measure of stability. A set of hydraulic geometry equations was assembled for gravel-bed rivers and the use was demonstrated on an actual river (see Chapter 8, Section 8.3.1). The observed width and depth of a stream reach were compared to the regime width and depth calculated from the hydraulic geometry equations. Significant differences can then be assumed to imply that the observed channel is either in regime or it is not. Although this is a reasonable approach, hydraulic geometry equations must be used cautiously since they are empirically derived.

Montgomery and MacDonald (2002) suggest a diagnostic approach in which the system and system variables are defined, observations are made to characterize the condition of the system, and an evaluation is made to assess the causal mechanisms producing the current condition. Observations are based on characterizing both the valley bottom and the active channel based on a set of field indicators. Valley bottom indicators include the channel slope, confinement, entrenchment, riparian vegetation, and overbank deposits. Indicators for the active channel include the channel pattern, bank conditions, gravel bars, pool characteristics, and bed material.

Rosgen (2001) proposed a channel stability assessment method that is based on assessing stability for a stable reference reach and then the departure from the stable conditions on an unstable reach of the same stream type. The stability analysis consists of 10 steps that assess various components of stability. The steps include measurement or description of: (1) the condition or "state" categories (riparian vegetation, sediment deposition patterns, debris occurrence, meander patterns, stream size or order, flow regime, and alterations); (2) vertical stability in terms of the ratio of the lowest bank height in a cross section divided by the maximum bankfull depth; (3) lateral stability as a function of the meander width ratio and the bank erosion hazard index (BEHI); (4) channel pattern; (5) river profile and bed features; (6) width-depth ratio; (7) scour and fill potential in terms of critical shear stress; (8) channel stability rating using a modification of the Pfankuch (USDA 1978) method; (9) sediment rating curves; and (10) stream type evolutionary scenarios. This is a very data-intensive assessment method and not one that bridge inspectors or hydraulic engineers will likely use due to the time and expense of data collection. However, one of the more interesting components of this method is the procedure for step 8. Like the Pfankuch method, a rating of good, fair, or poor is obtained based on a numerical rating. However, Rosgen modified the method to account for differences across different stream types so that each stream type has a separate definition for good, fair, and poor. For example, a rating of 60 would be considered poor in a B1 stream, fair in a C1 stream, and good in a F1 stream. Although the approach is interesting and has merit, the basis for the separate rating schemes is not given.

The U.S. Corps of Engineers (USACE 2001, USACE 1994a, b) suggests a three-level stability analysis for the purpose of stream restoration design. Level 1 is a geomorphic assessment, Level 2 is a hydraulic geometry assessment, and Level 3 is an analytical stability assessment that includes a sediment transport study. As part of the geomorphic assessment, they recommend collecting the following field data: descriptions of watershed development and land use, floodplain characteristics, channel planform, and stream gradient; historical conditions; channel dimensions and slope; channel bed material; bank material and condition; bed forms, such as pools, riffles, and sedimentation; channel alterations and evidence of recovery; debris and bed and bank vegetation; and photographs. Indicators of channel degradation are given as terraces, perched channels or tributaries, headcuts and knickpoints, exposed pipe crossings, perched culvert outfalls, undercut bridge piers, exposed tree roots, leaning trees, narrow and deep channel, undercut banks on both sides of the channel, armored bed, and hydrophytic vegetation located high on the banks. Indicators of a stable channel include vegetated bars and banks, limited bank erosion, older bridges, culverts and outfalls with inverts at or near grade, no exposed pipeline crossings, and tributary mouths at or near existing main stem stream grade. Copeland et al. further suggest that spatial bias in assessing stability can be reduced by walking a distance well upstream and downstream of the project reach while temporal bias can be reduced by revisiting the site at different times of year. The USACE manual on assessing channel stability for flood control projects (USACE 1994a) provides a detailed example of a quantitative stability analysis, based primarily on critical and design flow shear stresses.

Annandale (1994; 1999) developed a two-level procedure to determine the risk of bridge failure that included river instability. The first level is a hazard assessment and procedure for rating hazards. The hazard assessment is comprised of river instability, potential for morphological change, fluvial hydraulics in the immediate vicinity of the river crossing, and the structural integrity of the river crossing. He provides four tables of values to assign for each of these factors. The values were based on river crossing failures in South Africa, New Zealand, and the US. The hazard rating is then the product of the four values. The table for assessing the hazard rating for the river stability factor is based on channel type from Schumm (1977) (see Table 5.2). A second table provides the ratings for the potential for morphological changes (degradation, bank erosion, and aggradation) due to extraneous factors. Annandale accounts for the location of the bridge with respect to stream meanders as a separate factor in his method. If the bridge is between meanders or on a tight bend, the factor value is increased. The hazard rating, based on the product of the four factors, is categorized as significant, moderate, or low.

Individual states have also developed protocols and methods for assessing stream stability. For example, the state of Vermont has put together an extensive manual on stream stability assessment (Vermont Water Quality Division 2001). Their method basically follows that of Pfankuch (USGS 1978). They include a field form for bridges and culverts; however, it is primarily an inventory for habitat disruptions, rather than part of the stability assessment.

The U.S. Geological Survey developed a method to determine a "potential scour index" for assessment and estimates of maximum scour at selected bridge sites in Iowa (USGS 1995).

This work was completed in cooperation with the Iowa Highway Research Board and the Iowa Department of Transportation and was also based on the Western Tennessee study by Simon et al. (FHWA 1989b).

The potential scour index used in the USGS study is comprised of 11 principal stream stability and scour components. A value is assigned to each component according to the results of an onsite evaluation, and the potential scour index is the sum of the component values. Larger values of the index suggest a greater likelihood for scour-related problems to occur. Evaluation of several of the index components is somewhat subjective and assigned values may vary depending on the inspector's judgment and experience. However, no single component dominates the potential-scour index, and variations in the assigned values probably tend to cancel each other out when the components are summed to produce the index (USGS 1995).

The 11 principal index components are:

- Bed material
- Bed protection (i.e., riprap)
- Stage of channel evolution
- Percentage of channel constriction
- Number of bridge piers in the channel
- Percentage of blockage by debris
- Bank erosion
- Proximity of river meander impact point to bridge
- Pier skew
- Mass wasting (bank failure) at bridge
- Angle of approach of high flows

A potential scour assessment is used to help determine whether a bridge may be vulnerable to scour. Although a potential scour assessment cannot predict actual scour during a flood, it provides a measure of the likelihood of scour-related problems occurring, both during a flood and over time as the channel-evolution processes work on the stream. The assessment is accomplished by an onsite evaluation using a scour-inspection form.

Using the USGS method, potential scour assessments were performed at 130 highway bridges throughout Iowa. The drainage areas upstream from the bridges range from 23 to 7,785 mi$^2$ (60 to 20,163 km$^2$). All of the bridges were structures supported by abutments and possibly one or more piers. The ages of the bridges ranged from less than 5 to more than 70 years. The results of the assessments are summarized in USGS Water-Resources Investigation Report 95-4015 (USGS 1995).

In addition to specific indicators listed for various methods, Shields (1996) determined that factors in the watershed should be examined as part of assessing current and future channel stability. Watershed characteristics include:

- Physical characteristics and the channel network. Shields suggests using multiple classification methods, such as Schumm (1977), Rosgen (1994), Harvey and Watson (1986), and the US Army Corps of Engineers (USACE 2001) methods to classify these physical characteristics. Using multiple methods is useful in that they each give different results and information.

- Nature of existing and future hydrologic response and sediment yield. Water and sediment discharges are affected by urbanization, deforestation, mining, logging, and other disturbances. Changes will cause a response in the stream channel, possibly causing instability.

- Existing instability in the overall system and the causes. Depending on the problem, channel instabilities can move upstream or downstream, possibly moving into a project area and causing destabilization there.

Bank stability can be determined as a function of either fluvial erosion or mass wasting (geotechnical). Factors that influence fluvial bank erosion include bank material, stream power, shear stress, secondary currents, local slope, bend morphology, vegetation, and bank moisture content (see Appendix B). Factors that influence mass wasting include bank height, angle, material, and moisture content.

Many stream channel stability indicators are common to multiple assessment methods discussed above. These indicators are summarized in Table 5.4. Characteristics of those indicators are also provided. Additional information is given in the references provided.

Johnson (2005) developed a rapid stability assessment method based on geomorphic and hydraulic indicators to provide a semi-quantitative Level 1 analysis and to determine whether it is necessary to conduct a more detailed Level 2 analysis. Thirteen qualitative and quantitative stability indicators are rated, weighted, and summed to produce a stability rating for gravel bed channels. This method was based largely on prior assessment methods and the indicators shown in Table 5.4. This method was extended to different physiographic regions in 2006 and is described in detail in the following section.

## 5.4 RAPID ASSESSMENT OF CHANNEL STABILITY

### 5.4.1 Overview

Given the time constraints for field reconnaissance and bridge inspections and the expense of conducting lengthy geomorphic studies, it would be desirable to have a technique for rapid channel stability assessments. Johnson et al. (1999) reviewed existing methods and concluded that there are a number of methods currently available for assessing channel stability. Some require the expertise of an experienced geomorphologist while others require only some period of training. All of these methods are, at least in part, based on subjective observations of a variety of parameters that describe the characteristics of the channel and surrounding floodplain.

Based on existing methods described in Section 5.3.3 and data collected at bridges in 13 physiographic regions of the continental United States (FHWA 2006), Johnson developed a rapid assessment method for the preliminary documentation and rating of channel stability near bridges. This method provides an assessment of channel stability conditions as they affect bridge foundations. It is intended for a quick assessment of conditions for the purpose of documenting conditions at bridges and for judging whether more extensive geomorphic studies or complete hydraulic and sediment transport analyses are needed to assess the potential for adverse conditions developing at a particular bridge in the future. In this section, the method is described in detail and illustrated with several examples.

Table 5.4. Summary of Common Indicators Used in Channel Stability Assessment Methods (Johnson 2005).

| | Characteristics | References |
|---|---|---|
| Boundary conditions | Changes in loads to the stream, including water and sediment inputs. Causes channel to adjust. Amount and type of adjustment primarily based on channel type. | Lane, 1955; Montgomery and MacDonald, 2002 |
| Flow habit and flashiness | Ephemeral, intermittent, or perennial; flashiness; stream order. Flashy urban or ephemeral streams tend to be unstable relative to non-flashy perennial streams. Large ephemeral streams more unstable than first-order. | USACE 1994a, Chang 1988, HEC-20 Figure 2.6 |
| Valley setting, confinement | Confinement by narrow valley, rock walls, or revetments limits lateral migration and may promote downcutting. | HEC-20 Figure 2.6 |
| Channel pattern and type | Indicates energy in the system, propensity for lateral migration, processes on larger scale. | HEC-20 Figure 2.6, Montgomery and MacDonald 2002, Schumm 1977, Annandale 1999, USGS 1966, Brookes 1987, Chang 1988, Rosgen 2001 |
| Entrenchment, incision | Indicated by levees, terraces, low width-depth ratio, confinement, exposed infrastructure. Indicates processes that have been occurring. | Thorne 1998, Montgomery and MacDonald 2002 |
| Bed material | Size, uniformity, and packing are typical measurements and observations that indicate movement, armoring, and energy of the system. Fraction of sand and gravel can also indicate sediment transport characteristics (Wilcock, 1997; Wilcock and Crowe, 2003). | USDA 1978, Wilcock 1997, Wilcock and Crowe 2003, Parker et al. 1982, USACE 1994, USGS 1966, Brookes 1987, Chang 1988, Andrews 1983, Carson and Griffiths 1987, Gordon et al. 1992 |
| Bar development | Indicators of bar stability include the relative bar size, bar material, and vegetation. For streams on higher slopes (greater than about 0.02) and low width-depth ratios < 12, bars are not typically evident. | HEC-20 Figure 5.2, Montgomery and MacDonald 2002 |
| Bed and bank obstructions | Obstructions include rock outcrops, cohesiveness, grade control, bridge bed paving, debris jams, dikes or vanes, revetments. They can cause flow diversions that result in erosion of beds and banks. | USDA 1978, Thorne 1998, Montgomery and MacDonald 2002 |
| Bank material | Layers/lenses, material size and sorting, and cohesiveness indicate the coherence of the material and resistance to erosion. | Osman and Thorne 1988, USACE 1994, Wilcock 1997, ASCE 1998; HEC-20 Appendix B |
| Bank angle and height | Critical heights can be determined, greater than which the bank is susceptible to failure. Stable banks have low angles and heights. | Wilcock 1997, USDA 1978, Thorne and Osman 1988, ASCE 1998; HEC-20 Appendix B |
| Bank and riparian vegetation | Woody vegetation helps to maintain stability. The type, diversity, density, health, vertical orientation and maturity of woody vegetation are indicators of stabilization effects. Rosgen (2001) includes root density. | USDA 1978, Thorne et al. 1996, Rosgen 2001; HEC-20 Figure 2.6 |
| Bank cutting | The relative amount of bare banks and root exposure along a bank are indicative of stability. | USDA 1978, Montgomery and MacDonald 2002; HEC-20 Appendix B |
| Mass wasting | Indicated by scalloping of banks, irregular channel width, tension cracks, slumping. | USDA 1978; HEC-20 Figures 2.8 and 2.9, Appendix B |

A rapid stability assessment method should have the following characteristics: (1) it should be brief such that it can be completed quickly; (2) it should be simple in that extensive training is not required (although some training will be required); (3) it should be based on sound indicators as discussed in Section 5.3.3; and (4) it should be based on the needs of the bridge engineering community.

One way to assure that all aspects of channel stability are included is to start at the watershed or regional level and focus in on vertical and lateral aspects of the channel, following the concepts of Thorne (1998) (see Appendix C) and Montgomery and MacDonald (2002). Thus, at the broader level, watershed and floodplain activities and characteristics, flow habit, channel pattern and type, and entrenchment are selected as appropriate indicators. At the channel level, indicators such as bed material consolidation and armoring, bar development, and obstructions are used. Indicators of bank stability include bank material, angle, bank and riparian vegetation, bank (fluvial) cutting, and mass wasting (geotechnical failure). Finally, the position of the bridge relative to the channel can be indicated by meander impact point and alignment.

### 5.4.2 Rapid Assessment Method

The rapid channel stability assessment method uses a set of indicators, as determined from the literature and field observations. For each indicator, a rating is selected, the ratings are summed for a total score, and the score compared to stability definitions. In order to provide an appropriate level of sensitivity, the stability is based on stream type. Given that the Montgomery–Buffington classification method (Table 5.3) is based on processes as well as physical characteristics, this scheme was used to provide additional sensitivity to the method.

There are several assumptions implicit in this method of obtaining an overall rank. First, all indicators are weighted equally. This assumption was tested by assigning weights to each of the indicators and creating a weighted score for every bridge where observations were made. The results showed that the weighted indicators yielded the same results as the equally weighted indicators. Thus, there was no advantage in using weights. Second, this method implies that each indicator is independent of all others. While it is possible that some correlation exists between several of the indicators, an attempt was made to select indicators that independently describe various aspects of channel stability; thus, correlation effects were judged to be insignificant. Third, the summing of the ratings implies a linear scheme. The impact of this is not precisely known; however, given that weighted ratings provided no change in the overall results (Johnson 2005; FHWA 2006), it can be assumed that the linearity will also not affect the results significantly.

In collecting the data and observations for this method, it is desirable for the engineer or other inspector to view aerial photos of the bridge crossing and surrounding area and to walk some distance upstream and downstream from the bridge, rather than making all observations of the channel from the bridge itself. The appropriate distance, however, depends on several factors, such as uniformity of stream conditions; magnitude of disturbances along the banks, in the floodplain, or in the watershed; time available; and accessibility. Ideally, the observer should walk at least 10 channel widths upstream and downstream of the bridge. Although it is possible to establish stability conditions in a lesser distance, the more of the stream that is observed, the better the observer will understand the causes, processes, and rates of change, assuming that such observations are repeated at different times. Roads and bridges often divide property and sometimes divide geomorphic features or regions. Thus, conditions upstream and downstream of the bridge may be significantly different. In this case, it may be necessary to conduct separate analyses upstream and downstream.

### 5.4.3 Stability Indicators

The 13 indicators identified for this study are listed in Table 5.5 (FHWA 2006). For each indicator, a rating of poor, fair, good, or excellent can be assigned based on descriptors listed in the table. After a rating is assigned for each of the indicators, an overall score is obtained by summing the 13 ratings. This total score provides the overall relative stability of the channel. Table 5.6 provides the range of scores for Excellent, Good, Fair, and Poor ratings of stability for each of the three divisions of stream channels. The simplified data collection sheets of Appendix E assist in obtaining information necessary to score the stability indicators.

Occasionally, rating of each of the thirteen factors for a particular bridge will result in one factor which stands out as being much higher (worse) than the others. This situation is worth noting and making additional observations during future inspections.

### 5.4.4. Lateral and Vertical Stability

The indicators in Table 5.5 can be divided into those that indicate vertical stability and those that indicate lateral stability; vertical stability is described by indicators 4–6, while lateral stability is indicated by indicators 8 –13. Each of the lateral and vertical stability scores, based on summing the appropriate ratings, were normalized by the total number of points possible in each category so that they could be represented as a fraction and more readily compared. If the lateral fraction is greater than the vertical fraction, then it can be expected that channel instability is expressed primarily in the lateral direction. Lateral and vertical processes may be ongoing simultaneously or they may be occurring differentially; this is not indicated by the assessment method. If both fractions are relatively low, this suggests minimal instability in either direction.

As an example, if the lateral score is significantly higher than the vertical score (say for example, 0.93 versus 0.67), indicating that lateral instability is dominant. If, on the other hand, the vertical score fraction is greater than the lateral, then bed degradation is the dominant source of instability. If both scores are high, then the channel is unstable due to both lateral and vertical processes. For example, if a channel has lateral and vertical fractions of 0.86 and 0.92, this indicates that the channel is both degrading and widening.

### 5.4.5. Examples

In this section, examples are provided using the stream stability assessment method. Figures 5.11 to 5.14 show four streams, along with their overall ratings. The streams represent a wide range of conditions, stream types, and physiographic regions, including Pacific Coastal (Figure 5.11), Great Plains (Figure 5.12), Atlantic Coastal (Figure 5.13), and New England (Figure 5.14). Tables 5.7 and 5.8 provide details of the ratings based on the stream stability assessment method. Indicators 1 and 3 were primarily based on a view of the area surrounding the bridge as seen from satellite imagery or aerial photos. Viewing the bridge and surrounding area from above provides a "big picture" view that cannot be easily obtained in a short visit to a bridge. The overall stability was obtained from Table 5.6, in each case using the category of "pool-riffle, plane-bed, dune-ripple, and engineered channels." The lateral and vertical fractions of stability indicate that in each case neither lateral nor vertical problems predominate.

Table 5.5. Stability indicators, descriptions, and ratings. Range of values in ratings columns provide possible rating values for each factor.

| Stability Indicator | Excellent (1-3) | Good (4-6) | Fair (7-9) | Poor (10-12) |
|---|---|---|---|---|
| 1. Watershed and floodplain activity and characteristics | Stable, forested, undisturbed watershed. | Occasional minor disturbances in the watershed, including cattle activity (grazing and/or access to stream), construction, logging, or other minor deforestation. Limited agricultural activities. | Frequent disturbances in the watershed, including cattle activity, landsliding, channel sand or gravel mining, logging, farming, or construction of buildings, roads, or other infrastructure. Urbanization over significant portion of watershed. | Continual disturbances in the watershed. Significant cattle activity, landsliding, channel sand or gravel mining, logging, farming, or construction of buildings, roads, or other infrastructure. Highly urbanized or rapidly urbanizing watershed. |
| 2. Flow habit | Perennial stream with no flashy behavior | Perennial stream or ephemeral $1^{st}$ order stream with slightly increased rate of flooding | Perennial or intermittent stream with flashy behavior | Extremely flashy; flash floods prevalent mode of discharge; ephemeral stream other than $1^{st}$ order stream |
| 3. Channel pattern | Straight to meandering with low radius of curvature; primarily suspended load | Meandering, moderate radius of curvature; mix of suspended and bed loads; well maintained engineered channel | Meandering with some braiding; tortuous meandering; primarily bed load; poorly maintained engineered channel. | Braided; primarily bed load; unmaintained engineered channel. |
| 4. Entrenchment / channel confinement | Active floodplain exists at top of banks; no sign of undercutting infrastructure; no levees | Active floodplain abandoned, but is currently rebuilding; minimal channel confinement; infrastructure not exposed; levees are low and set well back from the river | Moderate confinement in valley or channel walls; some exposure of infrastructure; terraces exist; floodplain abandoned; levees are moderate in size and have minimal setback from the river. | Knickpoints visible downstream; exposed water lines or other infrastructure; channel width to top of banks ratio small; deeply confined; no active floodplain; levees are high and along the channel edge. |
| 5. Bed material Fs = approximate portion of sand in the bed | Assorted sizes tightly packed, overlapping, and possibly imbricated. Most material > 4 mm. Fs < 20% | Moderately packed with some overlapping. Very small amounts of material < 4 mm. 20 < Fs < 50% | Loose assortment with no apparent overlap. Small to medium amounts of material < 4 mm. 50 < Fs < 70% | Very loose assortment with no packing. Large amounts of material < 4 mm. Fs > 70% |
| 6. Bar development<br><br>S = Slope<br>W/Y = Width to Depth ratio | For S < 0.02 and W/Y > 12, bars are mature, narrow relative to stream width at low flow, well vegetated, and composed of coarse gravel to cobbles. For S > 0.02 and W/Y < 12, no bars are evident. | For S < 0.02 and W/Y > 12, bars may have vegetation and/or be composed of coarse gravel to cobbles, but minimal recent growth of bar evident by lack of vegetation on portions of the bar. For S > 0.02 and W/Y < 12, no bars are evident. | For S < 0.02 and W/Y > 12, bar widths tend to be wide and composed of newly deposited coarse sand to small cobbles and/or may be sparsely vegetated. Bars forming for S > 0.02 and W/Y < 12. | Bar widths are generally greater than ½ the stream width at low flow. Bars are composed of extensive deposits of fine particles up to coarse gravel with little to no vegetation. No bars for S < 0.02 and W/Y > 12. |
| 7. Obstructions, including bedrock outcrops, armor layer, large woody debris jams, grade control, bridge bed paving, revetments, dikes or vanes, riprap | Rare or not present. | Occasional, causing cross currents and minor bank and bottom erosion. | Moderately frequent and occasionally unstable obstructions, cause noticeable erosion of the channel. Considerable sediment accumulation behind obstructions. | Frequent and often unstable causing a continual shift of sediment and flow. Traps are easily filled causing channel to migrate and/or widen. |

Table 5.5. Stability indicators, descriptions, and ratings. Range of values in ratings columns provide possible rating values for each factor.

| Stability Indicator | Excellent (1-3) | Good (4-6) | Fair (7-9) | Poor (10-12) |
|---|---|---|---|---|
| 8. Bank soil texture and coherence | Clay and silty clay; cohesive material | Clay loam to sandy clay loam; minor amounts of noncohesive or unconsolidated mixtures; layers may exist, but are cohesive materials. | Sandy clay to sandy loam; unconsolidated mixtures of glacial or other materials; small layers and lenses of noncohesive or unconsolidated mixtures | Loamy sand to sand; noncohesive material; unconsolidated mixtures of glacial or other materials; layers or lenses that include noncohesive sands and gravels |
| 9. Average bank slope angle (where 90° is a vertical bank) V = Vertical H = Horizontal | Bank slopes < 3H:1V (18°) for noncohesive or unconsolidated materials to < 1:1 (45°) in clays on both sides | Bank slopes up to 2H:1V (27°) in noncohesive or unconsolidated materials to 0.8:1 (50°) in clays on one or occasionally both banks | Bank slopes to 1H:1V (45°) in noncohesive or unconsolidated materials to 0.6:1 (60°) in clays common on one or both banks. | Bank slopes over 45° in noncohesive or unconsolidated materials or over (60°) in clays common on one or both banks |
| 10. Vegetative or engineered bank protection | Wide band of woody vegetation with at least 90% density and cover. Primarily hard wood, leafy, deciduous trees with mature, healthy, and diverse vegetation located on the bank. Woody vegetation oriented vertically. In absence of vegetation, both banks are lined or heavily armored. | Medium band of woody vegetation with 70-90% plant density and cover. A majority of hard wood, leafy, deciduous trees with maturing, diverse vegetation located on the blank. Woody vegetation oriented 80-90° from horizontal with minimal root exposure. Partial lining or armoring of one or both banks. | Small band of woody vegetation with 50-70% plant density and cover. A majority of soft wood, piney, coniferous trees with young or old vegetation lacking in diversity located on or near the top of bank. Woody vegetation oriented at 70-80° from horizontal often with evident root exposure. No lining of banks, but some armoring may be in place on one bank. | Woody vegetation band may vary depending on age and health with less than 50% plant density and cover. Primarily soft wood, piney, coniferous trees with very young, old and dying, and/or monostand vegetation located off of the bank. Woody vegetation oriented at less than 70° from horizontal with extensive root exposure. No lining or armoring of banks. |
| 11. Bank Cutting | Little or none evident. Infrequent raw banks, insignificant percentage of total bank. | Some intermittently along channel bends and at prominent constrictions. Raw banks comprises minor portion of bank in vertical direction. | Significant and frequent on both banks. Raw banks comprise large portion of bank in vertical direction. Root mat overhangs. | Almost continuous cuts on both banks, some extending over most of the banks. Undercutting and sod-root overhangs. |
| 12. Mass wasting or bank failure | No or little evidence of potential or very small amounts of mass wasting. Uniform channel width over the entire reach. | Evidence of infrequent and/or minor mass wasting. Mostly healed over with vegetation. Relatively constant channel width and minimal scalloping of banks. | Evidence of frequent and/or significant occurrences of mass wasting that can be aggravated by higher flows, which may cause undercutting and mass wasting of unstable banks. Channel width quite irregular and scalloping of banks is evident. | Frequent and extensive mass wasting. The potential for bank failure, as evidenced by tension cracks, massive undercuttings, and bank slumping, is considerable. Channel width is highly irregular and banks are scalloped. |
| 13. Upstream distance to bridge from meander impact point and alignment | More than 115 ft (35 m); bridge is well aligned with river flow | 66-115 ft (20 - 35 m); bridge is aligned with flow | 33 - 66 ft (10 - 20 m); bridge is skewed to flow or flow alignment is otherwise not centered beneath bridge | Less than 33 ft (10 m); bridge is poorly aligned with flow |

Table 5.6. Overall Scores for Three Classifications of Channels.

| Category | Score, R | | |
|---|---|---|---|
| | Pool-Riffle, Plane-Bed, Dune-Ripple, and Engineered Channels | Cascade and Step-Pool Channels | Braided Channels |
| Excellent | R < 49 | R < 41 | N/A |
| Good | 49 ≤ R < 85 | 41 ≤ R < 70 | R < 94 |
| Fair | 85 ≤ R < 120 | 70 ≤ R < 98 | 94 ≤ R < 129 |
| Poor | 120 ≤ R | 98 ≤ R | 129 ≤ R |

Table 5.7. Stability Ratings for Streams in Figures 5.11 – 5.14.

| Stream | Indicator | | | | | | | | | | | | | Total | Rating Based on Table 5.5 |
|---|---|---|---|---|---|---|---|---|---|---|---|---|---|---|---|
| | 1 | 2 | 3 | 4 | 5 | 6 | 7 | 8 | 9 | 10 | 11 | 12 | 13 | | |
| Figure 5.11 | 12 | 4 | 6 | 12 | 11 | 10 | 10 | 12 | 12 | 11 | 12 | 12 | 3 | 127 | Poor |
| Figure 5.12 | 9 | 12 | 10 | 7 | 11 | 8 | 3 | 11 | 8 | 10 | 6 | 7 | 8 | 110 | Fair |
| Figure 5.13 | 8 | 2 | 4 | 5 | 3 | 5 | 5 | 8 | 5 | 2 | 2 | 2 | 6 | 57 | Good |
| Figure 5.14 | 3 | 2 | 3 | 3 | 4 | 3 | 3 | 4 | 5 | 4 | 4 | 1 | 5 | 44 | Excellent |

Table 5.8. Lateral and Vertical Stability for Streams in Figures 5.11 – 5.14.

| Stream | Lateral | Vertical | Lateral Fraction | Vertical Fraction |
|---|---|---|---|---|
| Figure 5.11 | 62 | 33 | 0.86 | 0.92 |
| Figure 5.12 | 50 | 26 | 0.69 | 0.72 |
| Figure 5.13 | 25 | 13 | 0.35 | 0.36 |
| Figure 5.14 | 23 | 10 | 0.32 | 0.28 |

Figure 5.11. Significant cattle activity, agricultural activity; poorly maintained engineered channel; deeply confined with no active floodplain; very loose assortment of sediment with no packing; frequent obstructions; noncohesive bank material (loess); steep banks; no vegetation; bank cutting and mass wasting. R = 127 = Poor (FHWA 2006).

Figure 5.12. Cattle and farming activity; flashy flows; braiding; small loose bed material; minor obstructions; noncohesive bank material; some woody vegetation; limited mass wasting downstream from channel bend; meander bend 66 ft (20 m) upstream. R = 110 = Fair (FHWA 2006).

Figure 5.13. Development in the watershed; perennial stream; low levees; well-packed bed material; minor obstructions from concrete, riprap, roots; sandy clay banks; moderately steep banks; heavily vegetated or protected banks; no mass wasting; moderately well aligned (downstream from bend). R = 57 = Good (FHWA 2006).

Figure 5.14. Stable watershed; perennial stream; mild meanders; no entrenchment; large, packed bed material; no obstructions; cohesive bank material; heavily vegetated banks with vertically oriented trees; little bank cutting; no mass wasting; well aligned with bridge. R = 44 = Excellent (FHWA 2006).

## 5.5 QUALITATIVE EVALUATION OF CHANNEL RESPONSE

### 5.5.1 Overview

The major complicating factors in river mechanics are: (1) the large number of interrelated variables that can simultaneously respond to natural or imposed changes in a stream system, and (2) the continual evolution of stream channel patterns, channel geometry, bars and forms of bed roughness with changing water and sediment discharge. In order to understand the responses of a stream to human activities and nature, a few geomorphic concepts are presented here. Quantitative techniques for analysis of channel stability are introduced in Chapter 6.

The dependence of stream form on slope, which may be imposed independent of other stream characteristics, is illustrated schematically in Figure 5.15 (FHWA 2001, Schumm and Kahn 1972). Any natural or artificial change which alters channel slope can result in modifications to the existing stream pattern. For example, a cutoff of a meander loop increases channel slope. Referring to Figure 5.15 this shift in the plotting position to the right could result in a shift from a relatively tranquil, meandering pattern toward a braided pattern that varies rapidly with time, has high velocities, is subdivided by sandbars, and carries relatively large quantities of sediment. Conversely, it is possible that a decrease in slope could change an unstable braided stream into a meandering one.

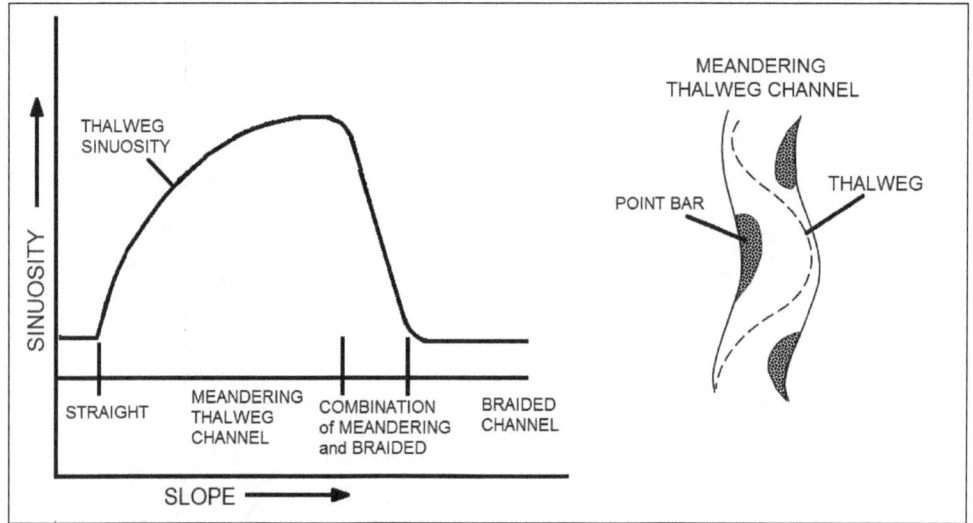

Figure 5.15. Sinuosity vs. slope with constant discharge (after FHWA 2001).

### 5.5.2 Lane Relation and Other Geomorphic Concepts

The significantly different channel dimensions, shapes, and patterns associated with different quantities of discharge and amounts of sediment load indicate that as these independent variables change, major adjustments of channel morphology can be anticipated. Further, a change in hydrology may cause changes in stream sinuosity, meander wave length, and channel width and depth. A long period of channel instability with considerable bank erosion and lateral shifting of the channel may be required for the stream to compensate for the hydrologic change. The reaction of a channel to changes in discharge and sediment load may result in channel dimension changes contrary to those indicated by many regime equations.

For example, it is conceivable that a decrease in discharge together with an increase in sediment load could cause a decrease in depth and an increase in width.

Figures 5.16a and b illustrate the dependence of sand-bed stream form on channel slope and discharge. According to Lane (1955), a sand-bed channel meanders where:

$$SQ^{0.25} \leq K_u \tag{5.1}$$

where:

$K_u$ = 0.0017  English
$K_u$ = 0.00070 SI

and:

S = channel bed slope, ft/ft (m/m)
Q = mean annual discharge, ft³/s (m³/s)

Similarly, a sand-bed channel is braided where:

$$SQ^{0.25} \geq K_u \tag{5.2}$$

where:

$K_u$ = 0.010  English
$K_u$ = 0.0041 SI

The zone between the lines defining braided streams and meandering streams in Figures 5.16a and b is the transitional range, i.e., the range in which a stream can change readily from one stream form to the other.

Many rivers in the United States are classified as intermediate sand-bed streams and plot in this zone between the limiting curves defining meandering and braided stream. If a stream is meandering but its discharge and slope borders on the transitional zone, a relatively small increase in channel slope may cause it to change, with time, to a transitional or braided stream.

Leopold and Wolman (1960) plotted slope and discharge for a variety of natural streams. They observed that a line could separate meandering from braided streams. The equation of this line is:

$$SQ^{0.44} = K_u \tag{5.3}$$

where:

$K_u$ = 0.06    English
$K_u$ = 0.00125 SI

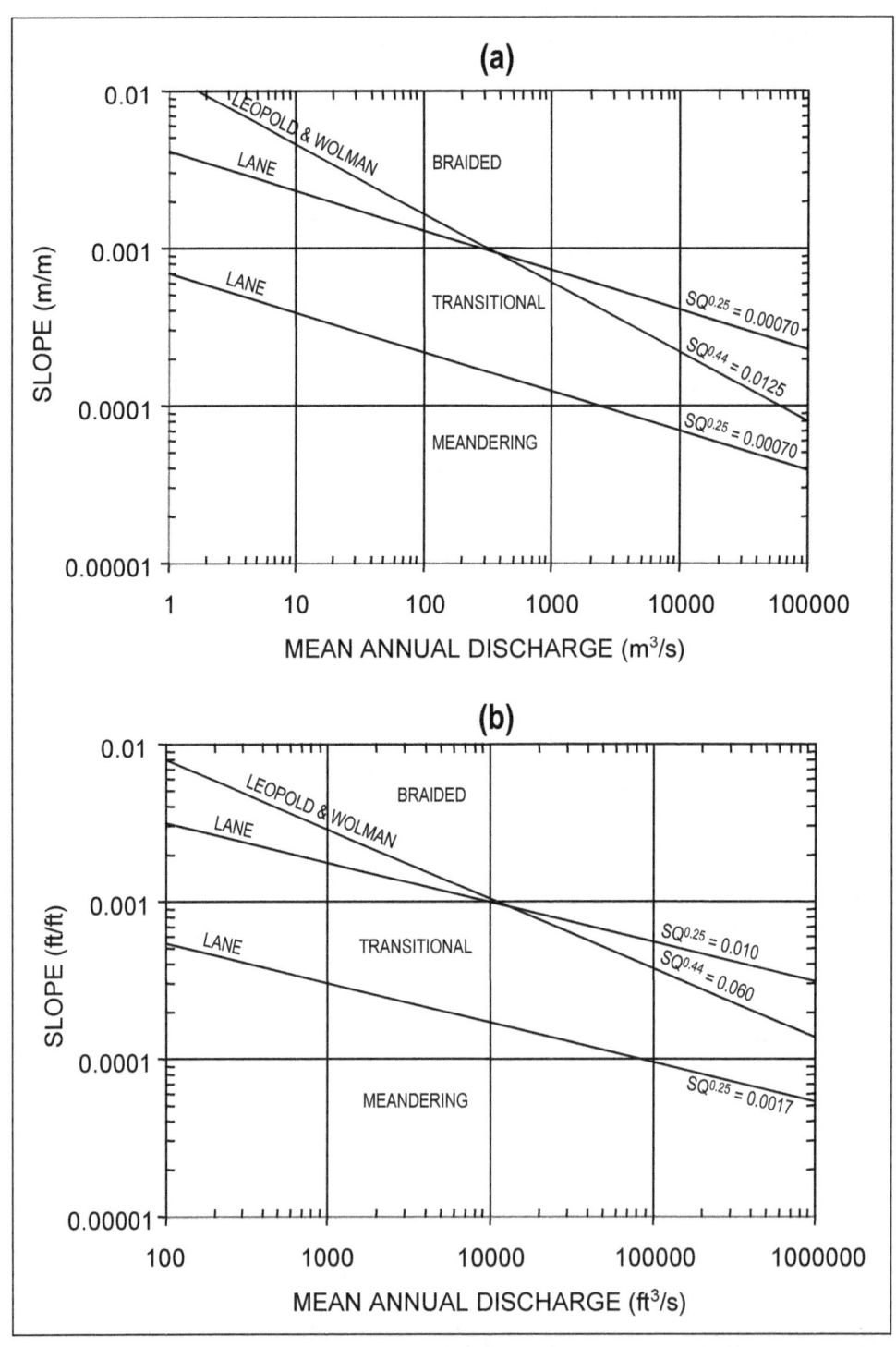

Figure 5.16a, b. Slope-discharge relationship for braiding or meandering in sand-bed streams (after Lane 1955).   a = SI Units   b = English Units

Streams classified as meandering by Leopold and Wolman are those whose sinuosity is greater than 1.5. Braided streams are those which have numerous alluvial islands and, therefore, two or more channels. They note that sediment size is related to slope and channel pattern but do not try to account for the effect of sediment size on the morphology of streams. They further note that braided and meandering streams can be differentiated based on combinations of slope, discharge, and width/depth ratio, but regard width as a variable dependent mainly on discharge.

Long reaches of many streams have achieved a state of equilibrium, for practical engineering purposes. These stable reaches are called "graded" streams by geologists and "poised" streams by engineers. However, this condition does not preclude significant changes over a short period of time or over a period of years. Conversely, many streams contain long reaches that are actively aggrading or degrading (see Section 2.4). These aggrading and degrading channels pose definite hazards to highway crossings and encroachments, as compared with poised streams.

Regardless of the degree of channel stability, human activities may produce major changes in stream characteristics locally and throughout an entire reach. All too frequently, the net result of a stream "improvement" is a greater departure from equilibrium than existed prior to "improvement." Designers of stream channel modifications should invariably seek to enhance the natural tendency of the stream toward equilibrium and a stable condition. This requires an understanding of the direction and magnitude of change in channel characteristics which will result from the actions of man and nature. This understanding can be obtained by:

- Studying the stream in a natural condition
- Having knowledge of the sediment and water discharge
- Being able to predict the effects and magnitude of future human activities
- Applying to these a knowledge of geology, soils, hydrology, and hydraulics of alluvial rivers

Predicting the response to channel modifications is a very complex task. There are large numbers of variables involved in the analysis that are interrelated and can respond to changes in a stream system in the continual evolution of stream form. The channel geometry, bars, and forms of bed roughness all change with changing water and sediment discharges. Because such a prediction is necessary, useful methods have been developed to qualitatively and quantitatively predict the response of channel systems to changes.

Quantitative prediction of response can be made if all of the required data are known with sufficient accuracy (see Chapter 6). Often, however, available data are not sufficient for quantitative estimates, and only qualitative estimates are possible. For example, Lane (1955) studied the changes in stream morphology caused by modifications of water and sediment discharges and developed simple qualitative relationships among the most important variables indicating stream behavior. Similar but more comprehensive treatments of channel response to changing conditions in streams have been presented by Leopold and Maddock (USGS 1953), Schumm (1971), and Santos-Cayado (1972). All research results support the relationship originally proposed by Lane:

$$QS \propto Q_s D_{50} \tag{5.4}$$

where:

$Q$ = Discharge of water
$S$ = Energy slope
$Q_s$ = Sediment discharge
$D_{50}$ = Median sediment size

This proportional relationship (Equation 5.4) is very useful to predict qualitatively channel response to climatological changes, stream modifications, or both. The geomorphic relation expressed is only an initial step in analyzing long-term channel response problems. However, this initial step is useful because it warns of possible future difficulties related to channel modifications. Examples of its use are given in the next section and in HDS 6 (FHWA 2001).

### 5.5.3 Stream System Response

Streambed aggradation or degradation affects not only the stream in which a bed elevation change is initiated, but also tributaries to the stream and the stream to which it is tributary. Thus, the stream system is in an imbalanced condition regarding sediment supply and sediment transport capacity, and it will seek a new state of equilibrium. A few examples are cited to illustrate the system-wide response to natural and human-induced changes. These examples also illustrate the use of several geomorphic concepts introduced in Section 5.5.2 and in the discussion of Section 2.4.

Example 1. A degrading principal stream channel will cause its tributaries to degrade, thus contributing additional sediment load to the degrading stream. This larger sediment load will slow the rate of degradation in the principal stream channel and may halt or reverse it for a period of time if the contribution is large enough or if a tributary transports material which armors the bed of the degrading stream.

Using Equation 5.4, the basic response of the principal stream can be expressed as:

$$QS^+ \alpha Q_s^+ D_{50}$$

Here, it is assumed that water discharge (Q) and sediment size ($D_{50}$) remain unchanged. **(Note: When neither + or - appears as a superscript in the Lane relationship, conditions remain unchanged).** Thus, the increase in sediment discharge ($Q_s^+$) derived from the tributary stream must result in an increase in slope ($S^+$) on the principal stream if the geomorphic balance expressed by the Lane relationship is to hold. The increase in slope then slows or reverses the original degradation of the principal stream which initiated the stream system response.

Example 2. The sediment supply available for transport by a reach of stream may be reduced by changes in the watershed which reduce erosion, mining of sand and gravel from the streambed upstream of the reach, or the construction of a dam to impound water upstream of the reach. In general, for the two latter cases, sediment transported by the stream is trapped in the mined areas (or removed from the system) or trapped in the reservoir and mostly clear water is released downstream. Figure 5.17 illustrates the principle by use of the example of a dam. Referring to Equation 5.4, a decrease in sediment discharge will cause a decrease in slope, if the discharge and median sediment size remain constant, or:

$$QS^- \alpha Q_s^- D_{50}$$

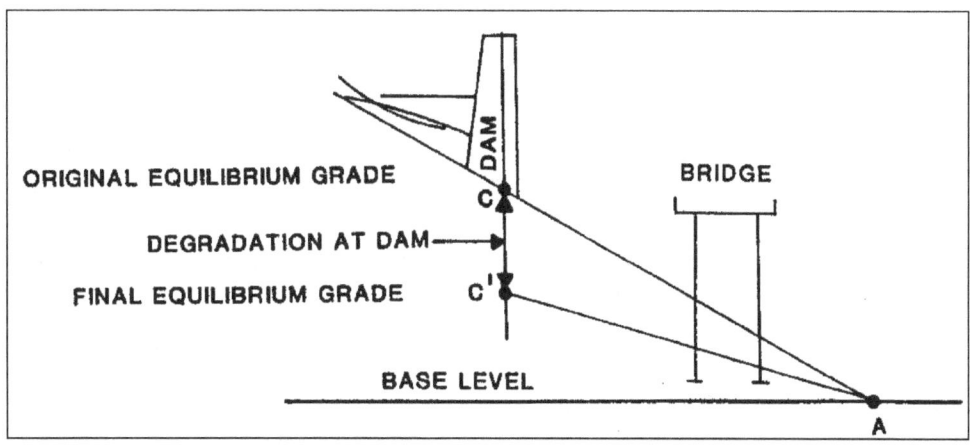

Figure 5.17. Changes in channel slope in response to a decrease in sediment supply at point C.

The original equilibrium channel gradient (Figure 5.17) is represented by the line CA. A new equilibrium grade represented by C'A will result from a decrease in sediment supply. The dam is a control in the channel which prevents the effects from extending upstream. Except for the channel control formed by the dam, similar effects are experienced at any location which undergoes a reduction in sediment supply.

Referring to Figure 5.15, for a low sinuosity braided stream, this decrease in slope below the dam could result in an increase in sinuosity and a change in planform toward a combination meandering/braided stream (see (1) on Figure 5.18a). If the stream below the dam were initially a meandering stream at near maximum sinuosity for the original slope, the decrease in slope below the dam could shift the planform of the stream toward a reduced sinuosity, meandering thalweg channel (see 2 on Figure 5.18a).

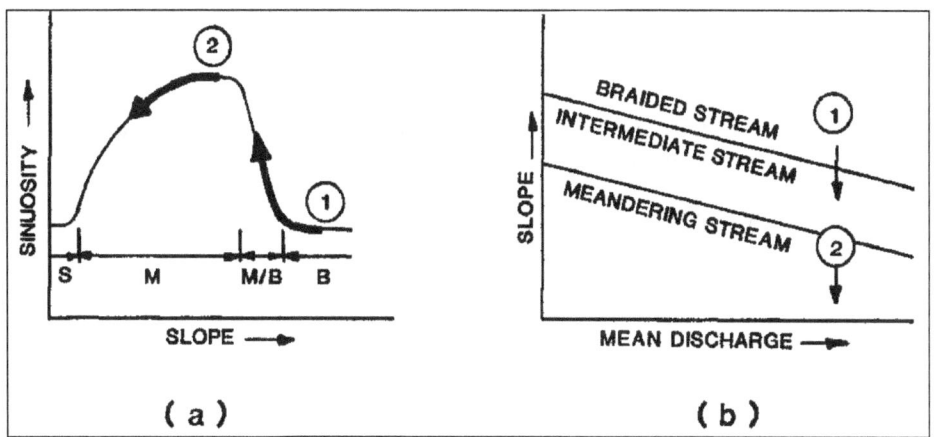

Figures 5.18a and b. Use of geomorphic relationships of Figures 5.15 and 5.16 in a qualitative analysis.

A similar result can be derived from Figure 5.16. For an initially braided channel pattern below the dam [(1) on Figure 5.18b], a decrease in slope below the dam could indicate a tendency to shift the stream's plotting position downward, possibly into the intermediate stream range (i.e., a combination of meandering and braided as on Figure 5.18a). For an initially meandering stream [(2) on Figure 5.18b], the decrease in slope below the dam could indicate a tendency toward a less meandering channel (as on Figure 5.18a). It should be noted that both of these cases have assumed a constant discharge (Q).

As discussed in Section 2.4, the effects downstream of a dam are more complex than a simple reduction in sediment supply. If the reservoir is relatively small and water flow rates downstream are not significantly affected, degradation may occur downstream initially and aggradation may then occur after the reservoir fills with sediments. Except for local scour downstream of the dam, the new equilibrium grade may approach line CA (Figure 5.17) over the long term. This could apply to a diversion dam or other small dam in a stream.

Dams constructed to impound water for flood control or water supply usually have provisions for sediment storage. Over the economic life of the project, essentially clear water is released downstream. For practical purposes, the sediment supply to downstream reaches is permanently reduced. Reservoirs developed for these purposes, however, also reduce the water flow rates downstream. Referring to Equation 5.4, a reduction in discharge (Q) may have a moderating effect on the reduction in slope S and, consequently, on degradation at the dam CC' in Figure 5.17. If sediment discharge or sediment size remain constant below the dam (e.g., a tributary downstream continues to bring in a large sediment discharge), this would be expressed as:

$$Q^- S^+ \alpha \ Q_s D_{50}$$

Considering the more likely scenario of stream response to a dam, both water discharge (Q) and sediment discharge ($Q_s$) would decrease. It is also possible that sediment size ($D_{50}$) in the reach below the dam would increase due to armoring or tributary sediment inflow. Using Equation 5.4, this complex result could be expressed as:

$$Q^- S^{\pm} \alpha \ Q_s^- D_{50}^+$$

Here, the resulting response in slope ($S^{\pm}$) would depend on the relative magnitude of changes in the other variables in the relationship.

### 5.5.4 Regime Equations for Sand-Bed Channels

Hydraulic geometry is a general term applied to alluvial channels to denote relationships between discharge Q and the channel morphology, hydraulics, and sediment transport (FHWA 2001). As defined in Chapter 2, channels forming in their own sediments are called alluvial channels. In alluvial channels the morphologic, hydraulic, and sedimentation characteristics of the channel are determined by a large variety of factors. While the mechanics of such factors are not fully understood, alluvial streams do exhibit some quantitative hydraulic geometry relations. In general, these relations apply to channels within a physiographic region and can be obtained from data available on gaged rivers. It is understood that hydraulic geometry relations express the integral effect of all the hydrologic, meteorologic, and geologic variables in a drainage basin for in-bank flows.

The hydraulic geometry relations of alluvial streams are useful in river engineering for evaluating river response. The forerunner of these relations are the "regime" theory equations of stable alluvial canals (see for example, Kennedy 1895, Lacy 1930, and Leliavsky 1955), hence the term regime equations. Hydraulic geometry relations were developed by Leopold and Maddock (USGS 1953) for different regions in the United States and for different types of rivers. In general, the hydraulic geometry relations are stated as power functions of the discharge.

Work by Julien and Wargadalam (1995) updated the theory and applications of hydraulic geometry relationships for alluvial channels. In their study, the hydraulic geometry of alluvial channels, in terms of bankfull width, average flow depth, mean flow velocity, and friction slope was examined from a 3-dimensional stability analysis of noncohesive particles under 2-dimensional flows. The analytical formulations were tested with a comprehensive data set consisting of 835 field channels and 45 laboratory channels. The data set covered a wide range of flow conditions from meandering to braided sand-bed, and gravel-bed rivers.

A more recent study by Lee and Julian (2006) extends the earlier work of Julien and Wargadalam (1995). A larger database for the downstream hydraulic geometry of alluvial channels was examined through a nonlinear regression analysis. The data used for validation included sand-bed, gravel-bed, and cobble-bed streams with meandering to braided planform geometry. The parameters describing downstream hydraulic geometry include: channel width W, average flow depth d, and mean flow velocity V. The three independent variables are discharge Q, median bed particle diameter $D_{50}$ and channel slope S.

These are combined in the following relationships:

$$W = 2.189 \, Q^{0.426} \, D_{50}^{-0.002} \, S^{-0.153} \tag{5.5}$$

$$d = 0.237 \, Q^{0.336} \, D_{50}^{-0.025} \, S^{-0.060} \tag{5.6}$$

$$V = 4.624 \, Q^{0.198} \, D_{50}^{-0.007} \, S^{-0.242} \tag{5.7}$$

These equations are for Q in ft$^3$/s and $D_{50}$ in mm. The coefficients for equations 5.5 through 5.7 should be 3.046, 0.201, and 2.996 for SI units (Q in m$^3$/s and $D_{50}$ in mm). While the Lee and Julian equations extend into the gravel and cobble bed range, additional regime equations specific to gravel bed rivers are presented in Chapter 8. These equations, developed by Hey and Thorne (1986) include consideration of bankline vegetation type and composite bank material with cohesive fine sand, silt, and clay overlying gravel.

### 5.5.5 Complex Response

Generally, a simple qualitative evaluation of stream system response assumes that the stream is "graded" or "poised," that is, in a condition of steady-state equilibrium (Schumm 1977). This condition in a process-response system is maintained by self-regulation or negative feedback, which operates to counteract or reduce the effects of external change on the system so that it returns to an equilibrium condition. For a qualitative evaluation of stream stability at highway structures this is a reasonable initial assumption; however, stream system response can be much more complex.

The fluvial system can be viewed either as a physical system, where the workings of and relations among the components of the system are the major concern, or as a historical system, where change is viewed from a longer term perspective (see, for example, Figures

2.1 and 2.2). As Schumm (1977) points out, in actuality the fluvial system is a physical system with a history. When viewed over the longer span of an erosion cycle (geologic time scale) the characteristics of the system progressively change; however, a state of "dynamic" equilibrium may exist if there is a long-term balance (on an engineering time scale) between the water and sediment supplied by the watershed and that transported by the stream system. When dynamic equilibrium exists, bed scour and fill and bankline migration may occur, but on an engineering time scale, reach averaged characteristics and the balance between sediment inflow and sediment outflow are maintained.

To add to the complexity of stream system response, some change may be episodic rather than progressive. Fluvial systems may change through time to a condition of incipient instability, without a change of external influences. Exceeding such a geomorphic threshold during, for example, a major flood can lead to rapid, non-progressive (episodic), and potentially catastrophic change in a stream system. Schumm (1977) cites as an example, a valley in a semi-arid region where sediment storage through time progressively increases valley slope until failure (erosion) occurs. Another common example of a geomorphic threshold is the progressive increase in channel sinuosity and meander amplitude until a cutoff of a meander loop or avulsion occurs. This is due to channel lengthening, increased resistance, and gradient reduction that accompany increases in sinuosity. A history of meander growth and cutoff is shown for two rivers in Figure 2.5. Meander migration is discussed in more detail in Section 6.2.1.

Figure 2.2 provides an example of complex response on a time scale of interest to the highway engineer if bridges cross or highway structures encroach on the affected stream system. Another example is the response of a drainage basin to "rejuvenation," that is, the lowering of the base level of the main channel by long-term degradation. When base level (bed elevation) is lowered, erosion and channel adjustments occur near the mouth of the basin and, in fact, the main channel probably will be adjusted to the change long before the tributaries have responded (Schumm 1977). However, when the tributaries are in turn rejuvenated as they adjust to the new base level, the increased sediment production is delivered to a main channel that has already adjusted to the base level change, but not to the increased sediment loads from upstream. Thus, the original incision would not be the only adjustment made by the main channel. In fact, a complex sequence of responses could be expected before the stream system attains a new condition of dynamic equilibrium; and all bridges or highway structures, not only on the main channel, but also on the tributaries, could be affected.

An example of analyzing river pattern thresholds and complex response using Figure 5.16 is presented in HDS 6, Chapter 5. Design considerations for highway encroachments and river crossings for both simple and complex situations are presented in HDS 6, Chapter 9 (FHWA 2001).

# CHAPTER 6

# QUANTITATIVE TECHNIQUES FOR STREAM STABILITY ANALYSIS

## Application of Level 2 Analysis Procedures

## 6.1 INTRODUCTION

Highway and bridge design, scour and stream stability analyses, bridge rehabilitation and countermeasure design, and channel restoration projects are all affected by changes in the morphologic characteristics of a stream. While qualitative techniques (Chapters 2 and 5) and classification and reconnaissance (Chapter 5) provide insight on channel processes, the application of quantitative geomorphic and engineering techniques may be necessary to evaluate the potential impact of changes in channel morphology for highway planning, design, and rehabilitation.

In general, the highway engineer needs to address three questions in regard to stream stability:

- What is the bank stability at the highway structure?
- What is the lateral (planform) stability of the stream channel?
- What is the vertical (profile) stability of the streambed?

The effects of lateral (planform) instability of a stream on a highway encroachment or bridge crossing are dependent on the extent of the bank erosion or channel migration and the design of the encroachment or bridge. Bank erosion can undermine piers and abutments located outside the channel and erode abutment spill slopes or breach approach fills. Migration of a bend through a bridge opening changes the direction of flow (angle of attack) through the opening, accentuating local and contraction scour. This chapter provides quantitative techniques to evaluate the lateral stability of a channel including: meander characteristics and prediction of the effects of lateral channel migration.

The typical effects associated with bed elevation (vertical) changes at highway bridges are erosion at abutments and the exposure and undermining of piers from degradation or scour, or a reduction in flow area from aggradation under bridges. Aggrading and degrading channels can also change planform, potentially eroding floodplain areas and highway embankments on the floodplain. In this chapter, specific quantitative procedures for estimating incipient motion and armoring characteristics of the streambed are presented. An indication of relative channel stability can be obtained from an application of an equation for incipient motion particle size developed from the Shields diagram. Determining the critical or threshold conditions at which hydrodynamic forces are sufficient to move a sediment particle provides insight on what flow conditions might mobilize the bed and affect channel vertical stability. A simple procedure developed by the U.S. Bureau of Reclamation for determining the depth of degradation necessary to produce an armor layer sufficient to arrest vertical instability is also presented.

Going beyond a comparison of historic streambed profiles or simple quantitative techniques to assess streambed vertical stability requires considerable expertise in sediment transport analyses. However, sediment continuity analysis and equilibrium slope concepts provided in this chapter offer a relatively straight forward approach to more detailed vertical stability analyses. If a more rigorous analysis of channel vertical dynamics is desired, application of the USACE HEC-RAS computer model (USACE 2010) can be considered.

Finally, the highway engineer must be cognizant of the potential need to restore or rehabilitate environmentally degraded stream channels when designing, constructing, and maintaining highway stream crossings. Chapter 9 provides an introduction to channel restoration concepts and reference to recently published guidelines for channel restoration design.

## 6.2 LATERAL CHANNEL STABILITY

Under ideal circumstances, a stable channel is one that does not change in size, form, or position over time. However, all alluvial channels change to some degree and, therefore, have some degree of inherent instability. An unstable channel is one with a rate or magnitude of change that is sufficiently large to be a significant factor in the design and maintenance of engineered structures within the river environment.

Although a stream or river may appear unstable, this does not necessarily indicate that it is not an equilibrium or regime channel. Based on the relationship of channel width, depth, and slope to discharge, most natural alluvial channels have probably attained or approached a state of equilibrium at one time or another. Yet, these channels migrate laterally at rates ranging from imperceptible to very rapidly. Thus, equilibrium or regime channels may not necessarily be stable in the practical engineering sense. An actively migrating channel may maintain its equilibrium slope and cross section while posing a threat or hazard to engineered structures. Some types of lateral instability are shown in Figure 6.1.

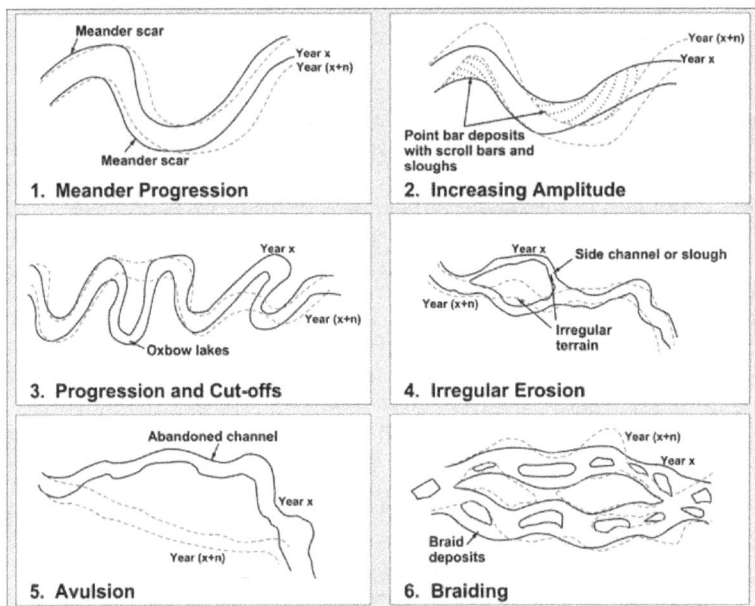

Figure 6.1. Types of lateral activity and typical associated floodplain features (Thorne 1998).

Bank retreat and active meander migration produce lateral instability and channel widening. As discussed in Appendix B, there are two mechanisms by which banks retreat: fluvial entrainment and mass failure. The specific failure mechanisms at a given location are related to the characteristics of the bank material. Commonly, mass failure and fluvial entrainment act in concert; fluvial erosion scours the bank toe followed by oversteepening and failure of the bank. Removal of the failed bank material from the base of the bank occurs through fluvial erosion and the process is repeated.

The bank erosion process can result from channel incision (degradation), flow around bends, flow deflection due to local deposition or obstructions, aggradation, or any combination of these. Flow around a bend can cause erosion at the toe of the outside or convex bank and subsequent bank failure due to increased shear stress on the outside of the bend. Fluvial entrainment through grain detachment can be a significant process in areas of concentrated flow and high shear stress (e.g., on the outside of bends). However, studies of bank erosion processes indicate that mass failure and subsequent fluvial transport of the failed material is the primary mechanism by which the lateral adjustment occurs.

It is important to note that fluvial erosion of previously failed bank material plays a significant role in determining the rates of bank retreat. Fluvial activity controls the state of basal endpoint control and removal of the failed material results in the formation of steeper banks and may induce toe erosion by removing the material along the toe that buttresses the bank slope (see Appendix B). These factors rejuvenate the process of bank erosion by mass failure. Without basal erosion, mass failure of the bank material would lead to bank slope reduction and stabilization within a relatively short period of time (Thorne 1981, Lohnes and Handy 1968).

### 6.2.1 Meander Migration

Meandering streams are classified as either actively or passively meandering. An actively meandering stream has sufficient stream power to deform its channel boundaries through active bed scour, bank erosion, and point bar growth. Conversely, while a passively meandering stream is sinuous, it does not migrate or erode its banks.

*Initiation of Meanders*. Although there is no completely satisfactory explanation of how or why meanders develop (Knighton 1998), it is known that meanders are initiated by localized bank retreat which alternates from one side of the channel to the other in a more or less regular pattern. In addition, deformation of the channel bed may be an important prerequisite that modifies the pattern of flow prior to meandering. It is believed that secondary helicoidal flow develops spontaneously in straight channels as a result of vortices generated at the boundary walls (Figure 6.2) (Einstein and Shen 1964, Shen and Komara 1968). A pair of surface-convergent helical cells will form if vortices develop along both banks. Inequalities in bank roughness may induce asymmetry in these cells and periodic reversal of the dominant cell. This periodically reversing helicoidal flow has an important influence on the pattern of erosion and deposition through meanders, and more specifically by forming a meandering thalweg and alternating bars (Einstein and Shen 1964). In addition, macroturbulent flow and the bursting process (i.e., streamwise fluctuations in the velocity field) are also important components in bank deformation (Yalin 1971 and 1992).

*Flow Pattern Through Meander Bends*. The primary features of the flow pattern through meander bends are:

- Superelevation of the water surface against the outside (convex) bank (Figure 6.3)

- Transverse current directed towards the outer bank at the surface and towards the inner bank at the bed producing a secondary circulation additional to the main downstream flow

- Maximum-velocity current which moves from near the inner bank at the bend entrance to near the outer bank at the bend exit, crossing the channel at the zone of maximum bend curvature

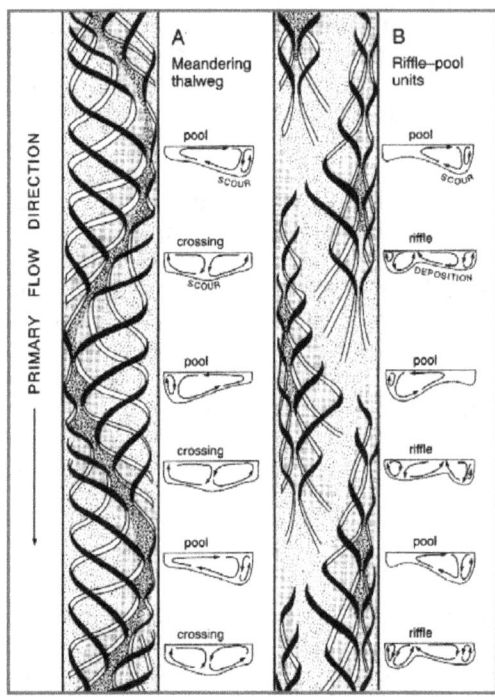

Figure 6.2. Models of flow structure and associated bed forms in straight alluvial channels: (A) Einstein and Shen's (1964) model of twin periodically reversing, surface-convergent helical cells, (the dark stippled line shows the trace of the thalweg) (B) Thompson's (1986) model of surface-convergent flow produced by interactions between the flow and a mobile bed, creating riffle-pool units of alternate asymmetry. Black lines indicate surface currents, and white lines represent near-bed currents (Knighton 1998).

The interaction between centrifugal force acting outwardly on the water as it flows around the bend and an inward-acting pressure gradient force driven by the cross stream tilting of the water surface is reflected in the above characteristics. The transverse current and the primary downstream flow component combine to produce the helicoidal motion to the flow. The superelevation of the water surface against the outer bank of a bend produces a locally steep downstream energy gradient and, in turn, a zone of maximum boundary shear stress in close proximity to the outer bank just downstream of the bend apex (Figure 6.3A). The maximum shear stress zone shifts outward further upstream as a result of the bar-pool topography and cross-sectional asymmetry characteristic of meander bends.

Secondary currents, which are usually weaker than primary ones, influence the distribution of velocity and boundary shear stress. Markham and Thorne (1992) divided the bend cross-section into three regions relative to the pattern of secondary flow (Figure 6.3B):

- Mid-channel region, helicoidal flow is well established passing nearly 90 percent of the flow
- Cell of opposite circulation develops in the *outer bank region*: the strength of this cell increases with discharge, the steepness of the bar, and the acuteness of the bend
- Inner bank region where shoaling over the point bar induces a net outward flow, forcing the core of maximum velocity more rapidly toward the outer bank (Dietrich and Smith 1983, Dietrich 1987); increasing stage tends to reduce the shoaling, allowing an inward component of near-bed flow over the bar top

The location and timing of the flow pattern varies with discharge, bend tightness, and cross-sectional form. Primary currents are dominant at high discharges because the main flow follows a straighter path, but secondary currents are relatively strong at intermediate discharges (Bathurst et al. 1979). The degree of superelevation and the strength of the secondary circulation increase in tighter bends (i.e., low $R_c/W$). In bends where $R_c/W < 2$, flow impinges on the outer bank at a sharper angle causing flow separation and generating a strong back-eddy along the outer bank near the bend apex, possibly inducing sedimentation along the outer bank upstream of the bend apex (Hicken and Nanson 1975 and 1984). The width/depth ratio exerts a major influence on flow pattern (Markham and Thorne 1992). Point bar development is more extensive and the shoaling effect over the bar directs the inner-bank flow radially outward when the width/depth ratio is relatively large. However, in narrow, deep channels, especially where $W/y < 10$, bars are less likely to form, thus reducing the shoaling effect and allowing an inward movement of near-bed flow.

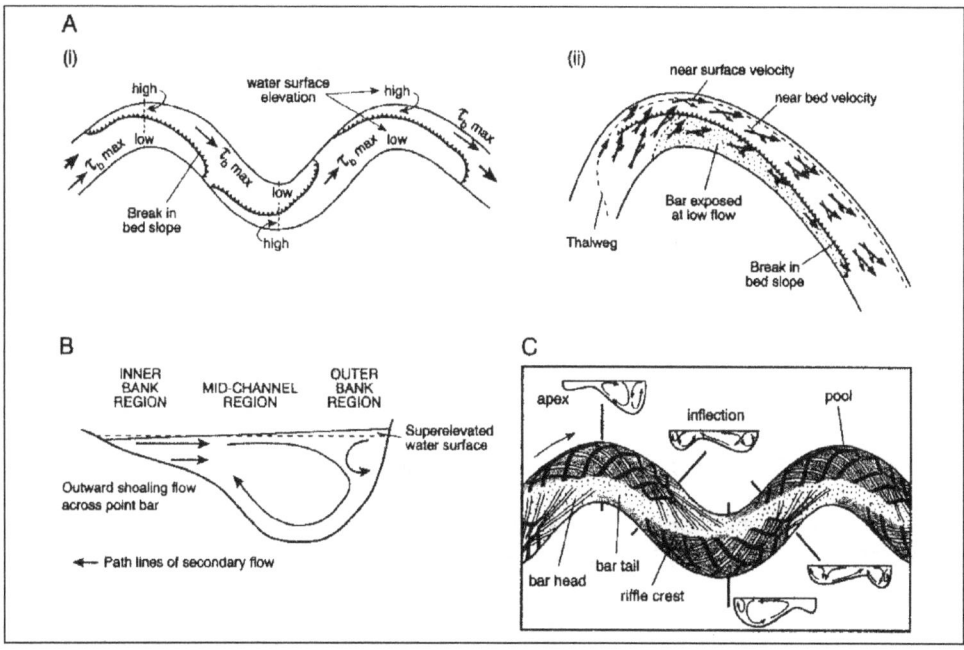

Figure 6.3. Flow patterns in meanders: (A) (i) Location of maximum boundary shear stress ($\tau_o$), and (ii) flow field in a bend with a well-developed point bar (after Dietrich 1987), (B) Secondary flow at a bend apex showing the outer bank cell and the shoaling-induced outward flow over the point bar (after Markham and Thorne 1992), (C) Model of the flow structure in meandering channels (after Thompson 1986). Black lines indicate surface currents and white lines represent near-bed currents (Knighton 1998).

The pattern of primary and secondary currents influences the distribution of erosion and deposition in meanders. In general, erosion in the bend is concentrated along the outer bank downstream of the bend apex where the currents are strongest, while point bar building predominates in a parallel position along the opposite bank, with material supplied by longitudinal and transverse currents. This produces a largely downvalley component to meander migration.

Meander Geometry. Meanders are defined by their geometry; specifically by their shape, bend radius, and wavelength (Figure 6.4). Consistent relationships exist between these meander parameters and channel width. Relative to the planform of a sinuous channel, pools are located at meander bends and riffles are situated at the crossings between bends. The riffle spacing, which is generally 5 to 7 times the width (W), is approximately half the meander wavelength ($\lambda$), which is 10 to 14 times the width (see Figure 2.4). The radius of curvature ($R_c$) of bend is generally 2 to 3 times the width.

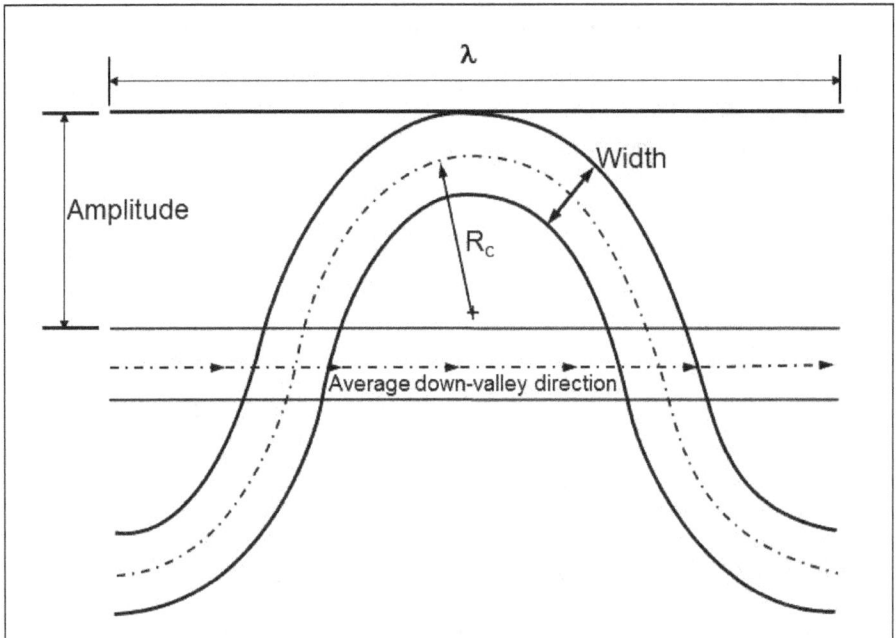

Figure 6.4. Schematic of an idealized meander bend illustrating geometric variables.

Langbein and Leopold (USGS 1966a) characterized meander geometry in terms of a sine-generated curve defined by the following equation:

$$\theta = \omega \sin\left(\frac{2\pi x}{\lambda}\right) \tag{6.1}$$

where:

$\theta$ = Channel direction
$\omega$ = Maximum angle between a channel segment and the mean downvalley axis
$x$ = Sinusoidal function of distance, ft (m)
$\lambda$ = Meander wave length, ft (m)

This curve closely approximates the curve of minimum variance, or least work in turning around the bend, and describes the form of symmetrical meander paths relatively well. However, real meanders are asymmetrical and deviate significantly from idealized, perfectly symmetrical, sine-generated curves. Bend asymmetry occurs because the point of deepest scour and maximum attack on the outer (convex) bank in a bend is usually located downstream of the geometric apex of the bend. This causes the bend to migrate downstream through time, becoming skewed in the downvalley direction as they shift.

Leopold and Wolman (USGS 1957a, 1960) established a link between meander wavelength and channel width over several orders of scale of flow in a variety of natural environments. Their equations (referred to as hydraulic geometry relationships) were developed from meander characteristics of free-flowing regime channels as follows (English units):

$$\lambda = 10.9 \, W^{1.01} \tag{6.2}$$

$$A_m = 2.7 \, W^{1.1} \tag{6.3}$$

$$\lambda = 4.7 \, R_c^{0.98} \tag{6.4}$$

$$R_c = 2.3 \, W \tag{6.5}$$

where:

$\lambda$ = Meander wavelength (ft or m) measured along the axis of the channel
W = Channel topwidth (ft of m) at the dominant discharge
$A_m$ = Meander amplitude (ft of m)
$R_c$ = Bend centerline radius of curvature (ft or m)

and the constants in Equations 6.2 through 6.4 are 11.0, 3.0, and 4.6 in SI units, respectively.

Subsequent reanalysis of the Leopold and Wolman data by Richards (1982) has resulted in an acceptable linear relationship of meander wavelength of the form:

$$\lambda = 12.34 \, W \tag{6.6}$$

The coefficient in this equation is very close to being twice the systematic riffle-pool spacing in a straight channel as defined by $2\pi W$ or 5-7 channel widths (Figure 2.4b). Hydraulic geometry relationships show that channel topwidth should be closely related to discharge in an alluvial channel. Thus, there should be a relationship between discharge and wavelength. Although it is well established that width is approximately proportional to the square root of discharge, the determination of the discharge that is physically most significant in shaping meanders is still under debate. It is likely that meander geometry is related to a range of discharges, the competence of which varies with boundary materials. Therefore, the degree and magnitude of lateral instability on a meandering stream is most likely dependent on a combination of its bank material composition and discharge.

Schumm analyzed large empirical data sets for sand bed channels in an attempt to account for the effect of boundary materials on meander wavelength explicitly by using a weighted silt-clay index of the bed and bank sediments (USGS 1968). As seen in the following equations, as the proportion of fine material in the bed and banks increases, the meander wavelength decreases (English units):

$$\lambda = 1890 \, Q_m^{0.34} \, M^{-0.74} \tag{6.7}$$

$$\lambda = 438 \, Q_b^{0.43} \, M^{-0.74} \tag{6.8}$$

$$\lambda = 234 \, Q_{ma}^{0.48} \, M^{-0.74} \tag{6.9}$$

where:

$Q_m$ = Mean annual discharge (ft³/s or m³/s)
$Q_b$ = Bankfull discharge (ft³/s or m³/s)
$Q_{ma}$ = Mean annual flood (ft³/s or m³/s)
M = Percent silt-clay in the channel boundary

and the constants in Equations 6.7 through 6.9 are 1935, 618, and 395 for SI units, respectively.

This indicates that the greater erosion resistance of silt-clay banks results in a narrow cross-section with steeper banks and tighter, shorter wavelength bends than those channels with noncohesive or less cohesive, easily eroded banks. In addition, Fisk's work on the Lower Mississippi River indicates that the form of most meanders is influenced by variations in the erodibility of the materials in the outer bank (USACE 1944, 1947).

Schumm (USGS 1968) also demonstrated the relationship of channel sinuosity to the weighted silt-clay index and the form ratio (width/depth) using the following:

$$P = 0.94 \, M^{0.25} \tag{6.10}$$

$$P = 3.50 \, F^{-0.27} \tag{6.11}$$

where:

P = Planform sinuosity
F = Width/depth ratio

These equations link the characteristic wavelength of meandering channels to the formative flow in the channel, its width, and the nature of the boundary materials.

The combination of the wavelength relations, width, and bend radius described above yields the following relationship:

$$R_c \approx 2 \text{ to } 3 \, W \tag{6.12}$$

For radius-to-width ratios ($R_c/W$) of 2 to 3, Bagnold (USGS 1960) showed that energy losses due to the curving of flow in the bend were minimized. Plots of both meander migration rate and bend scour depth as a function of bend tightness also peak sharply at a $R_c/W$ of between 2 and 3, indicating that these bends are the most effective at eroding their bed and banks (see Figure 6.3 and related discussion). The fact that many bends in nature develop and retain an $R_c/W$ value of 2 to 3 while migrating across the floodplain may be consistent with their conformance to the most efficient hydraulic shape, which also maximizes their geomorphic effectiveness (Thorne 1997).

Thorne examined the distribution of bend scour with bend geometry in a study of the Red River and determined that in very long radius bends ($R_c/W > 10$) mean scour pool depth is about 1.5 times the mean riffle (crossing) depth, and the maximum scour depth is between 1.7 and 2 times the mean crossing depth (Thorne 1992). Scour depths ranged from 2 to 4 times the mean crossing depth for bends with $R_c/W$ values between 2 and 4, with the greatest scour associated with $R_c/W$ of about 2. Evidence suggests that maximum scour depths decrease with decreasing bend radius for extremely tight bends with $R_c/W < 2$.

Care should be taken when using these relationships for the rehabilitation or restoration of meanders to an optimal form at specific sites. As Knighton states: "These various relationships indicate a self-similarity of meander geometry over a wide range of scales and environmental conditions. However, the regularity which they imply is not everywhere apparent, and the use of single parameters provides only a partial and often subjective characterization of meander form" (Knighton 1998). Chapter 9 provides an introduction to channel restoration concepts.

### 6.2.2 Bank Failure

As noted in Chapter 2 (Section 2.3.9) the appearance of a stream bank is a good indicator of relative stability, and field inspection of a channel will often identify characteristics which are associated with erosion rates and modes of bank failure. Resistance of a stream bank to erosion is closely related to several characteristics of the bank material which can be broadly classified as cohesive, noncohesive, and composite (see for example Figure 2.9).

Causes of bank failure and width adjustments can be viewed at several different spatial scales. Bank erosion, for example, may occur because high discharges cause increased shear forces on the banks. However, the high discharges themselves may be a result of changes in land use or climatic changes that are controlled by processes external to a particular river reach. For example, the increase in impervious surfaces as a result of urbanization throughout the watershed could generate increased stormwater runoff, which in turn causes bank erosion and increases in channel width and cross-sectional area (ASCE 2008).

It is important for engineers to understand both local and larger-scale causes of bank erosion and width adjustment. In some cases, bank protection or other small-scale engineering structures may provide the best solution to a bank erosion problem. In other cases, however, trying to mitigate erosion at a particular reach may be futile and wasteful because the problem is ultimately caused by land use practices throughout a watershed (ASCE 2008).

Thus, the erosion, instability, and/or retreat of a stream bank is dependent on the processes responsible for the erosion of material from the bank and the mechanisms of failure resulting from the instability created by those processes. Bank retreat is often a combination of these processes and mechanisms varying at seasonal and sub-seasonal timescales. Bank retreat processes may be grouped into three categories: weakening and weathering processes, direct fluvial entrainment, and mass failure. Appendix B provides a more detailed geotechnical discussion of factors influencing bank failure focusing on local processes that may affect individual river cross sections. However, the broader context of the entire watershed should always be considered in trying to understand and solve bank failure and width adjustment problems (see also Section 6.2.3).

### 6.2.3 Channel Width Adjustments

Bankfull channel width is an important measure of stream size and is a key parameter of interest for many applications in hydrology, fluvial geomorphology, and stream ecology (Faustine et al. 2009) (see also Section 2.3.2). A change in channel width (increase or decrease) often implies some degree of lateral channel instability.

As noted in a summary of the causes of stream bank erosion and channel width adjustments (ASCE 2008), river width adjustments have varied causes and can occur in different geomorphic settings. Widening can occur by erosion of one of both banks without substantial incision. Widening in sinuous channels may occur when outer bank retreat exceeds the rate of advance of the opposite bank. In braided rivers, bank erosion by flows deflected around growing braid bars is a primary cause of widening. In degrading streams, widening often follows incision of the channel where the increased height and steepness of the banks cause them to become unstable. Bank failures can cause very rapid widening under these circumstances. Widening in coarse-grained, aggrading channels can occur when flow acceleration due to a decreasing cross-sectional area, coupled with current deflection around growing bars, generates bank erosion.

Processes of channel narrowing are equally diverse (ASCE 1998). Rivers may narrow through the formation of in-channel berms or benches at the margins. The growth of berms or benches often occurs when bed levels stabilize following a period of degradation and can eventually lead to the creation of a new, low-elevation floodplain and establishment of a narrower, quasi-equilibrium channel. Encroachment of riparian vegetation into the channel often contributes to the growth, stability and, in some cases, to the initiation of berm or bench features. Narrowing in sinuous channels occurs when the rate of alternate or point bar growth exceeds the rate of retreat of the cut bank. In braided channels, narrowing may result when a marginal anabranch is abandoned. Sediment is deposited in the abandoned channel until it merges into the floodplain. Also, braid bars or islands may become attached to the floodplain, especially following a reduction in discharge. Attached islands and bars may, in time, become part of the floodplain bordering a much narrower, often single-threaded channel.

One method for evaluating bank erosion and width adjustment involves the development of qualitative conceptual models such as the Channel Evolution Model (CEM) introduced in Chapter 2 (see Figure 2.2). Although applicable only to incised channels, this and similar evolution models (Harvey and Watson 1986) have been of value in developing an understanding of watershed channel dynamics and in characterizing whether or not a reach is stable (ASCE 2008). This model (Figure 2.2) was originally based on observations of Oaklimiter Creek in northern Mississippi (Schumm et al. 1984). The sequence described a systematic response of a channel to base level lowering and encompasses conditions that range from disequilibrium (Figure 2.2B) to a new state of dynamic equilibrium (Figure 2.2E). Stages B and C in Figure 2.2 illustrate the widening that can accompany incision. These stages are only conceptual and variations may be encountered in the field; however, the sequence enables the evolutionary state of the channel to be determined from field observations that record the characteristic channel forms associated with each stage of evolution. The morphometric characteristics of the channel reach types can also be correlated with hydraulic, geotechnical, and sediment transport characteristics (ASCE 2008).

Somewhat more quantitative relationships for channel width adjustment have been developed. Faustini et al. (2009) present a set of hydraulic geometry relationships for bankfull channel width as a function of drainage area. These are similar in concept to the Leopold and Wolman hydraulic geometry relationships presented in Section 6.2.1 that establish a link between meander wavelength and channel width. Faustini et al. (2009) developed regression plots and predictive equations for bankfull width in nine large ecoregions that comprise the conterminous United States based on a nationwide sample of streams with bankfull widths between 3 and 250 ft (1 and 75 m) and draining watersheds between 0.4 and 3860 sq/mi (1 and 10,000 km$^2$). These equations provide a useful first-order estimate of channel width (although the authors note that they are no substitute for more detailed investigation for projects requiring accurate site-specific information). This analysis was also able to show that bankfull width exhibited a detectable response to human disturbance in several regions.

As noted in Chapter 2 (Section 2.3.9), mature trees on a graded bank slope are convincing evidence of bank stability. A detailed study of bank erosion on streams in southern British Columbia provides a quantitative assessment of the role of riparian vegetation in limiting channel width adjustments in bends during major flood events (Beeson and Doyle 1995). A total of 748 bends in four stream reaches were assessed by comparing pre- and post-flood aerial photography. Bends without riparian vegetation were found to be nearly five times as likely as vegetated bends to have undergone detectable erosion during the flood events. The likelihood of erosion on semi-vegetated bends was between that of the vegetated and non-vegetated category of bends. Most of the non-vegetated bends experienced major erosion (widening) in excess of 150 ft (45 m).

## 6.3 PREDICTING MEANDER MIGRATION

In general, most streams are sinuous to some degree and the majority of bank retreat and lateral migration occurs along meander bends. As such, the following discussion on evaluating and predicting lateral migration will focus on meander bends. Two related approaches for determining lateral stability and meander migration rates will be discussed: (1) the analysis of sequential historic aerial photographs, maps, and surveys; (2) a simple, practical overlay comparison technique for maps and aerial photographs that has been tested and proven to be useful.

### 6.3.1 Map and Aerial Photograph Comparison

The most accurate means of measuring changes in channel geometry and lateral adjustments is through repetitive surveys of the channel cross section. However, this data is rarely available. The next easiest and relatively accurate method of determining migration rates and direction is through the comparison of sequential historical aerial photography (photos), maps, and surveys. Brice provides a comprehensive methodology for conducting a stream stability and meander migration assessment using a comparative analysis of aerial photos, maps, and channel surveys (USARO 1975, FHWA 1982).

Accuracy in such an analysis is greatly dependent on the period over which migration is evaluated, the amount and magnitude of internal and external perturbations forced on the system over time, and the number and quality of sequential aerial photos and maps. The analysis will be much more accurate for a channel that has coverage consisting of multiple data sets (aerial photos, maps, and surveys) covering a long period of time (several tens of years to more than 100 years) versus an analysis consisting of only two or three data sets covering a short time period (several years to a few tens of years). Predictions of migration for channels that have been extensively modified or have undergone major adjustments attributable to extensive land use changes will be much less reliable than those made for channels in relatively stable watersheds.

Historical aerial photos and maps can be obtained from a number of federal, state, and local agencies (Tables 4.2 and 6.1). Extensive topographic map coverage of the United States at a variety of scales can be obtained from the local or regional offices of the U.S. Geological Survey. In general, both air photos and maps are required to perform a comprehensive and relatively accurate meander migration assessment. Since the scale of aerial photography is often approximate, contemporary maps are usually needed to accurately determine the true scale of air photos. Distortion of the image on aerial photos is also a common problem and becomes greater as one moves further away from the center of the photo. Expensive equipment, which is generally needed to rectify and eliminate aerial photo distortion, is often unavailable, so distortion and scale differences must be accounted for by some other means. The scale problem is easily rectified through the use of multiple distance measurements taken between common reference points on the photos and maps. The measurements of distance between several reference-point pairs common to both the photos and maps are then averaged to define an average scale for the photos. Common reference points can include cultural features such as building corners, roads or fences and their intersections, irrigation channels and canals, or natural features such as isolated rock outcrops, large boulders, and trees, drainages and stream confluences, and the irregular boundaries of water bodies.

The accurate delineation of a bankline on aerial photos is primarily dependent on the density of vegetation at the top of the bank. Top bank is easily defined if stereo-pairs of photos are available. However, single photos can be used relatively easily if one knows what to look for. For banks with little or no vegetation, the top of the bank is easily identified. The abrupt change between the water and the top of the bank along the convex bank in a bend or an eroding cutbank is defined by an abrupt change in the contrast and color (color photo) or gray tone (black and white photo). Usually the water is significantly darker than the top of bank. Along the concave or inner bank of a bend, exposed bar sediment is lighter colored than the river or the top of bank. The top bank along a point bar is usually defined by persistent vegetation such as mature trees and shrubs.

Where vegetation becomes increasingly dense along a bank, small sections of the top of the bank may be visible such that a line can be drawn connecting the sections. Often, the top of the bank may be completely obscured by vegetation and one may be required to locate the top of the bank by approximation. In this case, one can assume that the trunks of the largest trees growing along the river are nearly vertical and are located just landward of the top of the bank. Therefore, a line that approximates the top of the bank may be drawn just riverward of the center of the tree. The amount of error involved with this method increases with decreasing stream size.

### 6.3.2 An Overlay Comparison Technique

Background. Channel migration is typically an incremental process. On meandering streams, the problem at a bridge site may become apparent two or three decades after the bridge is constructed. Channel migration is a natural phenomenon that occurs in the absence of specific disturbances, but may be exacerbated by such basin-wide factors as land use changes, gravel mining, dam construction, and removal of vegetation. Remedial action such as constructing spurs or installing bank protection becomes increasingly expensive or difficult as the channel migrates. As noted in Section 6.2, channel migration includes lateral channel shift (expressed in terms of distance moved perpendicular to the channel center line, per year) and down valley migration (expressed in distance moved along the valley, per year).

Thus, predicting channel migration requires consideration of both local and system-wide factors. The morphology and behavior of a given river reach is strongly determined by the sediment and water discharge from upstream. Therefore, any significant modification of sediment load and water discharge, as a result of human or natural events, will impact local rates of channel change. Even without changes to the supply of water or sediment, lateral migration can occur and adversely impact highway structures.

As described in Section 6.2., the distribution of velocity and shear stress locally and the characteristics of bed and bank sediment will control channel behavior. Therefore, local channel morphology such as dimensions (width, depth, meander wavelength and amplitude), pattern (sinuosity, bend radius of curvature), shape (width-depth ratio), and gradient will not only reflect upstream controls, but will also provide information on the direction and rate of channel migration. For example, highly sinuous, equal-width channels are relatively stable, whereas less sinuous channels of variable width may migrate rapidly (see Section 2.3.13).

A practical methodology to predict the rate and extent of channel migration (i.e., lateral channel shift and down valley migration) in proximity to transportation facilities would enable practicing engineers to evaluate and determine bridge and other highway facility locations and sizes and ascertain the need for countermeasures considering the potential impacts of channel meander migration over the life of a bridge or highway river crossing. The methodology could also be applied to locate and design a new bridge or highway facility to accommodate anticipated channel migration or to evaluate the risk to existing facilities, and if necessary, to determine the need for and design countermeasures against the effects of channel migration. A prediction of channel migration could also be used to alert bridge inspection personnel to the potential for channel change that could affect the safety of a bridge.

These objectives were considered in a study conducted under the National Cooperative Highway Research Program (NCHRP 2004a and b). This research produced a stand-alone Handbook for Predicting Stream Meander Migration using aerial photographs and maps (NCHRP 2004b). The Handbook deals specifically with the problem of incremental channel shift and provides a methodology for predicting the rate and extent of lateral channel shifting and down valley migration of meanders. The methodology is based, primarily, on the analysis of bend movement using map and aerial photo comparison techniques.

Map and Aerial Photography Sources. Historical and contemporary aerial photos and maps can be obtained inexpensively from a number of Federal, State, and local agencies. Table 6.1 lists some of the main sources. The Internet provides numerous sites with links to data resources and sites having searchable data bases pertaining to maps and aerial photography. Often, just typing in a few key words relative to aerial photos or maps for a particular site into a search engine will generate a large number of links to related web sites, which can then be evaluated by the user. It is this ready availability of aerial photography resources that makes the methodologies presented in the Handbook powerful and practical tools for predicting meander migration.

Extensive topographic map coverage of the United States at a variety of scales can be obtained from the local or regional offices of the U.S. Geological Survey (USGS). In general, both aerial photos and maps are required to perform a comprehensive and relatively accurate meander migration assessment. Since the scale of aerial photography is often approximate, contemporary maps are usually needed to accurately determine the true scale of unrectified aerial photos.

Geo-referenced and rectified maps and aerial photos are the most desirable for use in the analysis of meander migration, but often can be expensive to obtain. Presently, aerial photos for the 1990s for most areas of the United States can be obtained from three major sources, the MSN TerraServer World Wide Web site, the USDA Farm Service Agency, and the U.S. Geological Survey (Table 6.1).

A major source of 1990s aerial photos available to the public is the TerraServer Web site operated by Microsoft Corporation. TerraServer, in partnership with the USGS and Compaq, provides free public access to a vast data store of maps and aerial photographs of the United States. Aerial photos and topographic maps at a wide variety of resolutions can be downloaded from the TerraServer Web site. The advantages of the TerraServer images are that they are rectified, geo-referenced, and are in digital format so that they are easily manipulated by a wide variety of software and can be used in GIS applications.

For sites where TerraServer photographic coverage from the 1990s is unavailable, aerial photos can be ordered from the USGS Earth Resources Observation System (EROS) Data Center in Sioux Falls, South Dakota, or from the USDA Farm Service Agency Aerial Photo Field Office (APFO) in Salt Lake City, Utah. Both agencies have World Wide Web sites (Table 6.1) with searchable catalogs of available contemporary and historic aerial photography.

Table 6.1. Sources of Contemporary and Historical Aerial Photographs and Maps.

| Source | Key Search Terms | Comments |
| --- | --- | --- |
| Google Earth | Google Earth<br>Google Earth Pro | Operated by Google. A highly functional virtual globe, map, and geographical information program that maps the Earth by the superimposition of images obtained from satellite imagery, aerial photography and GIS 3D globe. Also includes historic aerial photography (mostly post-1990) and topographic map overlay capability. |
| Microsoft Research Maps (formerly Microsoft TerraServer – USA) | msrmaps | Operated by MSN in conjunction with Hewlett-Packard and the USGS. Free downloads of contemporary digital topographic and aerial photo files. |
| USGS EROS Data Center<br>Sioux Falls, South Dakota | USGS EROS | Operated by the USGS. Interactive data base search for historic and contemporary topographic maps and aerial photos. Earth Explorer web site provides access to millions of land-related products, including satellite images, aerial photographs (NAPP, NHAP), digital cartographic data (DEM, DLG, DRG, and digital orthophoto quadrangles), and USGS paper maps |
| USDA Farm Service Agency, Aerial Photo Field Office (FSA - APFO)<br>Salt Lake City, Utah | USDA, APFO | Operated by the USDA Farm Service Agency. Catalog of historic and contemporary aerial photos for much of the United States. Sources include SCS (NRCS), Forest Service, BLM, Park Service, and other government agencies. |

| Table 6.1. Sources of Contemporary and Historical Aerial Photographs and Maps (cont). | | |
|---|---|---|
| Source | Key Search Terms | Comments |
| USGS Special Collections Library Denver, Colorado and Reston, Virginia | USGS Library USGS Special Collections | Operated by the United States Geological Survey. *Field Records Collection* is an archive of historic records including maps and aerial photography collected by USGS scientists dating back to 1879. *Special Collections* includes archived topographic maps for all states dating back to early 1880s. |
| National Archives and Records Administration (NARA) - Cartographic and Architectural Records, Washington D.C. | NARA Cartographic and Architectural Section | Operated by the National Archives and Records Administration. Archive of historic maps and pre-1941 aerial photos. |
| Western Association of Map Libraries (WAML) San Diego, California | WAML | Operated by the Western Association of Map Libraries - an independent association of map librarians and other people interested in maps and map librarianship. Provides references to information on obscure historic maps and where they can be found for reproduction. |
| U.S. Army Corps of Engineers | USACE | Operated by the U.S. Army Corps of Engineers. The various Corps Districts often have a wealth of historic photos, maps, navigation charts, and survey data. |

Aerial photographs from the EROS Data Center that were flown in the 1980s and 1990s are usually part of the National Aerial Photography Program (NAPP) or the National High Altitude Photography Program (NHAP) and are at scales of 1:40,000 (1 in = 3,333 ft) or 1:60,000 (1 in = 5,000 ft). Because of the scale of these photos, small objects may be difficult to see, the resolution of enlarged portions may be poor, and measurements made from the photos may be inaccurate. Historic aerial photos ordered from EROS or from APFO range in scale from 1:5,000 to 1:40,000 with most flights having optimal scales of 1:20,000 or 1:24,000. Although both agencies have the ability to enlarge any photo to specification, some resolution is lost with increasing enlargement.

Topographic maps in paper or electronic format can be obtained from a variety of sources. Paper copies of topographic maps can be obtained from the USGS or any commercial map supplier. Digital maps (DRGs, DEMs) can be downloaded free from the EROS Web site or purchased from commercial suppliers as well. Most digital maps are geo-referenced and can be loaded directly into GIS-based applications. Portions of geo-referenced topographic maps can be downloaded free from the TerraServer web site and pieced together to form a complete map of a given area or used to fill in gaps. The Handbook cautions that care should be taken when using digital maps and photos because the geo-referenced coordinates and dimensions are usually in metric (SI) units while the contours and spot elevations shown on the maps may be in English units.

Screening and Classification. As described in Section 5.3, a number of morphological classification schemes have been applied to alluvial rivers (USARO 1975; Schumm 1977 and 1981; Montgomery and Buffington 1997; Rosgen 1994). For predicting meander migration, a classification system should have the following attributes:

- It is simple and directly applicable to the meanders encountered in the field.
- The classification provides a rational basis for screening out stable and highly unstable patterns.
- It is in pictorial format, requiring only a map, aerial photograph, or visual inspection to apply.
- Its application does not require field data (e.g., sediment sampling).
- It requires minimal training and/or instruction for end users.

Based on these criteria, a method developed by Brice (USARO 1975) is the most appropriate classification approach for predicting meander migration (see Figure 5.6). However, not all the channel classes that Brice identified would commonly be encountered by hydraulic engineers working in the field. Hence, a modified method was developed for the Meander Migration Handbook. This consists of nine channel categories that were optimized for use in bend classification and screening for meander migration prediction as shown in Figure 6.5.

For an **initial screening**, alluvial streams can be classified according to their patterns as single-thread, braided, or anastomosing (Figure 5.6). The appearance of the stream when viewed on aerial photographs is used to screen out braided and anastomosing channels, which are beyond the scope of the methodologies presented in the Handbook. Only channels identified as meandering are analyzed beyond the initial screening.

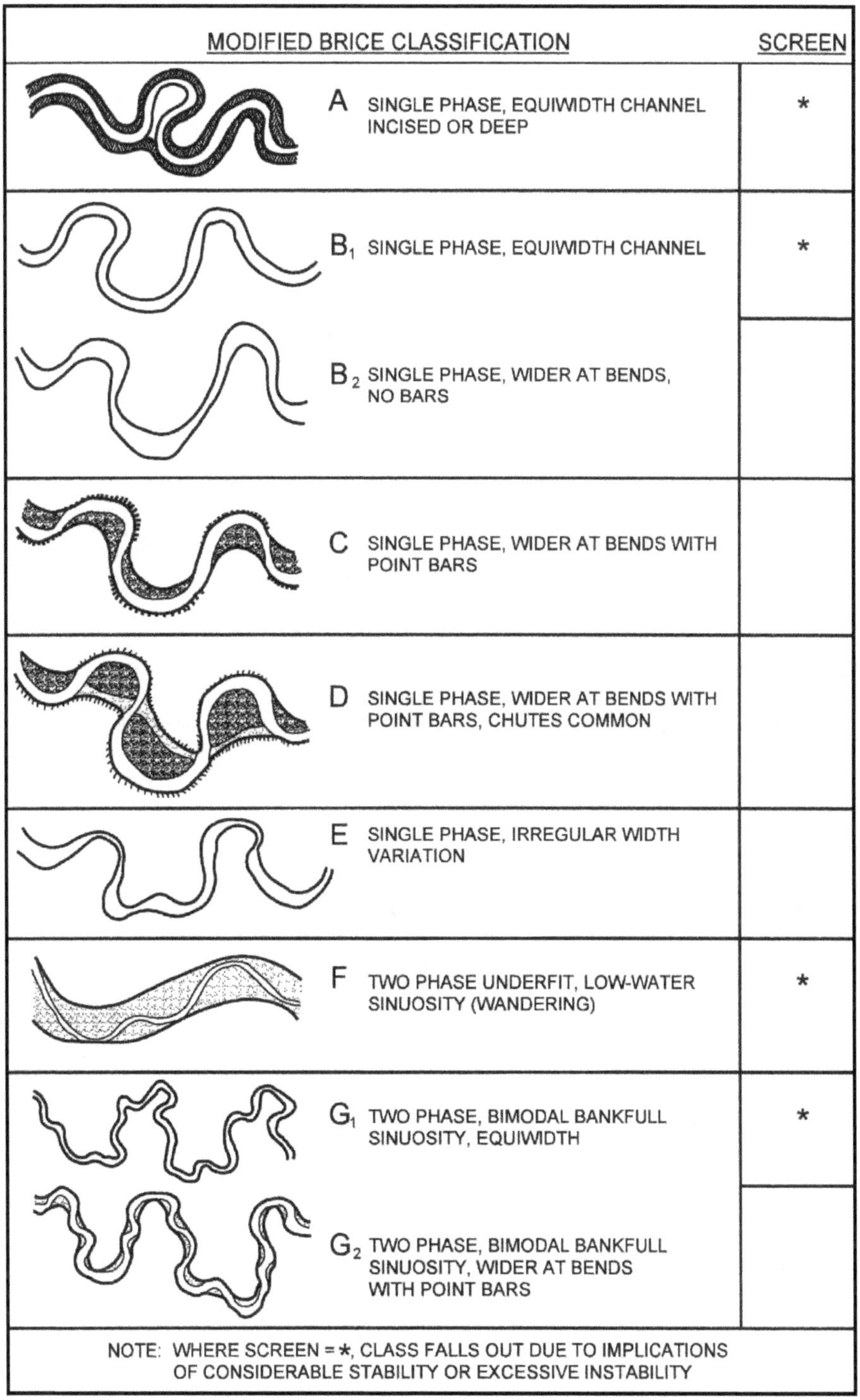

| MODIFIED BRICE CLASSIFICATION | | SCREEN |
|---|---|---|
| A | SINGLE PHASE, EQUIWIDTH CHANNEL INCISED OR DEEP | * |
| $B_1$ | SINGLE PHASE, EQUIWIDTH CHANNEL | * |
| $B_2$ | SINGLE PHASE, WIDER AT BENDS, NO BARS | |
| C | SINGLE PHASE, WIDER AT BENDS WITH POINT BARS | |
| D | SINGLE PHASE, WIDER AT BENDS WITH POINT BARS, CHUTES COMMON | |
| E | SINGLE PHASE, IRREGULAR WIDTH VARIATION | |
| F | TWO PHASE UNDERFIT, LOW-WATER SINUOSITY (WANDERING) | * |
| $G_1$ | TWO PHASE, BIMODAL BANKFULL SINUOSITY, EQUIWIDTH | * |
| $G_2$ | TWO PHASE, BIMODAL BANKFULL SINUOSITY, WIDER AT BENDS WITH POINT BARS | |

NOTE: WHERE SCREEN = *, CLASS FALLS OUT DUE TO IMPLICATIONS OF CONSIDERABLE STABILITY OR EXCESSIVE INSTABILITY

Figure 6.5. Modified Brice classification of meandering channels (NCHRP 2004a and b).

Before predicting migration rates for selected bends along a meandering stream, a **secondary screening** is necessary to identify whether meandering is stable or active. The method employed for this purpose is that of Brice (FHWA 1982) who was able to discriminate qualitatively between stable and laterally active channels from the degree of width variability along the course of the stream. Through extensive historical documentation and field observations of streams he discovered that channels that do not vary significantly in width are either static or relatively stable, whereas channels that are wider at bends migrate actively. In Brice's study, highly sinuous, equal-width streams were the most stable, equal-width streams of lower sinuosity were slightly active, and wider-at-bend streams had the highest migration rates.

As a result, secondary screening of meandering channels using the Brice width-variability criterion can be used to differentiate between stable meandering streams and channels with actively migrating bends. Bends of stable meandering streams can be considered sufficiently stable that bend migration is unlikely to pose a threat to bridges or highway structures and no further analysis is required.

In the modified classification scheme (Figure 6.5), equal-width rivers in classes $B_1$ and $G_1$ are static or relatively stable and are screened out on this basis. Equal-width, deep or incised channels of Class A are also screened out. Class A channels may not migrate because they are deeply inset into geologically competent materials such as bedrock, or they may be actively incising through erodible materials, but do not migrate laterally to any significant degree.

Class F streams are compound channels with a "wandering" low-water channel, wide bars, and back channels. Wandering streams feature lateral activity that is sporadic and spatially disorganized. Lateral migration in such situations is highly unpredictable and wandering streams should be screened out as potentially so unstable and unpredictable that further evaluation would not be likely to produce a meaningful prediction of meander migration.

Migration rates for meander bends on rivers classified in any of the remaining categories ($B_2$, C, D, E, $G_2$) are amenable to prediction by overlay comparison techniques.

Manual Overlay and Prediction. The following steps illustrate a simple overlay comparison of historic banklines and the process of predicting the potential future position of a bend based on past channel migration characteristics (NCHRP 2004a and b).

STEP 1 - The first step in conducting a meander migration analysis using an overlay technique is to obtain aerial photographs and maps for the study area using sources such as those listed in Table 6.1. Appendix A of the Handbook (NCHRP 2004a) provides general instructions on downloading digital aerial photographs and topographic maps from Microsoft's TerraServer Web site. Then, initial and secondary screening are used to determine if the channel can be classified as "predictable" using Figure 6.5.

STEP 2 - The maps and photos must be enlarged or reduced to a common working scale. The scale of the most recent map or photo should be used since it will be the basis for making and comparing historical meander pattern changes and predicting the position of a given bend in the future.

STEP 3 - After defining a working scale, the photos and maps are registered to a common base map or photo by identifying several features or points that are common to each photo/map being compared. The registration points do not need to be common to all the maps and photos, only to the subsequent map or photo to which it is being compared, since comparisons can be performed in pairs.

For example, Figures 6.6 and 6.7 show the 1937 and 1966 aerial photos, respectively, for a reach of the White River in Indiana (Note that this reach of the White River would be considered a "Class C" channel in Figure 6.5 and flow is from left to right). Four registration points have been identified on the 1937 photo that are also on the 1966 aerial photo. Two registration points are road intersections and two are isolated vegetation (trees or large shrubs). Registration points that bracket the site on both sides of the stream and at both ends of the reach are most useful because they reduce the amount of potential error within the bracketed area. Intermediate points between the end points are helpful in accurately registering the middle sections of the reach. More than five or six registration points can make registration difficult because of the difficulty in aligning all the registration points among the various aerial photos and maps used. However, there will be instances where there will be very few identifiable registration points common to both photos, and these sites may have the potential for significant error.

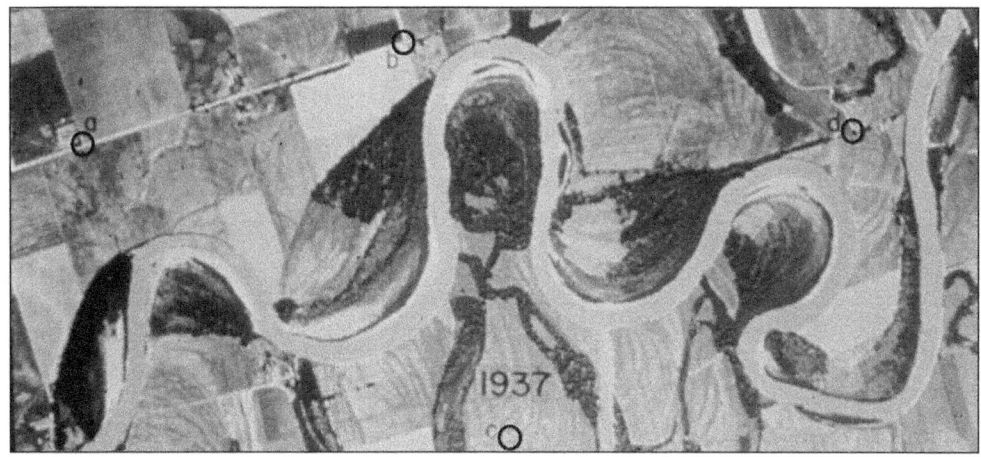

Figure 6.6. Aerial photograph of a site on the White River in Indiana showing four registration points (circles designated a through d) common to the 1966 aerial photo.

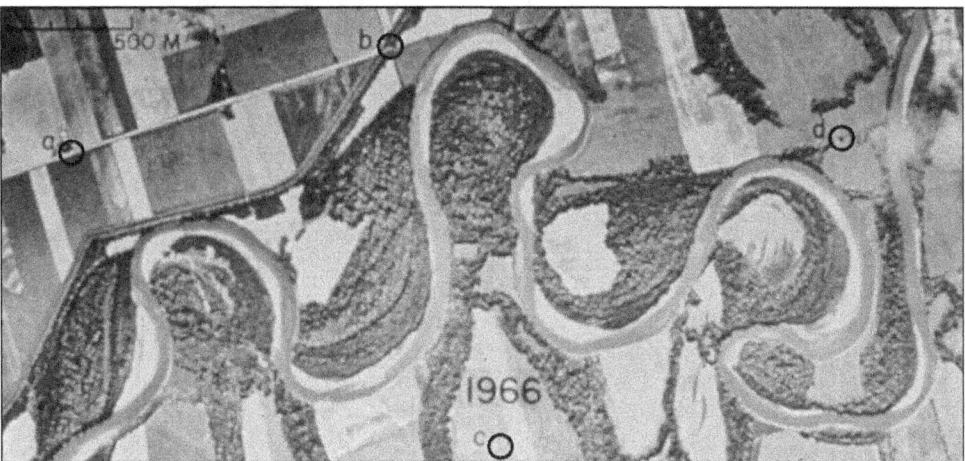

Figure 6.7. Aerial photo of a site on the White River in Indiana showing the four registration points common to the 1937 aerial photo in Figure 6.6.

STEP 4 - After identifying the registration points, banklines and registration points for each year are traced from the aerial photo onto a transparent overlay. The method for identifying and tracing the banklines is discussed in Section 6.3.1 and described in detail in Appendix B of the Handbook (NCHRP 2004a). Registration points are included on the overlay so that they can be easily plotted onto other aerial photos or maps for comparative purposes. The traced banklines and registration points of the White River for 1937 are plotted on the 1966 aerial photo in Figure 6.8 for comparative purposes.

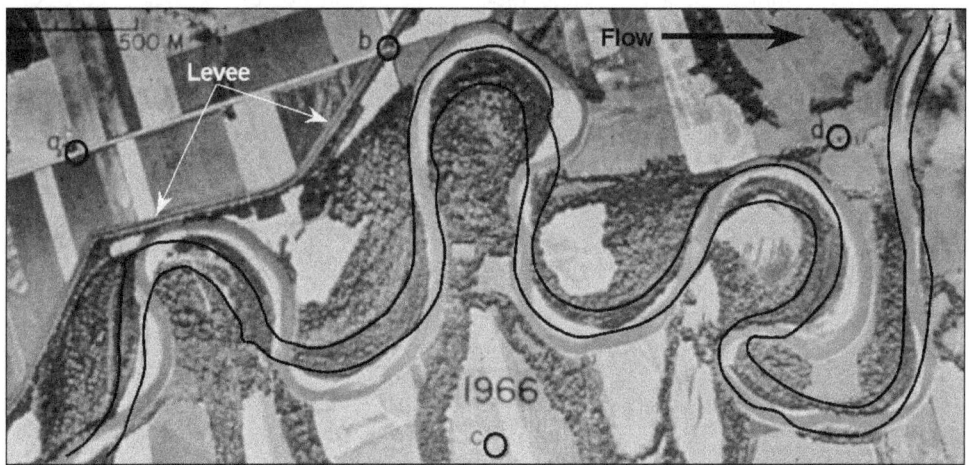

Figure 6.8. The 1966 aerial photo of the White River in Indiana with the 1937 bankline tracing and registration points.

Since most meander bends are not simple loops, the loop classification of Brice (USARO 1975) can be used to characterize the shape of each bend that is to be analyzed (Figure 6.9). Meander bends seldom form single symmetrical loops, but instead are comprised of one or more arcs combined to form either symmetrical or asymmetrical loops. Brice derived the classification scheme for meander loops from a study of the meandering patterns of 125 alluvial streams. The scheme consists of four main categories of loops (simple and compound symmetrical and asymmetrical) comprising 16 form types. Although compound loops are regarded as aberrant forms of indefinite radius and length, the meandering patterns can be divided into simple loops whose properties can be described, measured, and analyzed. The radius of curvature of most bends can be defined by fitting one or more circles or arcs to the bend centerline or outer bankline of a meander loop.

STEP 5 - Once the banklines for each of the historic aerial photos have been traced, circles are best-fit to the outer bank of each bend to define the average bankline arc, the radius of curvature ($R_C$) of the bend, and the bend centroid position (Figure 6.10). The number of circles required to define the bend is based on the loop classification described above and shown in Figure 6.9. A detailed description of the method used to fit a circle to the outer bankline of a meander bend is provided in Appendix B of the Handbook (NCHRP 2004a). The radius of curvature and centroid position of the circle used to describe the bend will be used to make comparison with the bend measurements of previous and subsequent years. These measurements can then be used to determine migration rates and direction and estimate future bend migration characteristics.

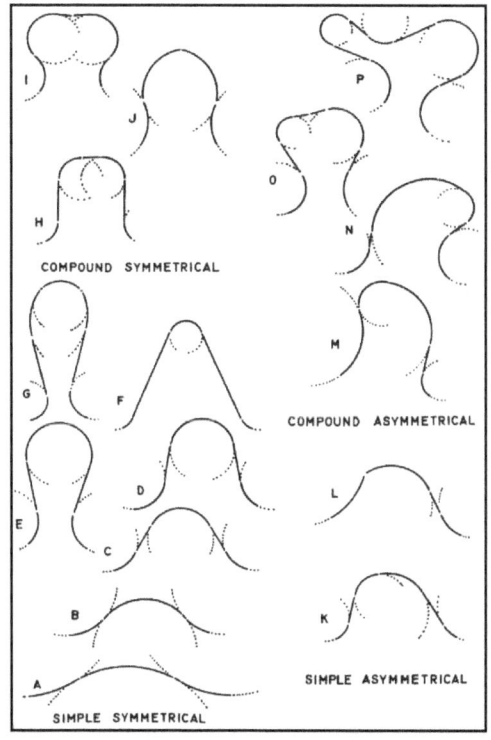

Figure 6.9. Meander loop evolution and classification scheme proposed by Brice (USARO 1975). Flow is left to right.

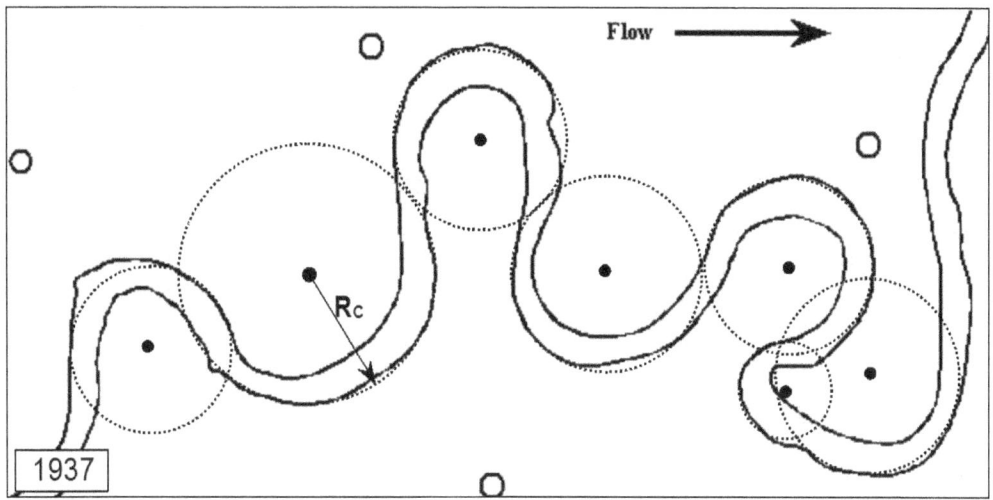

Figure 6.10. Circles that define the average outer banklines from the 1937 aerial photo of the White River site in Indiana. Also shown are the bend centroids and the radius of curvature ($R_C$) for one of the bends.

Figure 6.11 compares the best-fit circles and bend centroids for each bend traced from the aerial photographs for 1937 and 1966. The vector arrow at each bend shows the direction and magnitude of movement of the bend centroid between 1937 and 1966. For each bend, this vector may be resolved into cross- and down-valley components to determine the rates of meander migration. The change in radius of curvature of each bend is defined by the difference between the magnitudes of the vectors for 1937 and 1966.

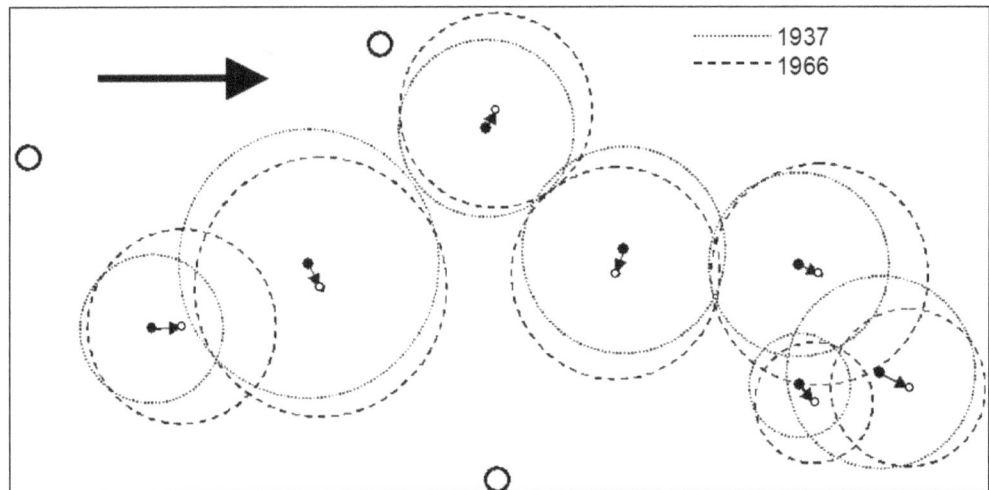

Figure 6.11. Depiction of the bends from 1937 (dotted line) and 1966 (dashed line) outer banklines as defined by best-fit circles. The movement of the bend centroids (arrows) defines migration of the bends.

STEP 6 - The position of the bend at a selected date in the future can be predicted by simple extrapolation if it is assumed that the bend will continue to move at the same rate and in approximately the same direction as it has in the past. To estimate the position of a bend centroid in 1998, for example, the distance the centroid would be expected to move during the 32 years between 1966 and 1998 can be determined by multiplying the annual rate of movement for the 1937 to 1966 period by 32. This distance is plotted along a line starting at the 1966 centroid point and extending in the direction defined by the 1937 to 1966 migration vector. The radius of curvature of the bend in 1998 can be defined by determining the rate of change of the bend radius from 1937 to 1966 relative to the 1966 radius and multiplying this value by number of years from 1966 to 1998. A circle with that radius, centered on the predicted location of the centroid, is plotted on the tracing to indicate the expected location and radius of the bend in 1998.

Figure 6.12 shows the expected outer bank circles for each of the bends of the White River in 1998, based on simple extrapolation of the rates and directions of change during 1937-66. Banklines for the 1998 channel can then be constructed on the tracing by joining the outer bank circles through interpolation, with the 1937 and 1966 banklines used to indicate the reach-scale configuration of the channel.

Figure 6.13 shows the banklines observed in 1937 and estimated for 1998, overlain on the 1966 aerial photo. Inspection of the estimated banklines reveals that Bend 1 would encroach into the levee to the north by 1998 while growth of Bend 5 would likely cutoff Bends 6 and 7.

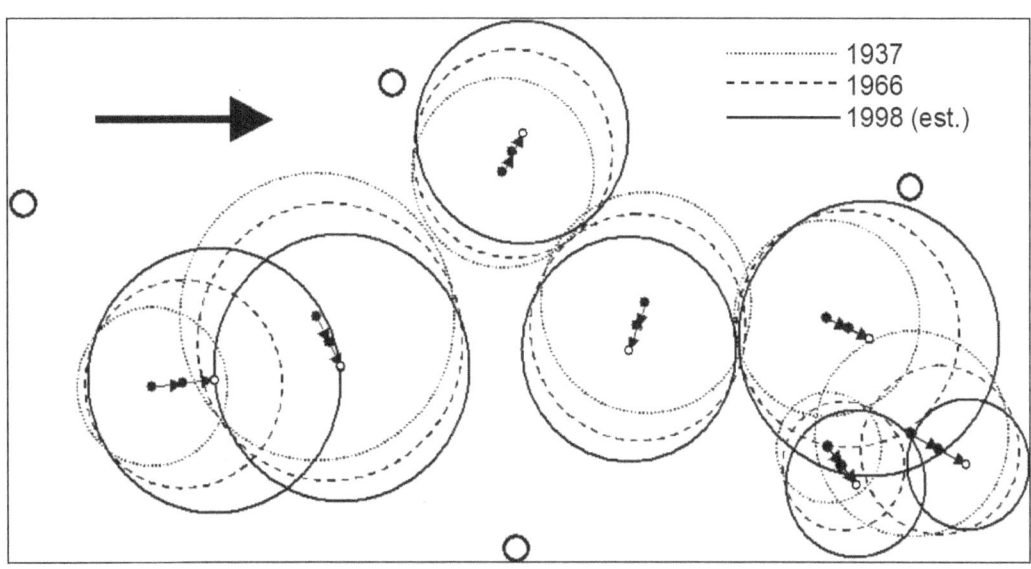

Figure 6.12. Depiction of the bends from the 1937 (dotted line) and 1966 (dashed line) outer banklines, as defined by best-fit circles, and the predicted location and radius of the 1998 outer bankline circle (solid line).

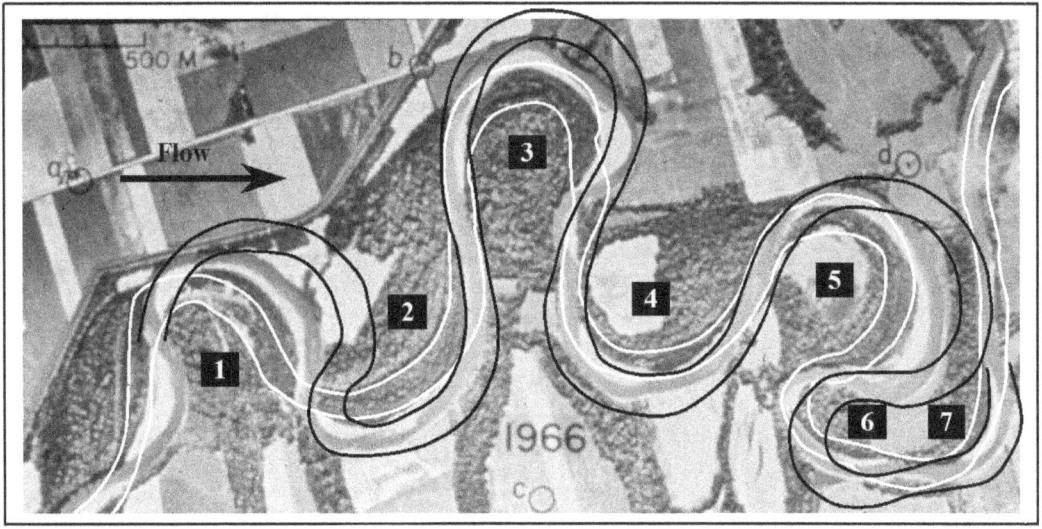

Figure 6.13. Aerial photo of the White River in 1966 showing the actual 1937 banklines (white) and the predicted 1998 bankline positions (black).

In Figure 6.14 the banklines predicted for 1998 by extrapolation of trends of change between 1937 and 1966 are superimposed on an aerial photograph taken in 1998. Two of the registration points used for this comparison are different because two of the original registration points from the previous aerial photos are no longer present on the 1998 aerial photo.

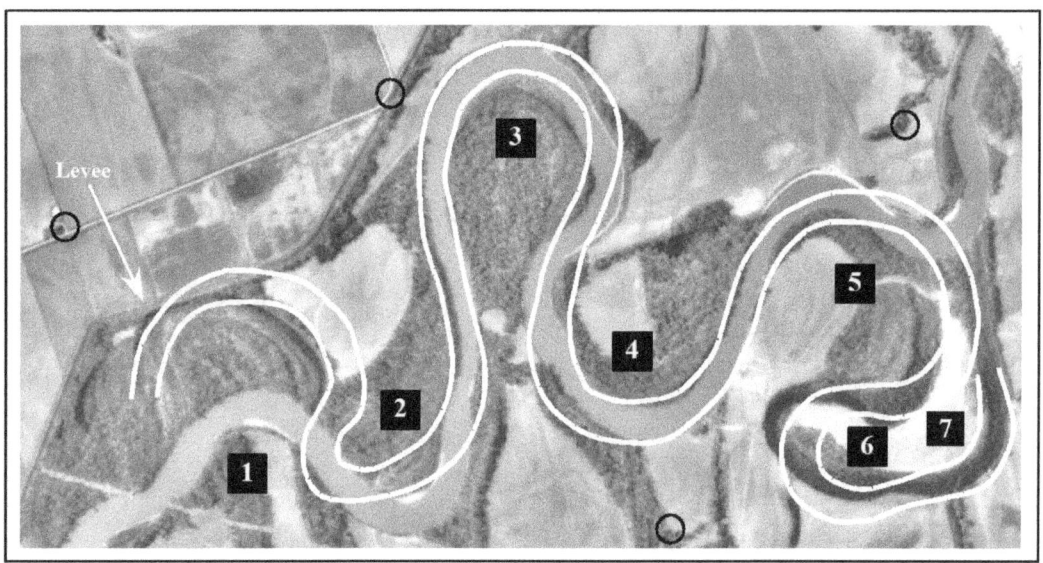

Figure 6.14. Aerial photograph of the White River site in Indiana in 1998 comparing the predicted bankline positions with the actual banklines.

Comparison of the actual and estimated banklines in Figure 6.14 illustrates that meander migration can be predicted relatively accurately using this simple approach. For example, the positions of Bends 3 and 4 and the cutoff at Bend 5 are accurately predicted. Errors in the predicted banklines can be accounted for by: (1) an artificial cutoff that affected Bends 1 and 2; (2) the natural cutoff at Bend 5 that led to Bends 6 and 7 being abandoned; and (3) construction of bank protection at Bends 3 and 5 during the period 1966-95. The artificial cutoff at Bend 1 may have been in response to the serious threat posed by bend migration toward the nearby levee. That cutoff caused Bend 2 to distort in a way that could not have been predicted from its previous behavior. Outer bank migration at Bends 3 and 5 appears to have been curtailed by bank revetments. The migration of Bends 3, 4, and 5, the cutoff of Bend 5, and the abandonment of Bends 6 and 7 were predicted with sufficient accuracy to meet the objectives of this study. It is likely that the positions of Bends 1 and 2, as well as the banklines in the revetted portions of Bends 3 and 4, would have been as predicted except for these engineering interventions.

The case study of the White River used a single period (1937 to 1966) to predict the position of the banklines in 1998. To improve the reliability and accuracy of predictions it is desirable to use multiple pairs of aerial photographs to generate more than one period of analysis. By evaluating multiple periods, meander migration analysis can detect trends of change in the rate and direction of bend migration as well as time-averaged values.

Limitations. As with any analytical technique, aerial photograph comparison technologies have limitations. The accuracy of photo comparison is greatly dependent on the period over which migration is evaluated, the magnitude of internal and external perturbations forced on the system over time, and the number and quality of sequential aerial photos and maps. The analysis will be much more accurate for a channel that has coverage consisting of multiple data sets (aerial photos, maps, and surveys) covering a long period of time (several decades to more than 100 years) versus an analysis consisting of only two or three data sets covering a short time period (several years to a decade). Predictions of migration for channels that have been extensively modified or have undergone major adjustments attributable to extensive land use changes will be much less reliable than those made for channels in relatively stable watersheds.

Overlay techniques require the availability of adequate maps and aerial photos that cover a sufficient period of time to be useful. In general, both air photos and maps will be required to perform a comprehensive and relatively accurate meander migration assessment. Since the scale of aerial photography is often approximate, contemporary maps are usually needed to accurately determine the true scale of air photos without the use of sophisticated photogrammetric instruments.

In addition to scale adjustment and distortion problems that are inherent in the use of aerial photography for comparative purposes, there are a number of physical characteristics of the river environment that can complicate the prediction of meander migration impacts on transportation facilities. Countermeasures to halt bank erosion or protect a physical feature within the floodplain can have an impact on the usefulness of the overlays and these features should be identified prior to developing the overlays. Anomalous changes in the bend or bankline configuration or a major reduction in migration rates may suggest that bank protection is present, especially in areas where the bankline is not completely visible or on images with poor resolution.

Geologic features, such as clay plugs or rock outcrops, in the floodplain can also limit the usefulness of the overlays because they can have a significant influence on migration patterns. Bends can become distorted as they impinge on these features and localized bankline erosion rates may decrease significantly as these erosion resistant features become exposed in the bank. In reaches where geologic controls are exposed predominantly in the bed of the channel, migration rates may dramatically increase because the channel bed is not adjustable, which may cause the channel to migrate rapidly across the feature. **A fundamental assumption of overlay techniques based on aerial photo or map comparison is that a time period sufficient to "average out" such anomalies will be available, making the historic meander rates a reasonable key to the future**.

## 6.4 VERTICAL CHANNEL STABILITY

### 6.4.1 Overview

Vertical channel stability was introduced in Chapter 2 (Section 2.4) and in Chapter 5 (Section 5.5) through discussion of the sediment continuity concept and the Lane relationship (QS α $Q_s D_{50}$). In the Lane relationship, the channel is assumed to be responding to a change in discharge or sediment supply and is moving from one equilibrium geometry to another, either by a change in slope or a change in sediment size. The sediment continuity concept compares the upstream sediment supply (inflow) with the channel's ability to convey sediment (transport capacity). A difference in the inflow of sediment and the transport capacity results in either aggradation or degradation of the channel bed. While these two concepts result in a prediction of channel response in the vertical (aggradation or degradation), they do not provide a prediction of the amount of aggradation or degradation required to reach a new equilibrium state or how quickly the channel will adjust.

Chapter 4 includes a discussion of three levels of analysis: Qualitative Geomorphic Analyses (Level 1), Basic Engineering Analyses (Level 2) and Mathematical or Physical Modeling Studies (Level 3). These three levels of analysis provide the engineer with an understanding of the likely direction of vertical instability and predictions of the amounts and rates of vertical adjustment.

In Level 1 (Qualitative Geomorphic Analyses), land use change, evaluation of vertical stability and prediction of channel response are discussed. Land use change is a common cause of vertical instability as it provides the change in flow or sediment supply causing the channel response. As discussed Section 4.5.5, historic bed elevation changes can be determined by comparing channel longitudinal profiles or comparing channel cross sections. Direct evidence of channel degradation includes (1) exposed utility crossings, (2) exposed bridge foundations, (3) channel banks failing due to excessive height and (4) comparison of channel profiles and cross sections. Bridge inspection reports, which should include soundings at each bent, are a valuable tool for assessing historic channel vertical stability and can be used to predict future trends. If a historic trend is identified, extrapolation can be used to estimate future aggradation or degradation over the life of the bridge. However, if the channel is reaching a new equilibrium condition, the extrapolation will over predict future change. Conversely, if the channel is responding to more recent conditions, extrapolation of historic rates may under predict future change.

In Level 2 (Basic Engineering Analyses), watershed sediment yield, incipient motion, armoring, and rating curve shifts are introduced as factors that influence vertical stability. Changing watershed sediment yield is one factor controlling sediment supply. In coarser bed materials, the channel bed may only be mobilized for relatively high flows and an incipient motion analysis provides insight on the frequency of bed mobilization and vertical stability. When a significant portion of the bed material cannot be moved even during extreme flows, an armor layer can arrest degradation. If a USGS stream gage is located near the bridge, review of historic rating curves (stage-discharge relationships) for the gage can be used to infer vertical stability (see Section 4.6.7). If the discharge increases for a particular stage (positive shift), then the channel has probably degraded.

Level 3 (Mathematical or Physical Modeling Studies) includes sediment transport modeling. Sediment routing using computer models is the most rigorous application of the sediment continuity concept and can be used to determine single event or long-term bed elevation changes in a river.

This section includes expanded discussion on the topics of predicting aggradation and degradation. For degradation, additional discussion on incipient motion analysis and armoring is presented. Expanding on the topic of channel response, stable slope analysis is included for estimation of a new equilibrium slope after the channel has adjusted to a new sediment supply. The topic of sediment continuity is also covered in more detail than in Chapter 2. Combining sediment continuity and transport relationships results in predictive tools for degradation and aggradation rates and amounts. These concepts, which can be used directly to estimate long-term aggradation or degradation, are the basis of sediment routing models.

### 6.4.2 Aggradation and Degradation Analysis

<u>Incipient Motion</u>. Incipient motion is the condition where the hydraulic forces acting on a sediment particle are equal to the forces resisting motion. The particle is at a critical condition where a slight increase in the hydraulic forces will cause the particle to move. The hydraulic forces consist of lift and drag and are usually represented in a simplified form by the shear stress of the flow acting on the particle. Incipient motion conditions can be analyzed using the Shields diagram or by the following equation developed from the diagram:

$$D_c = \frac{\tau_o}{K_s(\gamma_s - \gamma)} \qquad (6.13)$$

where:

$D_c$ = Diameter of the sediment particle at the critical condition, ft (m)
$\tau_o$ = Boundary shear stress, (lb/ft²) Pa
$\gamma$ = Specific weight of water, lb/ft³ (N/m³)
$\gamma_s$ = Specific weight of sediment, lb/ft³ (N/m³)
$K_s$ = Dimensionless coefficient often referred to as the Shields parameter

The Shields parameter can range from 0.03 to 0.10 for natural sediments based on particle shape, angularity, gradation and imbrication. The use of 0.047 for sand sizes provides reasonable results (Meyer-Peter and Muller 1948, Gessler 1971), but lower values (0.03) are commonly used for gravel and cobble sizes.

Equation 6.13 can be used to calculate a sediment particle size that will move for a particular hydraulic condition or to calculate the shear stress required to move a particular particle size. The average shear stress acting on the channel ($\gamma RS$) includes all the factors contributing to resistance to flow. Only the shear stress acting on the individual particles should be used for this calculation. For sand sizes, the base value of Manning roughness coefficient is representative of the grain resistance and the shear stress can be computed from:

$$\tau_o = \frac{\gamma n^2 V^2}{K_u^2 R^{(1/3)}} \tag{6.14}$$

where:

n = Manning roughness coefficient
V = Average channel velocity, ft/s (m/s)
R = Hydraulic radius, ft (m)
$K_u$ = 1.486 English
$K_u$ = 1.0 SI

For coarser grained materials (gravel and larger) the Manning roughness coefficient is a function of grain size and flow depth. The shear stress can be computed from:

$$\tau_o = \frac{\rho V^2}{\left[5.75 \log\left(\frac{12.27 R}{k_s}\right)\right]^2} \tag{6.15}$$

where:

$\rho$ = Density of water, slugs/ft³ (kg/m³)
$k_s$ = Grain roughness usually taken as 3.5 $D_{84}$ for gravel and coarser bed material, ft (m)

Equation 6.15 is essentially Equation 6.14 with the Limerinos equation (Equation 3.11) substituted for Manning roughness coefficient. In the Limerinos equation, the grain roughness is equivalent to 3.5 times $D_{84}$, although for poorly graded material grain roughness can be as low as 1.0 to 2.0 times $D_{84}$. The hydraulic depth (channel area divided by topwidth) can be substituted for hydraulic radius, R, in Equations 6.14 and 6.15 when the width-depth ratio exceeds 10. An incipient motion example problem is solved in Section 6.5.

Armoring. Armoring occurs when the hydraulic forces are sufficient to move a portion of the bed material but insufficient to move the larger sizes. Under these conditions, the smaller material is transported and removed from the bed leaving the coarse material or an armor layer. Armor layers often form in gravel bed rivers during the recession of floods. These armor layers may be disturbed during the next major flood and re-form during the flood recession. In a degrading stream with sufficient amounts of large particles, especially downstream of a dam, the large particles can form a permanent armor (pavement) which is stable under all flow conditions and arrests further degradation. The stability of an armor or pavement is relative to the armor forming discharge. If that discharge is exceeded, further degradation will occur.

The incipient motion equation can be used to determine the critical size of material that can resist a particular hydraulic condition. If at least five percent of the material is larger than the critical size ($D_{95}$ or smaller), armoring can occur. The following equation is used to predict the amount of degradation that would need to occur to form an armor layer (USBR 1984):

$$Y_s = y_a \left( \frac{1}{P_c} - 1 \right) \qquad (6.16)$$

where:

$Y_s$ = Depth of degradation or scour required to form the armor layer, ft (m)
$y_a$ = Thickness of the armor layer, ft (m)
$P_c$ = Percent of material coarser than the critical particle size expressed as a decimal fraction

Figure 6.15 illustrates armor layer development. The thickness of the armor layer ranges from one to three times the critical size ($D_c$) determined from the Shields incipient motion relation. A minimum of two times the critical size is required for a relatively stable armor layer. An armoring example problem is solved in Section 6.5.

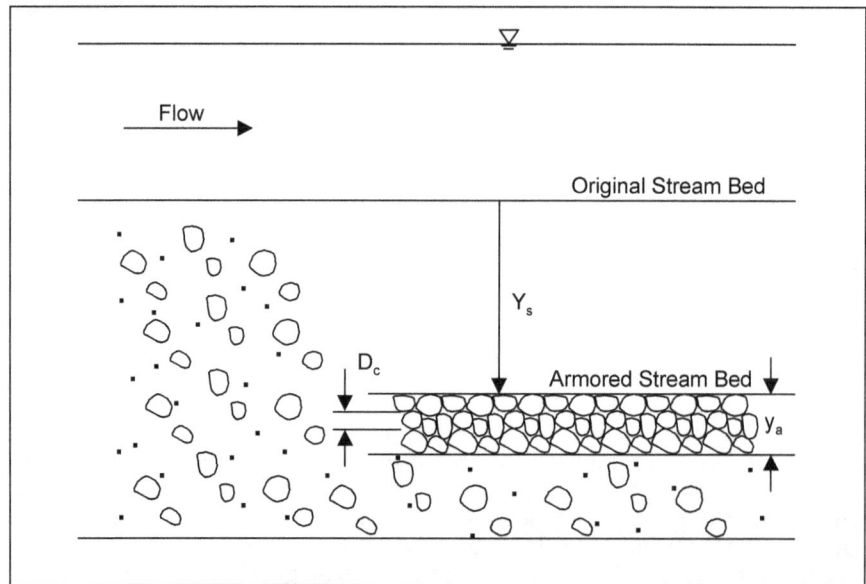

Figure 6.15. Channel armoring.

Equilibrium Slope Analysis. For clear-water releases of flow from dams or detention ponds, the channel immediately downstream would be expected to degrade until the reduction in slope results in a boundary shear stress too low to entrain the bed material. In a sand bed channel, the channel slope would have to be extremely low to reach incipient motion conditions and the amount of degradation could be significant. For a gravel bed channel, channel degradation would also occur, although, in addition to the reduction in slope, the formation of a pavement could arrest degradation. Depending on the bed and bank materials, the degrading channel can narrow as it deepens or the banks can become unstable and the channel can widen. Channel widening temporarily replenishes sediment supply.

For the case of **no sediment supply from upstream**, combining the incipient motion relation (Equation 6.13) and the Manning equation (Equation 3.3) results in an estimate of the equilibrium slope where bed material movement ceases:

$$S_{eq} = \left[ K_s D_c \left( \frac{\gamma_s - \gamma}{\gamma} \right) \right]^{(10/7)} \left( \frac{K_u}{qn} \right)^{(6/7)} \tag{6.17}$$

where:

$S_{eq}$ = Channel Slope at which particles $D_c$ will no longer move
$q$ = Channel discharge per unit width, ft²/s (m²/s)
$K_s$ = Shields parameter
$K_u$ = 1.486 English
$K_u$ = 1.0 SI
$n$ = Manning roughness coefficient
$D_c$ = Critical bed material size, ft (m)

This relationship assumes that the channel width remains constant for future conditions. The critical size ($D_c$) used in this equation should be $D_{90}$ because the bed will coarsen as degradation occurs.

Another approach to determining an equilibrium slope under conditions of no upstream sediment supply is presented by the USBR using the Meyer-Peter Muller equation for beginning of transport (USBR 1984). If adjustment of the hydraulic depth due to the reduction in channel slope is included in the equation, the USBR equation is:

$$S_{eq} = K_u \frac{(D_{50})^{10/7} n^{9/7}}{(D_{90})^{5/14} q^{6/7}} \tag{6.18}$$

where:

$K_u$ = 60.1 English
$K_u$ = 28.0 SI

The degradation computed from the reduction in slope could result in channel narrowing or bank failure and channel widening. Also, the appropriate discharge for use in the equation is difficult to select. A range of discharges are responsible for forming the channel. Given long periods of time, extreme discharges would ultimately be responsible for forming the channel under these conditions. An initial estimate for the clear-water condition is to use the bankfull discharge recognizing that as the channel degrades the dimensions will adjust.

A more typical situation involves a **reduction in sediment supply**. In this case, the equilibrium slope can be predicted using sediment transport relationships. As shown in the Lane relationship (Chapter 5), a reduction in sediment supply or an increase in discharge can cause a reduction in channel slope and degradation. The new equilibrium slope will produce hydraulic conditions where the channel sediment transport capacity matches the upstream sediment supply. This procedure can be performed using sediment transport equations directly or through simplified relationships. A detailed discussion of available sediment transport equations is presented in HDS 6 (FHWA 2001).

It is often useful to develop a sediment transport capacity relationship for a river reach in the form of:

$$q_s = aV^b Y^c \tag{6.19}$$

where:

$q_s$ = Sediment transport capacity per unit width, ft²/s (m²/s)
$V$ = Channel average velocity, ft/s (m/s)
$Y$ = Channel average depth, ft (m)
$a, b, c$ = Coefficient and exponents

The coefficient and exponents can be determined from fitting Equation 6.19 to observed data or a sediment transport equation appropriate to the stream conditions. If the coefficient and exponents are fit to Yang's sediment transport equation for sand (FHWA 2001, Yang 1996), reasonable results (generally within 25 percent) are produced by the following equations. In English units the coefficients are:

$$a = 0.025\, n^{(2.39 - 0.8 \log(D_{50}))} (D_{50} - 0.07)^{-1.4} \tag{6.20}$$

$$b = 4.93 - 0.74 \log(D_{50}) \tag{6.21}$$

$$c = -0.46 + 0.65 \log(D_{50}) \tag{6.22}$$

where:

$D_{50}$ = Mean sediment size, mm (for both SI and English applications)
$n$ = Manning roughness coefficient

For metric units, b and c are unchanged, but the coefficient, a, must be multiplied by a factor of $0.3048^{(2-b-c)}$ when using Equation 6.20. **The range of data used to develop Equations 6.20 through 6.22 is shown in Table 6.2**.

| Table 6.2. Range of Parameters. ||
| Parameter | Value Range |
|---|---|
| $D_{50}$, mm | 0.1 - 2.0 |
| Velocity, ft/s (m/s) | 2.0 – 8.0 (0.61 - 2.44) |
| Depth, ft (m) | 2.0 – 25 (0.61 - 7.62) |
| Slope | 0.00005 - 0.002 |
| Manning roughness coefficient | 0.015 - 0.045 |
| Froude Number | 0.07 - 0.70 |
| Unit Discharge, ft²/s (m²/s) | 1.0 – 200 (0.9 - 18.6) |

For specific values of a, b, and c, the equilibrium slope can then be computed from:

$$S_{eq} = \left(\frac{a}{q_s}\right)^{\frac{10}{3(c-b)}} q^{\frac{2(2b+3c)}{3(c-b)}} \left(\frac{n}{K_u}\right)^2 \qquad (6.23)$$

where:

$S_{eq}$ = Equilibrium slope for the channel to match the upstream sediment supply
$q_s$ = Upstream sediment supply per unit width, ft²/s (m²/s)
$q$ = Unit discharge, ft²/s (m²/s)
$K_u$ = 1.486 English
$K_u$ = 1.0 SI

In the case of a reduction in sediment supply to a reach that was previously in equilibrium and with all other characteristics remaining constant (discharge, roughness and channel width), the equilibrium slope can be related to the existing channel slope by simplifying Equation 6.23 to produce:

$$S_{eq} = S_{ex} \left(\frac{Q_{s(future)}}{Q_{s(existing)}}\right)^{\frac{10}{3(b-c)}} \qquad (6.24)$$

where:

$S_{ex}$ = Existing channel slope
$Q_s$ = Sediment supply, ft³/s (m³/s)

The sediment supply, $Q_s$, for existing conditions can be measured or computed. The sediment supply for future conditions must be computed using an applicable sediment transport relationship (FHWA 2001). Equations 6.23 and 6.24 also assume that channel width and bed material size remain constant as the channel degrades. The appropriate discharge for use in these equations is the effective discharge, which is defined as the discharge responsible for the greatest amount of sediment transport and, therefore, is considered to be responsible for channel formation. If the sediment rating curve is combined with a flow duration curve, the flow that is responsible for transporting the greatest quantity of sediment is the effective discharge.

Because Equations 6.23 and 6.24 use sediment transport capacity and sediment supply where each is determined from the same sediment transport relationship, the selection of the discharge does not greatly affect the equilibrium slope prediction. The bankfull discharge can be used as a reasonable estimate when additional information is unavailable. An equilibrium slope example problem is solved in Section 6.5.

Base Level Control. The equilibrium slope calculations provide an estimate for the slope adjustment inferred by the Lane relationship but do not yield a prediction of the extent or amount of degradation or the amount of time required to reach equilibrium. In a sediment deficient reach, degradation occurs first at the upstream end of the reach and progresses downstream. The downstream extent of degradation is limited by some vertical control to the channel base level (Figure 6.16). The base level control could be a geologic outcrop of erosion resistant material or extremely coarse material. In a tributary channel, the confluence with a much larger river could act as a downstream control. Lakes, reservoirs or the ocean can also act as controls. Grade control structures and culverts can also limit the extent of degradation downstream. If none of these controls exist, then degradation will continue until the channel reaches the equilibrium slope along the entire profile or until armoring takes place. As tributaries contribute sediment to the downstream channel, the effects of the reduced upstream sediment supply are diminished. The amount of ultimate degradation at a location upstream of the base level control can be estimated from the equilibrium slope computation as:

$$Y_s = L(S_{ex} - S_{eq}) \tag{6.25}$$

where:

$Y_s$ = Ultimate degradation amount, ft (m)
$L$ = Distance upstream of base level control, ft (m)

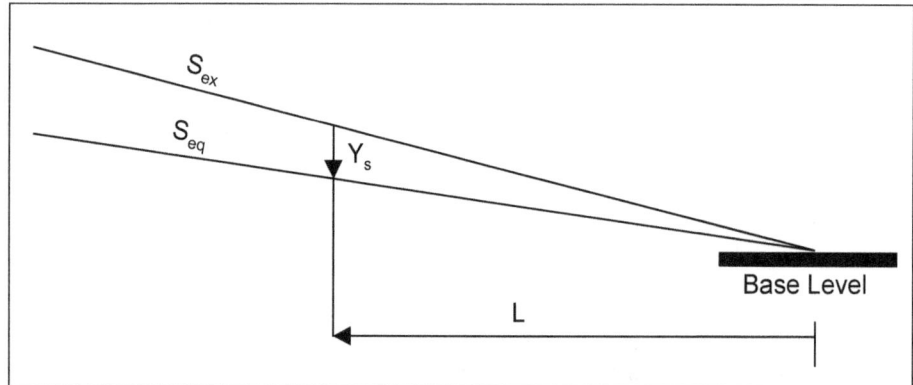

Figure 6.16. Base level control and degradation due to changes in slope.

Another consideration for base level control occurs when a control is removed or lowered on a primary channel and channel degradation progresses upstream. When a primary channel degrades, the base level control is also lowered for each of its tributaries and degradation can progress up these channels. Figure 6.17 illustrates two types of upstream migrating degradation. Headcuts form in cohesive sediment and often form vertical or near-vertical drops with plunge pools at the base of the drop. Headcuts can also be overhanging when weak layers are overlain by a more erosion resistant layer. Figure 6.18 shows a headcut that will migrate upstream and through the bridge crossing during future runoff events. The features of a headcut that can threaten a bridge include the long-term degradation that persists after the headcut has migrated upstream of the bridge, the plunge pool when the headcut is under the bridge, and channel widening that occurs when bed lowering destabilizes the channel banks.

Nickpoints form in noncohesive sediments in which the over-steepened reach translates upstream. For both headcuts and nickpoints, the cause of the degradation is a lowering of the downstream base level control. Headcuts and nickpoints are best identified though channel reconnaissance (Section 5.2). It is reasonable to assume that the amount of degradation will be consistent over the entire stream reach unless there is a longitudinal change in bed and bank materials. It is recommended that detailed monitoring of downstream headcuts be performed to assess threats to existing structures because there are no predictive equations for plunge pool depths, long-term degradation, channel widening, or rate of upstream migration for headcuts or nickpoints.

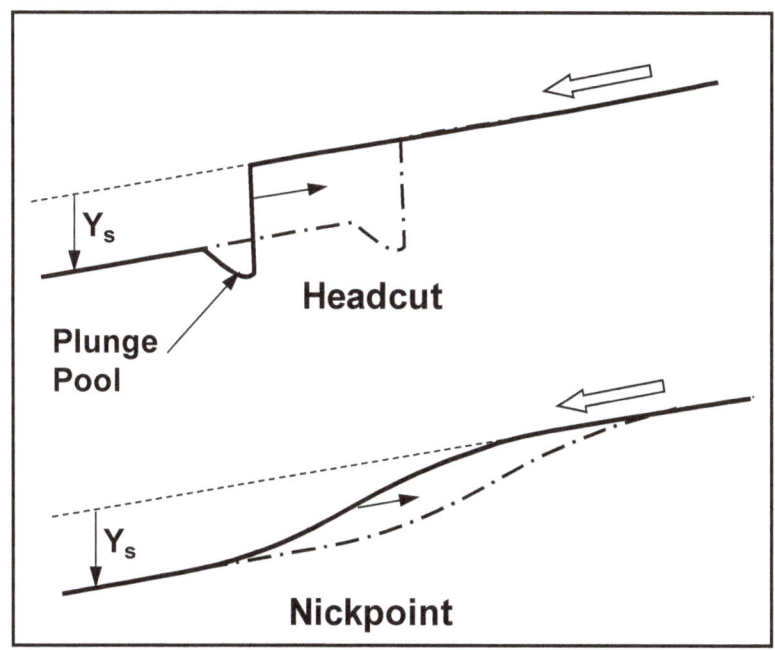

Figure 6.17. Headcuts and nickpoints.

Figure 6.18. Headcut downstream of bridge.

### 6.4.3 Sediment Continuity Analysis

<u>Sediment Transport Concepts</u>. Figure 6.19 shows the various modes of sediment transport. Sediment transport formula are developed to predict bed load, suspended bed material load or bed material load based on the sediment size and hydraulic conditions (see Chapter 7). Wash load is not hydraulically controlled, but is dependent on the supply of fine material from watershed and bank erosion. At high wash load concentrations the transport capacity of the bed material load can increase significantly. FHWA's HDS 6 - River Engineering for Highway Encroachments (FHWA 2001) includes an in-depth discussion of sediment transport processes, equations for predicting sediment transport, and recommendations on the selection of an appropriate equation.

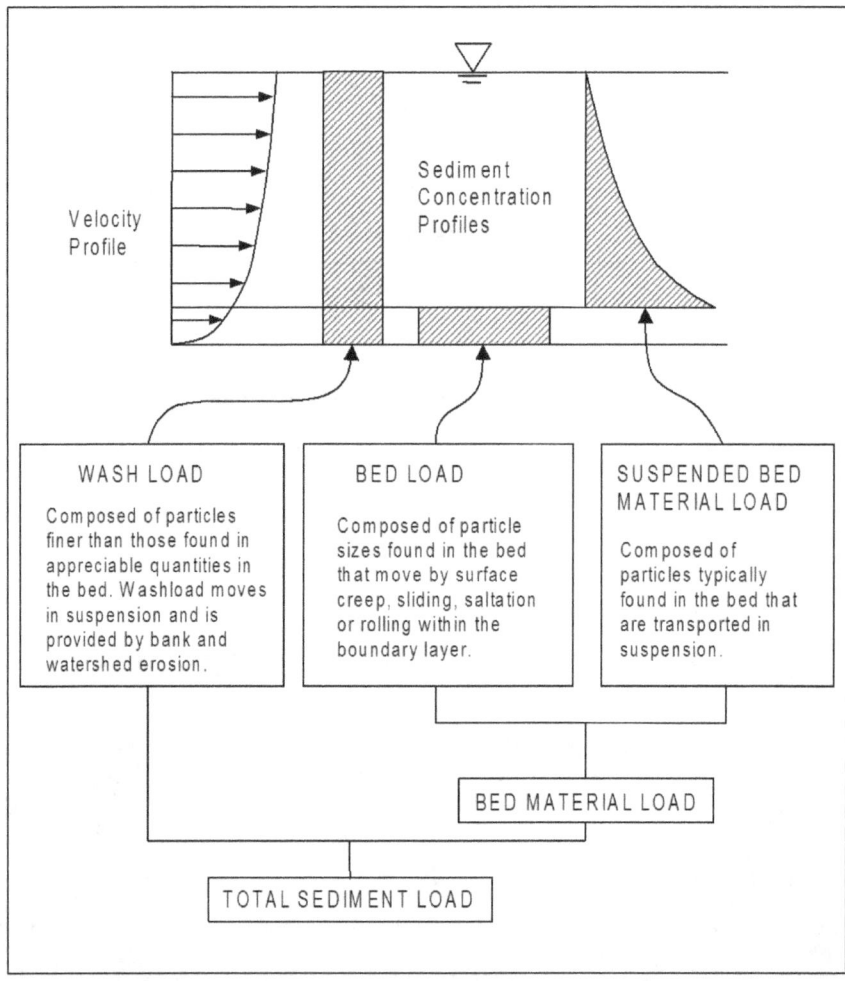

Figure 6.19. Definition of sediment load components.

Based on an appropriate sediment transport relationship, the sediment transport capacity of individual river cross sections in a channel reach can be predicted. The sediment transport can then be used to compute volumes of material being transported and, by comparing with sediment supply to a reach, aggradation and degradation rates can be predicted. For specific site conditions, simplified relationships in the form of Equation 6.19 can be fit to the results of the more rigorous sediment transport equations and estimates of equilibrium slope can be made.

Sediment Continuity Analysis for Aggradation or Degradation. Sediment transport rates can be determined for a range of discharges and combined with a flow duration curve to determine the effective channel discharge. The sediment transport rates can also be summed for a specific flood hydrograph to predict single event aggradation or degradation. In order to do this the sediment supply and the reach transport capacity must be computed. As shown in Figure 2.12, the difference between sediment inflow and outflow results in either bed aggradation or degradation. The volume of material either eroded or deposited is:

$$\Delta V = V_{s(inflow)} - V_{s(outflow)} \tag{6.26}$$

where:

$\Delta V$ = Volume of sediment stored or eroded, ft$^3$ (m$^3$)
$V_{s(inflow)}$ = Volume of sediment supplied to a reach, ft$^3$ (m$^3$)
$V_{s(outflow)}$ = Volume of sediment transport out of a reach, ft$^3$ (m$^3$)

The inflowing and outflowing sediment volumes are equal to:

$$V_s = q_s W \Delta t \tag{6.27}$$

where:

W = Channel width, ft (m)
$\Delta t$ = Time increment, s
$q_s$ = Unit sediment discharge, ft$^2$/s (m$^2$/s)

Equation 6.27 can be summed over a hydrograph to determine sediment volumes during a flood event or can be combined with a flow duration curve to predict long-term rates. The amount of aggradation or degradation is then computed with:

$$\Delta Z = \frac{\Delta V}{WL(1-\eta)} \tag{6.28}$$

where:

$\Delta Z$ = Change in bed elevation, ft (m)
$\eta$ = Porosity of the bed material (volume of the voids/total volume of a sample)
L = Reach length, ft (m)

Because channel aggradation or degradation are adjustments towards a new equilibrium condition, the hydraulic model should be adjusted by the amounts computed in Equation 6.28 before a new flood hydrograph is analyzed. Also, the stability of the new bank heights should be assessed to determine whether channel widening will occur.

A sediment continuity analysis example is solved in Section 6.5.

Sediment Transport Modeling. The sediment continuity analysis described above can be complex and labor intensive. Sediment transport models use the above procedures to route sediment down a channel and adjust the channel geometry to reflect imbalances in sediment supply and transport capacity. The HEC-RAS (USACE 2010) model is an example of a sediment transport model that can be used for single event or long-term degradation estimates. The information needed to run such models includes:

1. Channel and floodplain geometry
2. Structure geometry
3. Roughness
4. Geologic or structural vertical controls
5. Downstream water surface relationship
6. Event or long term inflow hydrographs
7. Tributary inflow hydrographs
8. Bed material gradations
9. Upstream sediment supply
10. Tributary sediment supply
11. Selection of appropriate sediment transport relationship
12. Depth of alluvium

These models perform hydraulic and sediment transport computations on a cross section basis and adjust the channel geometry prior to proceeding with the next time step. Because the actual flow hydrograph is input, the simplifying assumption of using an effective discharge is avoided.

## 6.5 EXAMPLE PROBLEMS - VERTICAL CHANNEL STABILITY

This section provides example problems that illustrate quantitative techniques for vertical channel stability analysis, which were discussed in detail in Section 6.4. The techniques utilized in these example problems include incipient motion, armoring, equilibrium slope and base level control, and sediment continuity.

### 6.5.1 Example Problem 1 - Incipient Motion and Armoring Analysis

A scour vulnerability assessment is being completed for a bridge on a river with well-graded bed material ranging from fine sand to course gravel and cobbles. Determine if the development of an armor layer on the streambed will limit contraction scour. Use principles of incipient motion and armoring to make this assessment and assume a wide channel.

**Given**:

Design discharge = 63,100 cfs (1787 m³/s)
Velocity = 10.7 ft/s (3.26 m/s) (determined from hydraulic modeling)
Depth = 19.0 ft (5.79 m) (determined from hydraulic modeling)
Gradation curves from two bed material samples and their average (see following page)
$D_{50}$ = 110 mm = 0.361 ft (small cobbles)

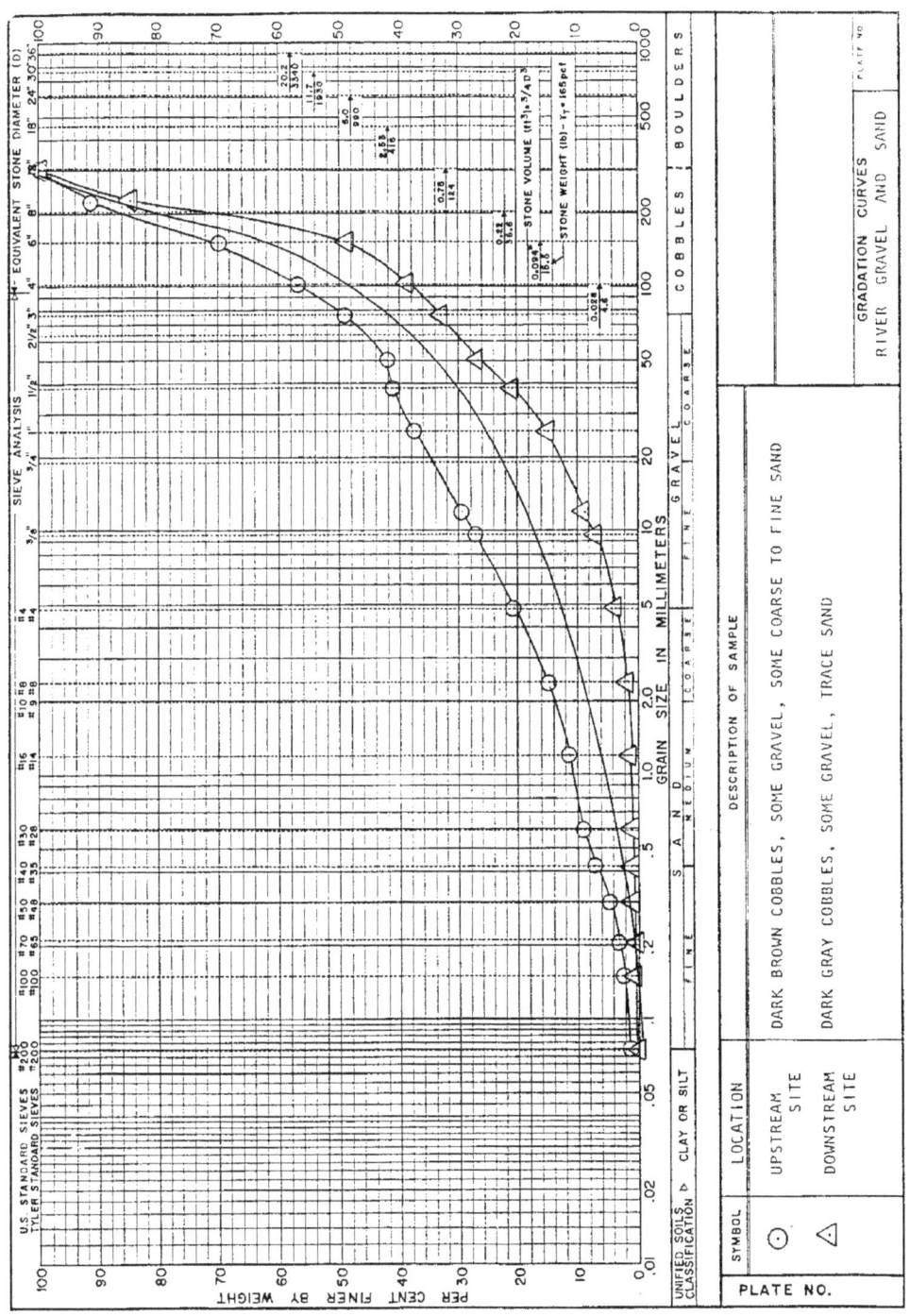

Bed material size gradation curves.

**Solution**:

1. Use Equation 6.15 to calculate the boundary shear stress acting on the bed. Since the bed material is well graded and coarse use $k_s = 3.5D_{84}$, where $D_{84}$ is determined to be 210 mm (0.689 ft) from the gradation curve. The hydraulic radius (R) is approximately equal to the flow depth (y) for many natural channels.

$$\tau_0 = \frac{\rho V^2}{\left[5.75 \log\left(\frac{12.27 R}{k_s}\right)\right]^2}$$

$$\tau_0 = \frac{1.94 \,(10.7)^2}{\left[5.75 \log\left(\frac{12.27(19.0)}{3.5(0.689)}\right)\right]^2} = 1.70 \,\text{lb/ft}^2 \,(81.4 \,\text{Pa})$$

2. Knowing the boundary shear stress calculate the bed material size for incipient motion. Use Equation 6.13 assuming Shields parameter = 0.03 for coarse bed material.

$$D_c = \frac{\tau_0}{K_s(\gamma_s - \gamma)}$$

$$D_c = \frac{1.70}{0.03[(2.65)(62.4) - (62.4)]} = 0.550 \,\text{ft} \,(0.168 \,\text{m})$$

The results indicate that during the design flood, hydraulic forces are adequate to transport bed material up to 0.550 ft (168 mm) in diameter. The gradation curve indicates that 70 percent of the bed material is less than or equal to this particle diameter. Therefore, 30 percent of the bed material is coarser than $D_c$.

3. More than 5 percent of the bed material is coarser than $D_c$. Therefore, armoring is possible. Use Equation 6.16 to estimate the depth of degradation at which an armor layer could form. Assume the armor layer thickness is 3 $D_c$

$$Y_s = y_a \left(\frac{1}{P_c} - 1\right)$$

$$Y_s = 3\,(0.550) \left(\frac{1}{0.30} - 1\right) = 3.8 \,\text{ft} \,(1.2 \,\text{m})$$

It is expected that the bed would armor after 3.8 ft (1.2 m) of degradation.

## 6.5.2 Example Problem 2 - Equilibrium Slope Analysis

The following example was adapted from the USBR (1984). A channel reach receives a majority of its sediment load from an upstream tributary. A small dam on the tributary is proposed to provide local farmers with water for irrigation. The agency responsible for a bridge on the main channel just downstream from the tributary confluence is concerned about the effects of the dam on channel stability. Given the existing hydraulic conditions in the channel reach, calculate the equilibrium slope that is expected to develop over time as a result of (1) removing 100 percent of the sediment supply and (2) reducing the existing supply to 35 percent of the existing value.

**Given:**

Dominant discharge = 780 cfs (22.1 m³/s)
Sediment Supply = 0.142 ft³/s (.004 m³/s)
Width = 350 ft (107 m)
Depth = 1.05 ft (0.32 m)
Slope = 0.0014
$D_{50}$ = 0.000984 ft (0.30 mm)
$D_{90}$ = 0.003150 ft (0.96 mm)
Manning roughness coefficient = 0.027 for the bed of the stream

**Solution for Part 1:**

Use Equation 6.17 to estimate the equilibrium slope assuming the sediment supply has been removed. Assume the Shields parameter is 0.047 and the specific gravity is 2.65 for this bed material.

$$S_{eq} = \left[K_s D_c \left(\frac{\gamma_s - \gamma}{\gamma}\right)\right]^{(10/7)} \left(\frac{K_u}{qn}\right)^{(6/7)}$$

$$S_{eq} = \left[0.047(0.003150)\left(\frac{2.65(62.4) - 62.4}{62.4}\right)\right]^{(10/7)} \left(\frac{1.486}{\left(\frac{780}{350}\right)0.027}\right)^{(6/7)}$$

$S_{eq} = 0.000108 \text{ ft / ft (m / m)}$

Use Equation 6.18 to estimate the equilibrium slope assuming the sediment supply has been removed.

$$S_{eq} = K_u \frac{(D_{50})^{10/7} n^{9/7}}{(D_{90})^{5/14} q^{6/7}}$$

$$S_{eq} = 60.1 \frac{0.000984^{(10/7)} 0.027^{(9/7)}}{0.003150^{(5/14)} \left(\frac{780}{350}\right)^{(6/7)}}$$

$S_{eq} = 0.000115 \text{ ft / ft (m / m)}$

**Solution for Part 2:**

Use Equation 6.23 to estimate the equilibrium slope given that the sediment supply has been reduced to 0.35 (0.142) = 0.050 ft³/s (0.00142 m³/s).

$$a = 0.025\, n^{(2.39-0.8\log(D_{50}))} (D_{50} - 0.07)^{-1.4}$$

$$a = 0.025\, (0.027^{(2.39-0.8\log(0.3))}) (0.3 - 0.07)^{-1.4} = 0.00000770$$

$$b = 4.93 - 0.74 \log(D_{50})$$

$$b = 4.93 - 0.74 \log(0.3) = 5.32$$

$$c = -0.46 + 0.65 \log(D_{50})$$

$$c = -0.46 + 0.65 \log(0.3) = -0.80$$

$$S_{eq} = \left(\frac{a}{q_s}\right)^{\frac{10}{3(c-b)}} q^{\frac{2(2b+3c)}{3(c-b)}} \left(\frac{n}{K_u}\right)^2$$

$$S_{eq} = \left(\frac{0.00000770}{\left(\frac{0.050}{350}\right)}\right)^{\frac{10}{3(-0.80-5.32)}} \left(\frac{780}{350}\right)^{\frac{2(2(5.32)+3(-0.80))}{3(-0.80-5.32)}} \left(\frac{0.027}{1.486}\right)^2$$

$$S_{eq} = 0.000789 \text{ ft/ft (m/m)}$$

Use Equation 6.24 to estimate the equilibrium slope given that the sediment supply has been reduced to 0.35 (0.142) = 0.05 ft³/s (0.00142 m³/s). This equation assumes the reach was previously in equilibrium.

$$S_{eq} = S_{ex}\left(\frac{Q_{s(future)}}{Q_{s(existing)}}\right)^{\frac{10}{3(b-c)}}$$

$$S_{eq} = 0.0014 \left(\frac{0.050}{0.142}\right)^{\frac{10}{3(5.32+0.80)}}$$

$$S_{eq} = 0.000793 \text{ ft/ft (m/m)}$$

## 6.5.3 Example Problem 3 - Base Level Control

From Example Problem 2 there is a concern that changes in bed elevation could threaten the bridge foundation of the bridge that crosses the main channel immediately downstream from the tributary. Given that a base level control exists approximately 8,000 ft (2,500 m) downstream of the bridge, calculate the degradation at the bridge assuming each of the equilibrium slopes from the Example Problem 2.

**Solution**:

Equation 6.25 is used to calculate degradation at the bridge assuming the equilibrium slope from Shields.

$$Y_s = L(S_{ex} - S_{eq})$$

$$y_s = 8,000(0.0014 - 0.000108) = 10.3\,\text{ft}\ (3.2\,\text{m})$$

Results from Equation 6.25 are presented for all the equilibrium slopes in the following table:

| Estimates of Degradation at the Bridge for Each Slope from Example 2. | | |
|---|---|---|
| Method of Calculating Slope | Equilibrium Slope | Degradation (ft) (m) |
| Shields (Equation 6.17) | 0.000108 | 10.3 (3.2) |
| MPM (Equation 6.18) | 0.000115 | 10.3 (3.2) |
| Regression (Equation 6.23) | 0.000789 | 4.9 (1.5) |
| Regression (Equation 6.24) | 0.000793 | 4.9 (1.5) |

The computed channel profiles from the bridge downstream to the base level control are presented in the following figure. Notice that the computed profiles are almost identical for a given inflowing sediment supply, regardless of the equation used; but the amount of inflowing sediment supply has a significant impact on the computed profile.

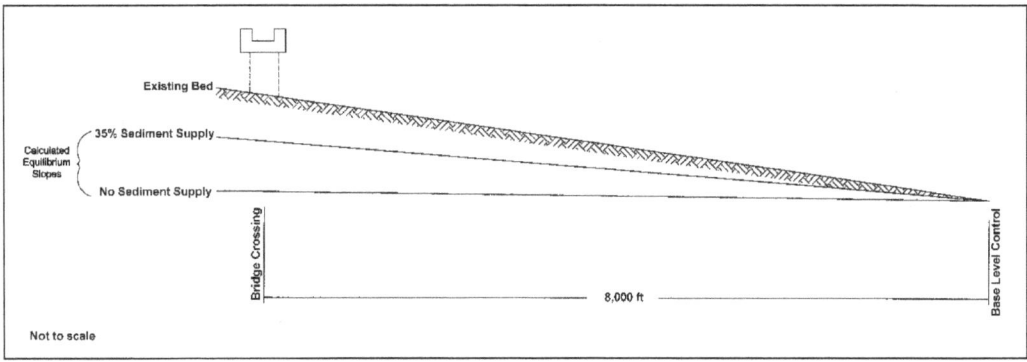

Computed channel profiles from the bridge downstream to the base level control.

### 6.5.4 Example Problem 4 - Sediment Continuity

For the channel of Example Problem 2, calculate the average change in bed elevation that is expected by the end of the first water year following construction of the dam. The data are also plotted in the following figure. The discharge hydrograph will not change, however, the sediment supplies are expected to be reduced to 35 percent of the existing value as a result of the dam. The following table shows monthly discharges and sediment supplies for a typical water year.

Assume the reach is 8,000 ft (2,500 m) long (the distance to downstream base level control) and that the channel properties: width, slope, particle size and Manning roughness coefficient are the same as presented in Example Problem 2. Use Equation 6.19 to calculate sediment transport capacity for the reach.

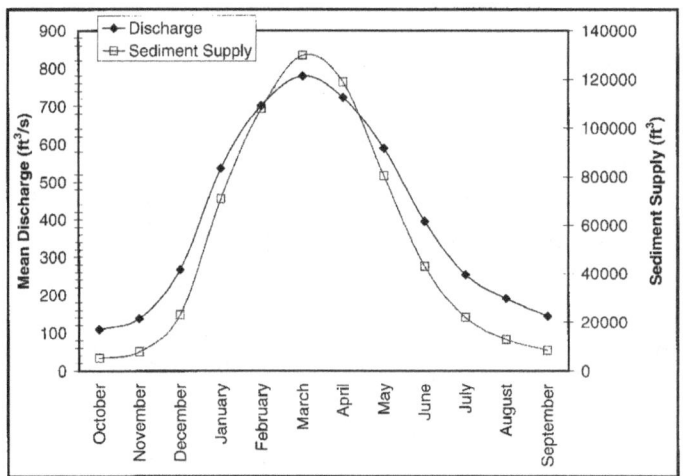

Plot of average discharge and total sediment supply distribution for year.

| Average Discharge and Total Sediment Supply Distribution for the Year. | | | | |
|---|---|---|---|---|
| Month | Average Discharge (ft³/s) | Average Discharge (m³/s) | Sediment Supply (ft³) | Sediment Supply (m³) |
| October | 109 | 3.1 | 5,240 | 148 |
| November | 138 | 3.9 | 7,880 | 223 |
| December | 268 | 7.6 | 23,060 | 653 |
| January | 537 | 15.2 | 70,970 | 2,010 |
| February | 703 | 19.9 | 108,050 | 3,060 |
| March | 780 | 22.1 | 129,840 | 3,677 |
| April | 724 | 20.5 | 118,910 | 3,367 |
| May | 590 | 16.7 | 80,440 | 2,278 |
| June | 396 | 11.2 | 42,940 | 1,216 |
| July | 254 | 7.2 | 21,990 | 623 |
| August | 191 | 5.4 | 12,940 | 366 |
| September | 145 | 4.1 | 8,500 | 241 |
| | | TOTAL | 630,760 | 17,861 |

**Solution:**

The first step is to calculate the hydraulic properties for each month. The values of depth and velocity are required to calculate sediment transport capacity using Equation 6.19. Assuming a wide rectangular channel, the Manning equation (Equation 3.3) can be rearranged to solve for flow depth. The following is a sample calculation for the month of March:

$$Y = \left[\frac{Qn}{1.486\, WS^{1/2}}\right]^{3/5}$$

$$Y = \left[\frac{(780)(0.027)}{1486(350)(0.0014)^{1/2}}\right]^{3/5} = 1.05\,\text{ft} \ (0.32\,\text{m})$$

Knowing discharge, width, and depth, velocity can be calculated using the continuity equation:

$$V = \frac{Q}{A}$$

$$V = \frac{780}{(350)(1.05)} = 2.12\,\text{ft/s} \ (0.65\,\text{m/s})$$

The following table presents the hydraulic properties corresponding to the mean discharge for each month. The channel properties given in Example Problem 2 were used.

| Hydraulic Properties for Each Month of the Year. | | | | | | |
|---|---|---|---|---|---|---|
| Month | Discharge (ft³/s) | Discharge (m³/s) | Depth (ft) | Depth (m) | Velocity (ft/s) | Velocity (m/s) |
| October | 109 | 3.1 | 0.32 | 0.10 | 0.97 | 0.29 |
| November | 138 | 3.9 | 0.37 | 0.11 | 1.07 | 0.33 |
| December | 268 | 7.6 | 0.55 | 0.17 | 1.39 | 0.42 |
| January | 537 | 15.2 | 0.84 | 0.25 | 1.83 | 0.57 |
| February | 703 | 19.9 | 0.99 | 0.30 | 2.03 | 0.62 |
| March | 780 | 22.1 | 1.05 | 0.32 | 2.12 | 0.65 |
| April | 724 | 20.5 | 1.00 | 0.31 | 2.07 | 0.62 |
| May | 590 | 16.7 | 0.89 | 0.27 | 1.89 | 0.58 |
| June | 396 | 11.2 | 0.70 | 0.21 | 1.62 | 0.50 |
| July | 254 | 7.2 | 0.53 | 0.16 | 1.37 | 0.42 |
| August | 191 | 5.4 | 0.45 | 0.14 | 1.21 | 0.36 |
| September | 145 | 4.1 | 0.38 | 0.12 | 1.09 | 0.33 |

The next step is to calculate the hydraulic capacity to transport sediment out of the reach for each month. Given that the coefficient and exponents from Equation 6.19 are functions only of $D_{50}$ and Manning roughness coefficient, their values will be unchanged from Example Problem 2 and will be the same for each month. From Example Problem 2, the values of a, b, and c are:

a = 0.00000770
b = 5.32
c = -0.80

Sediment transport capacity for each month can be calculated using Equation 6.19. The following is a sample calculation for the month of March:

$$q_s = aV^b Y^c$$

$$Q_s = Wq_s = W(aV^b Y^c)$$

$$Q_s = Wq_s = 350[0.00000770(2.12)^{5.32}(1.05)^{-0.80}] = 0.14116 \text{ ft}^3/\text{s} \ (.004 \text{ m}^3/\text{s})$$

The calculated sediment transport capacity for each month is presented in the following table. The total volume of sediment for each month was also calculated by multiplying the capacity and the number of seconds in a month (assuming an average of 2,628,000 seconds per month) as shown in the following table.

| Month | Sediment Transport Capacity for the Year. | | | |
|---|---|---|---|---|
| | Sediment Transport Capacity (ft³/s) | Sediment Transport Capacity (m³/s) | Total Volume of Sediment (ft³) | Total Volume of Sediment (m³) |
| October | 0.00570 | 0.000161 | 14,980 | 424 |
| November | 0.00856 | 0.000262 | 22,500 | 637 |
| December | 0.02507 | 0.000582 | 65,880 | 1,866 |
| January | 0.07716 | 0.002184 | 202,780 | 5,742 |
| February | 0.11747 | 0.003326 | 308,710 | 8,742 |
| March | 0.14116 | 0.004 | 370,970 | 10,505 |
| April | 0.12928 | 0.003366 | 339,750 | 9,620 |
| May | 0.08746 | 0.002476 | 229,840 | 6,508 |
| June | 0.04668 | 0.001321 | 122,680 | 3,474 |
| July | 0.02391 | 0.000677 | 62,840 | 1,779 |
| August | 0.01407 | 0.000398 | 36,980 | 1,047 |
| September | 0.00924 | 0.000262 | 24,280 | 688 |
| | | TOTAL | 1,802,190 | 51,032 |

The volume of degradation can be calculated for each month using Equation 6.26. The sediment inflows are the volume supplies reported in the Total Sediment Supply Table and the sediment outflows are the volume capacities reported in the Sediment Transport Capacity Table. The results of these calculations are shown in the following table. For the month of March:

$$\Delta V = V_{s(inflow)} - V_{s(outflow)} = 129,840 - 370,970 = -241,130 \text{ ft}^3 \ (-6828 \text{ m}^3)$$

| Change in Volume and Depth of Bed Material for Each Month. | | | | | | |
|---|---|---|---|---|---|---|
| Month | Inflow (ft³) | Inflow (m³) | Outflow (ft³) | Outflow (m³) | Change in Volume (ft³) | Change in Volume (m³) |
| October | 5,240 | 148 | 14,980 | 424 | -9,740 | -276 |
| November | 7,880 | 223 | 22,500 | 637 | -14,620 | -414 |
| December | 23,060 | 653 | 65,880 | 1,866 | -42,820 | -1,213 |
| January | 70,970 | 2,010 | 202,780 | 5,742 | -131,810 | -3,732 |
| February | 108,050 | 3,060 | 308,710 | 8,742 | -200,660 | -5,682 |
| March | 129,840 | 3,677 | 370,970 | 10,505 | -241,130 | -6,828 |
| April | 118,910 | 3,367 | 339,750 | 9,620 | -220,840 | -6,253 |
| May | 80,440 | 2,278 | 229,840 | 6,508 | -149,400 | -4,231 |
| June | 42,940 | 1,216 | 122,680 | 3,474 | -79,740 | -2,258 |
| July | 21,990 | 623 | 62,840 | 1,779 | -40,850 | -1,157 |
| August | 12,940 | 366 | 36,980 | 1,047 | -24,040 | -681 |
| September | 8,500 | 241 | 24,280 | 688 | -15,780 | -447 |
| TOTALS | 630,760 | 17,861 | 1,802,190 | 51,032 | -1,171,430 | -33,171 |

The following figure presents a plot of the inflow and outflow sediment volumes. By looking at the discrepancy between the two curves it is apparent that the reach will degrade by the end of the inflow hydrograph.

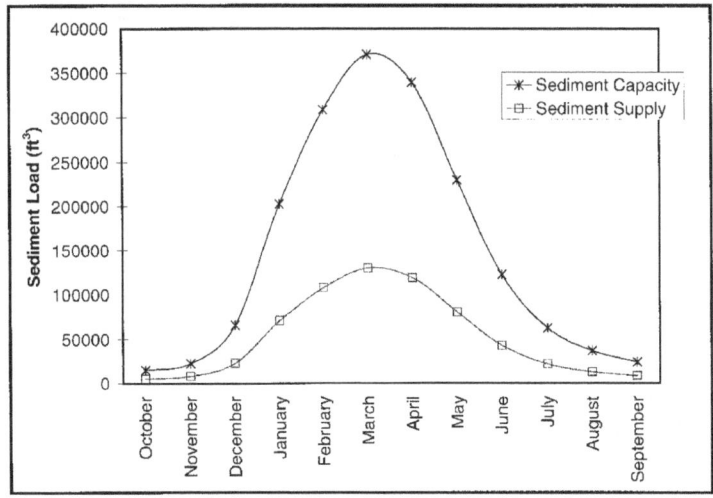

Inflow (Supply) and Outflow (Capacity) Plots Corresponding to the Inflow hydrograph.

The cumulative change in bed elevation expected to occur by the end of the water year can be computed using Equation 6.28, where $\Delta V$ is the total change in volume for the entire year. It is assumed for this bed material that the porosity is 40 percent.

$$\Delta Z = \frac{\Delta V}{WL(1-\eta)}$$

$$\Delta Z = \frac{-1171430}{350(8000)(1-0.40)} = -0.70 \text{ ft } (-0.21 \text{ m})$$

The average change in bed elevation for the 8,000 ft (2,500 m) reach is expected to be -0.70 ft (-0.21 m) by the end of the water year following installation of the dam. Bed lowering at the upstream end of the reach will be greater than the average because degradation begins upstream in a sediment deficient system. In the first water year, degradation could be zero or negligible at the lower portion of the reach.

### 6.5.5 Example Problem 5 - Sediment Continuity

For the channel in Example Problems 2 through 4, estimate the number of years it will take for the slope of the main channel to reach equilibrium given that the sediment supply will be reduced to 35 percent of the existing value. Use the total volume deficit from the Change in Volume table in Example Problem 4 as the annual volume of erosion for the reach. The total volume of sediment that will ultimately be eroded can be calculated from the results of Example Problem 3.

**Solution:**

1. Calculate the total volume of sediment that will have been eroded once equilibrium has been reached (see the Computed Channel Profiles Figure in Example Problem 3 and assume the degradational "wedge" is a right triangle and account for the porosity of the sediment).

Total Volume = W [1/2 $y_s$ L] = 350 [1/2 (4.9) (8,000)] = 6,860,000 ft³ (194,254 m³)

Sediment Volume = Total Volume (1 - η) = 6,860,000 (1 - 0.4) = 4,116,000 ft³ (116,552 m³)

2. Calculate the time to reach slope equilibrium.

$$\text{Time (years)} = \frac{4,116,000 \text{ ft}^3}{1,171,430 \text{ ft}^3 / \text{yr}} = 3.5 \text{ years (say 4 years)}$$

It is expected that the slope of the main channel will reach equilibrium in about 4 years after construction of the dam on the tributary.

# CHAPTER 7

# SEDIMENT TRANSPORT CONCEPTS

## 7.1 OVERVIEW

As discussed in Chapter 2, most channels and floodplains that roads cross are alluvial. Alluvial channels are formed by sediment that has been transported and deposited by flowing water and can be transported by the channel in the future. Channel adjustments include aggradation, degradation, width adjustment, and lateral shifting. Aggradation and degradation are the overall raising or lowering of a channel bed over time from sediment accumulation or erosion. Channel widening and shifting are the result of bank erosion due to hydraulic forces or by mass failure of the bank.

Sediment transport involves complex processes that interact to produce the existing channel form and future channel adjustments. The amount of material transported or deposited in a stream under a given set of conditions is the result of the interaction of two groups of variables. In the first group are those variables that influence the quantity and quality of the sediment brought down to that section of the stream. In the second group are variables that influence the capacity of the stream to transport that sediment. These groups of variables can be summarized as follows.

Group 1 - Sediment brought down to the stream depends on the geology and topography of watershed; magnitude, intensity, duration, distribution, and season of rainfall; soil moisture conditions; vegetal cover; cultivation and grazing; surface erosion and bank cutting.

Group 2 – The capacity of a stream to transport sediment depends on hydraulic properties of the stream channel. These are fluid and flow related properties including: slope, roughness, hydraulic radius, discharge, velocity, velocity distribution, turbulence, tractive force, viscosity and density of the fluid sediment mixture, and size and gradation of the sediment.

These variables are not all independent and, in some cases, their effect is not definitely known. The variables which control the amount of sediment brought down to the stream are subject to so much variation, not only between streams but at a given point of a single stream, that the quantitative analysis of any particular case is extremely difficult. It is practicable, however, to measure the sediment discharge over a long period of time and record the results, and from these records to determine a soil loss from the area. The variables that deal with the capacity of the stream to transport solids are subject to mathematical analysis. These variables are closely related to the hydraulic variables controlling the capacity of the stream to carry water.

From Group 1 it is apparent that many aspects of hydrology play a role in sediment transport analyses. These include not only peak flow rates, but also individual flood hydrographs, and the duration of flow. The entire range of flow may be significant because even though the highest flows have the highest rates of sediment transport, lower flows may have significantly longer durations and produce the greatest cumulative sediment transport. Channels respond and adjust to changes in flow and sediment supply. Therefore, changing watershed conditions often result in changes in channel geometry. Channel geometry, bed material, and vegetation determine hydraulic variables (velocity, depth, etc.), which in turn control sediment transport capacity. Therefore, sediment transport and channel stability depend not only on the specific physical processes, but also the history of natural and human-induced factors in the watershed.

The following sections provide a general overview of sediment transport concepts and processes. Other resources are available to provide the in-depth information required to perform sediment transport analyses. These resources include HDS 6 (FHWA 2001), Sedimentation Engineering (ASCE 2008), and textbooks (Simons and Senturk 1992, Yang 2003, Julien 2010).

## 7.2 SEDIMENT CONTINUITY

As noted in Chapter 2, the amount of material transported, eroded, or deposited in an alluvial channel is a function of sediment supply and channel transport capacity. Sediment supply is provided from the tributary watershed and from erosion occurring in the upstream channel bed and banks. Sediment transport capacity is primarily a function of sediment size and the hydraulic properties of the channel. When the transport capacity of the flow equals sediment supply from upstream, a state of equilibrium exists.

Application of the sediment continuity concept to a channel reach illustrates the relationship between sediment supply and transport capacity. The sediment continuity concept states that the sediment inflow minus the sediment outflow equals the rate of change of sediment volume in a given reach. More simply stated, during a given time period the amount of sediment coming into the reach minus the amount leaving the downstream end of the reach equals the change in the amount of sediment stored in that reach (see Figure 2.12). The sediment inflow to a given reach is defined by the sediment supply from the watershed and channel (upstream of the study reach plus lateral input directly to the study reach). The transport capacity of the channel within the given reach defines the sediment outflow. Changes in the sediment volume within the reach occur when the total input to the reach (sediment supply) is not equal to the downstream output (sediment transport capacity). When the sediment supply is less than the transport capacity, erosion (degradation) will occur in the reach so that the transport capacity at the outlet is satisfied, unless controls exist that limit erosion. Conversely, when the sediment supply is greater than the transport capacity, deposition (aggradation) will occur in the reach.

Controls that limit erosion may either be human induced or natural. Human-induced controls included bank protection works, grade control structures, and stabilized bridge or culvert crossings. Natural controls can be geologic, such as outcrops, or the presence of significant coarse sediment material in the channel. The presence of coarse material can result in the formation of a surface armor layer of larger sediments. For a discussion of armoring, base level control, and sediment continuity analysis for aggradation or degradation, see Chapter 6. Armoring in gravel-bed rivers is discussed in Chapter 8.

## 7.3 SEDIMENT PROPERTIES

A knowledge of the properties of sediment particles is important, as they indicate the behavior of the particles in their interaction with the flow. Several important sediment properties are discussed in the following sections.

### 7.3.1 Particle Size

Of the various sediment properties, physical size has by far the greatest significance to the hydraulic engineer. The particle size is the most readily measured property, and other properties such as shape, fall velocity and specific gravity tend to vary with size in a roughly predictable manner. In general, size represents a sufficiently complete description of the sediment particle for many practical purposes.

Particle size $D_s$ may be defined by its volume, diameter, weight, fall velocity, or sieve mesh size. Except for volume, these definitions also depend on the shape and density of the particle. The following definitions are commonly used to describe the particle size.

In general, sediments have been classified into boulders, cobbles, gravels, sands, silts, and clays on the basis of their nominal or sieve diameters. The size range in each general class is given in Table 2.1. The non-cohesive material generally consists of silt (0.004-0.062 mm), sand (0.062 - 2.0 mm), gravel (2.0 - 64 mm), or cobbles (64-250 mm).

The boulder class (250 - 4000 mm) is generally of little interest in sediment problems. The cobble and gravel class plays a considerable role in the problems of local scour and resistance to flow and to a lesser extent in bed load transport. The sand class is one of the most important in alluvial channel flow. The silt and clay class is of considerable importance in the evaluation of stream sediment loads, bank stability and problems of seepage and consolidation.

### 7.3.2 Particle Shape

Generally speaking, shape refers to the overall geometrical form of a particle. Sphericity is defined as the ratio of the surface area of a sphere of the same volume as the particle to the actual surface area of the particle. Roundness is defined as the ratio of the average radius of curvature of the corners and edges of a particle to the radius of a circle inscribed in the maximum projected area of the particle. However, because of simplicity and effectiveness of correlation with the behavior of particles in the flow, the most commonly used parameter to describe particle shape is the Corey shape factor, $S_p$, (see FHWA 2001):

### 7.3.3 Fall Velocity

The prime indicator of the interaction of sediments in suspension within the flow is the fall velocity of sediment particles. The fall velocity of a particle is defined as the velocity of that particle falling alone in quiescent, distilled water of infinite extent. In most cases, the particle is not falling alone, and the water is not distilled or quiescent. Measurement techniques are available for determining the fall velocity of groups of particles in a finite field in fluid other than distilled water. However, little is known about the effect of turbulence on fall velocity.

A particle falling at terminal velocity in a fluid is under the action of a driving force due to its buoyant weight and a resisting force due to the fluid drag. Fluid drag is the result of either the tangential shear stress on the surface of the particle or a pressure difference on the particle or a combination of the two forces (see FHWA 2001).

### 7.3.4 Sediment Size Distribution

Several methods of obtaining sediment size distribution are available for determining the size distribution of a sediment sample. Each method for size distribution analysis is appropriate for only a particular range of particle sizes. Two methods applicable for the non-cohesive sand size and coarser fraction are described briefly here.

<u>Sieve Size Distribution.</u> In the sand and gravel range size distribution is generally determined by passing the sample through a series of sieves of mesh size ranging from 32 mm to 0.062 mm. A minimum of about 100 grams of sand is required for an accurate sieve analysis. More is required if the sample contains particles of 1.0 mm or larger. Standard methods employed in soil mechanics are suitable for determining the sieve sizes of sand and gravel sediment samples (see Table 2.1).

Pebble Count Method. A pebble count can be used to obtain the size distribution of coarse bed materials (gravel and cobbles) which are too large to be sieved. Very often the coarser material is underlain by sands. Then, the underlying sands are analyzed by sieving. The two classes of bed material are either combined into a single distribution or used separately. The large material sizes are measured in situ by laying out a square grid. Within the grid, all the particle sizes are measured and counted by size intervals. For large samples a random selection of particles in the various classes is appropriate to develop frequency histograms of sediment sizes (see FHWA 2001).

### 7.3.5 Specific Weight

Specific weight is weight per unit volume. In the English system of units, specific weight is usually expressed in units of pounds per cubic foot and in the metric system, in grams per cubic centimeter. In connection with granular materials such as soils, sediment deposits, or water sediment mixtures, the specific weight is the weight of solids per unit volume of the material including its voids. The measurement of the specific weight of sediment deposits is determined simply by measuring the dry weight of a known volume of the undisturbed material.

### 7.3.6 Porosity

The porosity of granular materials is the ratio of the volume of void space to the total volume of an undisturbed sample. To determine porosity, the volume of the sample must be obtained in an undisturbed condition. Next, the volume of solids is determined either by liquid displacement or indirectly from the weight of the sample and the specific gravity of material. The void volume is then obtained by subtracting the volume of solids from the total volume.

### 7.3.7 Angle of Repose

The angle of repose is the maximum slope angle upon which non-cohesive material will reside without moving. It is a measure of the intergranular friction of the material and is different for dry versus submerged conditions.

### 7.4 SEDIMENT TRANSPORT CONCEPTS

### 7.4.1 Initiation of Motion

The initiation or ceasing of motion of sediment particles is involved in many geomorphic and hydraulic problems including stream stability and scour at highway bridges, sediment transport, erosion, slope stability, stable channel design, and design of riprap. These problems can only be handled when the threshold of sediment motion is fully understood.

Beginning of motion can be related to either shear stress on the grains or the fluid velocity in the vicinity of the grains. When the grains are at incipient motion, these values are called the critical shear stress or critical velocity. The choice of shear stress or velocity depends on: (1) which is easier to determine in the field; (2) the precision with which the critical value is known or can be determined for the particle size; (3) the type of problem. In sediment transport analysis most equations use critical shear stress. In stable channel design either critical shear stress or critical velocity is used; whereas, in riprap design critical velocity is generally used.

The initiation of motion of bed material particles exposed to flowing water is difficult to define precisely. The particles are subjected to drag and lift forces by the flowing water. The flow field near the boundary, turbulence fluctuations, and the particle size, shape and relative position with respect to other particles all contribute to these forces. The particle size, shape, and relative position to other particles also contribute to the forces that resist motion, which include gravitational and external support forces acting on the particle (friction and other point contacts between grains). This problem has been simplified and studied empirically by many scientists for laboratory conditions, dating back to Shields (1936). Detailed discussions are available from many sources including HDS 6 (FHWA 2001). Shields related the beginning of motion to particle size, particle submerged unit weight, and flow shear stress to predict the initiation of motion.

Standing water exerts hydrostatic pressure on the channel bed. For uniform flow with small slopes, the flowing water exerts a time-average shear stress in the direction of flow equal to the hydrostatic pressure times the channel slope:

$$\tau_0 = \gamma y S_o \tag{7.1}$$

where:

$\tau_0$ = Shear stress, lb/ft² (Pa)
$\gamma$ = Specific weight of water, lb/ft³ (N/m³)
$y$ = Flow depth (hydraulic radius, hydraulic depth for wide channels, or local depth) ft (m)
$S_o$ = Bed slope (or energy slope for gradually varied flow)

Another useful formula for estimating average shear stress for gradually varied flow conditions is:

$$\tau_0 = \frac{\gamma}{y^{(1/3)}} \left(\frac{nV}{K_u}\right)^2 \tag{7.2}$$

where:

$n$ = Manning roughness coefficient
$V$ = Flow velocity, ft/s (m/s)
$K_u$ = 1.486 English units
$K_u$ = 1.0 SI

Equation 7.2 shows the relationship between velocity and shear stress, i.e., shear stress is proportional to velocity squared. The Shields parameter relates critical shear stress to particle size and specific weight by.

$$\tau_c = k_s D_s (\gamma_s - \gamma) \tag{7.3}$$

where:

$\tau_c$ = Critical shear stress for beginning of motion, lb/ft² (Pa)
$k_s$ = Shields parameter
$D_s$ = Particle size, ft (m)
$\gamma_s$ = Specific weight of the particle, lb/ft³ (N/m³)

The Shields parameter ranges from 0.03 to 0.10 for natural sediments and depends on particle shape, angularity, gradation and imbrication. The use of 0.047 is common for sand sizes. When the shear stress of the flow exceeds the critical shear stress of the particle, the channel bed begins to mobilize and bed material is transported downstream. Particle motion begins as sliding and rolling of individual particles along the bed. It is important to recognize that the Shields equation is not a sediment transport equation because it does not provide any estimate of the amount of sediment in motion. It is also important to note that only the shear stress acting on the particles, or grain friction, should be used in applying this relationship.

### 7.4.2 Modes of Sediment Transport

Once the critical shear stress is exceeded, bed material begins to move (roll, slide, and saltate) along the bed surface. This material is referred to as bed load or contact load because it is in almost continuous contact with the bed. For small amounts of positive excess shear stress (defined as $\tau_o - \tau_c$), this is the only mode of bed material transport. As excess shear stress increases, turbulence begins to suspend some of the particles. The turbulence acts to mix the particles in the water column and gravity causes the particles to settle. Therefore, bed material can also transported downstream as suspended bed material load. The two types of bed material load are illustrated in Figure 6.19.

The suspended bed material load shown in Figure 6.19 depends on the interaction between gravity and turbulence. Because gravity is causing particles to settle, they are concentrated near the bed. Turbulence mixes the particles in the water column and, depending on the size and specific weight of the particles, relatively few particles may reach the surface. The suspension of particles is illustrated in Figure 7.1, which shows the concentration profile for various particle sizes in a turbulent flow field. The equation that describes the concentration profiles is:

$$\frac{c}{c_a} = \left[\left(\frac{y_o - y}{y}\right)\left(\frac{a}{y_o - a}\right)\right]^z \tag{7.4}$$

where:

| | | |
|---|---|---|
| $c$ | = | Sediment concentration at height y from the bed |
| $c_a$ | = | Sediment concentration at height a above the bed |
| $y_o$ | = | Total flow depth (from surface to bed) ft (m) |
| $z$ | = | Rouse number = $\omega/(\beta\kappa v_*)$ |
| $\omega$ | = | Fall velocity of the particle in quiescent water, ft/s (m/s) |
| $\beta$ | = | Parameter relating particle and momentum transfer due to turbulence, approximately equal to 1.0 for fine particles |
| $\kappa$ | = | Von Karman's constant of 0.4 |
| $v_*$ | = | Shear velocity = $\sqrt{\tau_o/\rho} = \sqrt{gRS}$ |
| $\rho$ | = | Water density, slugs/ft³ (kg/m³) |
| $g$ | = | Acceleration due to gravity, ft/s² (m/s²) |
| $R$ | = | Hydraulic radius, ft (m) |

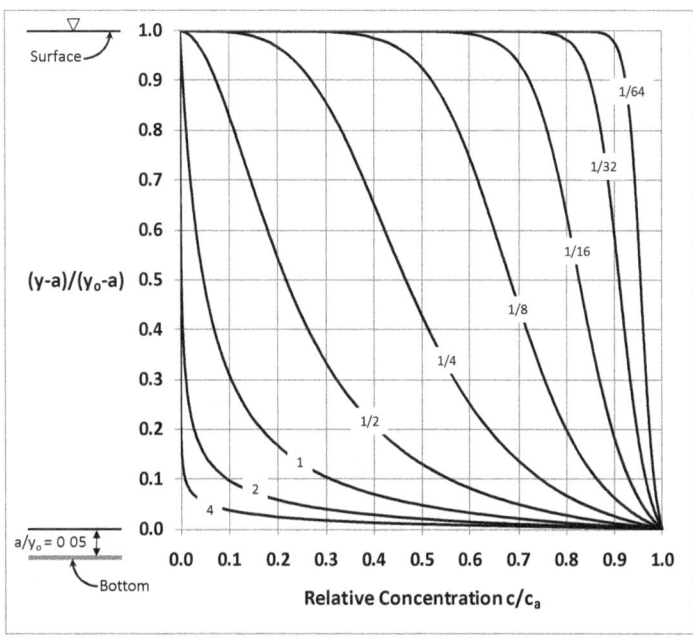

Figure 7.1. Suspended sediment concentration profiles (Rouse 1937).

Larger particles have greater fall velocities and larger Rouse numbers. Therefore, Figure 7.1 shows that for a given level of turbulence (as represented in the Rouse number by the shear velocity), large particles will remain close to the bed. Finer particles have smaller Rouse numbers, are mixed higher into the flow and have higher concentrations. Julien (2010) indicates that particle sizes with Rouse numbers less than 0.025 (1/40) will have essentially uniform concentration profiles. These particles are extremely fine, primarily silts and clays, and have very small fall velocities. They are defined as wash load, which are derived primarily from upland erosion and bank erosion of floodplain materials (see Figure 6.19). Wash load material is not found in appreciable quantities in the channel bed.

In summary, bed material is transported in contact with the bed (bed load) and in suspension (suspended bed material load). The total sediment load transported by the channel also includes wash load, which is supplied to the channel rather than derived from the bed. Wash load is also transported in suspension. In coarse bed channels, such as cobble-bed and boulder-bed streams, sand may act as wash load because it is not found in appreciable quantities in the bed and because the supply is far less than the channel capacity to transport this size.

### 7.4.3 Effects of Bed Forms at Stream Crossings

In sand-bed streams, sand material is easily eroded and is continually being moved and shaped by the flow. The interaction between the flow of the water-sediment mixture and the sand-bed creates different bed configurations which change the resistance to flow, velocity, water surface elevation, and sediment transport. At high flows, most sand-bed stream channels shift from a dune bed to a transition or a plane bed configuration (see Sections 3.4.3 and 3.4.4). The resistance to flow is then decreased by one-half to one-third of that preceding the shift in bed form. The increase in velocity and corresponding decrease in depth may increase scour around bridge piers, abutments, spurs or guide banks and may increase the required size of riprap.

Another effect of bed forms on highway crossings is that with dunes on the bed, there is a fluctuating pattern of scour on the bed. Methods for computing bed-form geometry can be found in Julien and Klaassen (1995) and Karim (1999). Karim included laboratory and field data where the crest-to-trough height, $\Delta$, for dunes ranged from less than 0.1y to up to 0.5y. Karim also showed a range of antidune heights between 0.1y and 0.4y. Bennett (USGS 1997) indicated an approximate upper limit as $\Delta < 0.4y$. The average dune height equation by Julien and Klaassen is:

$$\frac{\Delta}{y} = 2.5\left(\frac{D_{50}}{y}\right)^{0.3} \tag{7.5}$$

The lower and upper bounds on dune heights (95 percent) range from 0.3 to 3.2 times this average height. Dune lengths can be approximated as 6.25 times the flow depth. Care must be used in analyzing crossings of sand-bed streams in order to anticipate changes that may occur in bed forms and the impact of these changes on the resistance to flow, sediment transport, and the stability of the reach and highway structures. With a dune bed, the Manning n could be more than twice as large as a plane bed (see Figure 3.5). A change from a dune bed to a plane bed, or the reverse, can have an appreciable effect on depth and velocity. In the design of a bridge or a stream stability or scour countermeasure, it is good engineering practice to assume a dune bed (large n value) when establishing the water surface elevations, and a plane bed (low n value) for calculations involving velocity.

## 7.5 SEDIMENT TRANSPORT EQUATIONS

Equations for predicting bed material sediment transport differ depending on the mode of sediment transport. ASCE (2008) includes 16 bed load equations. The Meyer-Peter and Müller (1948) equation is considered to be a classic bed load equation (see FHWA 2001). The equation has the basic form of:

$$q_b = a(\tau_0 - \tau_c)^{3/2} \tag{7.6}$$

where:

$q_b$ = Bed load discharge per unit width of channel, ft²/s (m²/s)
$a$ = Empirical coefficient

As with the analysis of incipient motion, only grain friction should be included in the bed shear ($\tau_0$) variable. Many of the equations presented in ASCE (2008) include excess shear stress ($\tau_0 - \tau_c$) to the 1.5 power. Because bed shear is proportional to velocity squared (see Equation 7.2), bed-load dominated sediment transport, such as in gravel-bed rivers, is generally proportional to velocity cubed.

The Einstein (1950) suspended load equation is described in HDS 6 (2001) and is a solution to the general suspended load equation:

$$q_s = \gamma_s \int_a^{y_0} vc\,dy \tag{7.7}$$

Where the variables are defined as in Equation 7.4 and:

$q_s$ = Suspended load discharge per unit width
$\gamma_s$ = Weigh per unit volume of suspended sediment
v = Velocity at height y above the bed
c = Volumetric concentration at height y above the bed

The solution of the integral uses Equation 7.4 for sediment concentration and a logarithmic velocity profile equation. The concentration and velocity profiles are illustrated in Figure 7.2. This integration depends on a reference concentration at point "a" that is determined from the bed load (see Figure 7.1). Rouse assumed that "a" was approximately equal to 5% of the depth, $y_o$, while Einstein used 2 $D_{50}$ to estimate the reference concentration. ASCE (2008) presents nine equations for determining the reference concentration and an easily applied equation (Abad and Garcia 2006) to solve the integration of Equation 7.7. Because the rate of bed load transport and the concentration profile depend on grain size, the integration is performed for the range of grain sizes in the bed material and the total bed material is the sum of the proportionate transport rates computed for each size class. Julien (2010) used Equation 7.7 to show that bed load comprises 80 percent or more of the total load when shear velocity divided by fall velocity ($v^*/\omega$) is less than 0.5, and that suspended load comprises 80 percent or more of the total load when $v^*/\omega > 2$. For $0.5 < v^*/\omega < 2$ the sediment transport is considered to be mixed load.

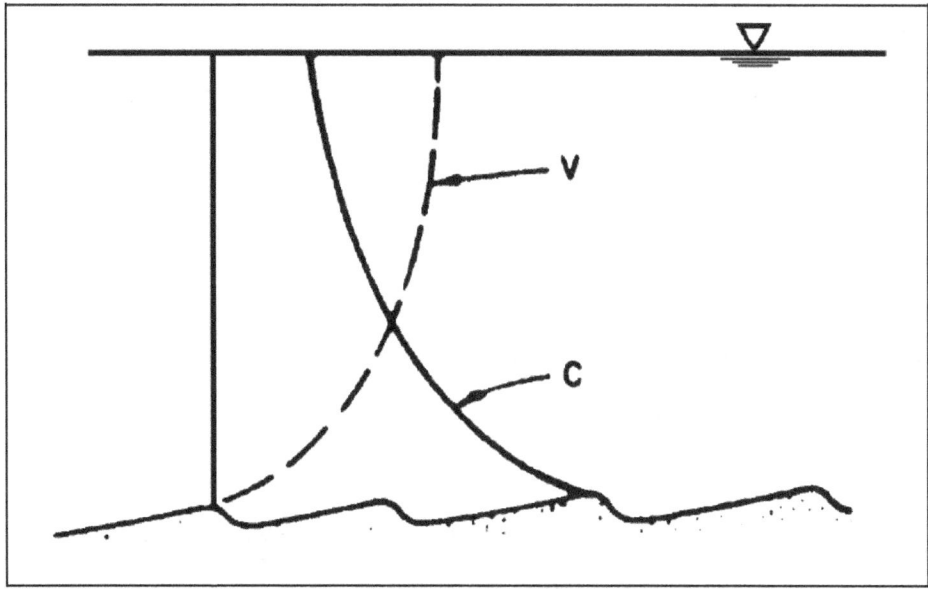

Figure 7.2. Velocity and sediment concentration profiles.

ASCE (2008) also presents six empirically based equations for determining total sediment load. These equations have the advantage of being more easily applied, but should only be used within the limits of the data used in their development. This concept applies to the use of any sediment transport equation. The HDS 6 manual (FHWA 2001) includes 20 sediment transport equations and discusses their applicability to various grain sizes. The HEC-RAS Reference Manual (USACE 2010) and the SAM reference manual (USACE 2002) include information on the range of data (particle size, specific gravity, velocity, depth, slope, channel width and temperature) used to develop many of the sediment transport equations used for sand and gravel sizes. Any equation that is considered for use should be evaluated for applicability to the specific conditions.

It is important to note that there are several ways of expressing and calculating rates of sediment transport. These include volumetric (ft$^3$/s, m$^3$/s), mass and weight (tons/day, metric-tons/day), and concentration (ppm, mg/l), sediment volume/total volume, and sediment weight/total weight. HDS 6 provides exact and approximate equations for converting between these expressions (FHWA 2001).

## 7.6 VARIABILITY IN SEDIMENT TRANSPORT ESTIMATES

A convenient method for estimating sediment transport is the Colby (1964) graphical method for bed material load in sand-bed rivers. Sand-bed channels are dominated by suspended sediment transport for most flow conditions. The Colby "curves" provide insight into the sensitivity of bed material load to various flow parameters (see HDS 6 (FHWA 2001)). The first step in the Colby method is to determine an uncorrected sediment discharge based on flow velocity. The Colby curves follow a trend of sediment discharge proportional to velocity to the power of between 3.5 and 6. These large powers indicate that suspension is more effective in transporting sediment in sand-bed channels. They also indicate that uncertainty in velocity generates extreme uncertainty in sediment transport calculations. Based on these observations, a 10 percent change in velocity can result in a 40 to 80 percent change in sediment transport rate. The Colby method also includes correction factors for water temperature and wash load concentration because these factors affect the fluid viscosity and particle fall velocity.

As flow velocity and shear stress increase bed load increases, but suspended load increases rapidly and can easily dominate the sediment transport process. This is because bed load is transported in a small fraction of the flow depth (often considered as twice the median sediment diameter ($D_{50}$)) and because the flow velocity (and bed load velocity) is low near the bed. Suspended load is carried through more, and potentially, all of the flow depth (see Figures 7.1 and 7.2). Velocity quickly increases with distance above the bed so suspended load is carried downstream at a much higher velocity than bed load.

The relative influence of viscosity, slope, bed sediment size and depth on bed sediment and water discharge can be examined using Einstein's (1950) bed-load function and Colby's (1964) relationships. Einstein's bed-load function is chosen because it is the most detailed and comprehensive treatment from the point of fluid mechanics. Colby's relations are chosen because of the large amount and range of data used in their development (FHWA 2001).

To study the relative influence of variables on bed material and water discharges, data taken by the U.S. Geological Survey on the Rio Grande near Bernalillo, New Mexico, was used (FHWA 2001). The width of the channel reach was 270 ft (82.3 m). In the analysis, the energy slope was varied from 0.7S to 1.5S, in which S is the average bed slope assumed to be equal to the average energy slope. Further, the kinematic viscosity was varied to correspond with variations in temperature from 39.2° to l00°F inclusive. The average bed material distribution for this reach of the Rio Grande given by Nordin (USGS 1964) was used. The average water temperature was assumed to be equal to 70°F and the average energy gradient of the channel was assumed to be equal to 0.00095. The water and sediment discharges were computed independently for each variation of the variables and for three subreaches of differing width of the Rio Grande near Bernalillo. The applicability of the results depends on the reliability of the modified Einstein bed-load function and Colby's relationships used in the analysis rather than on the choice of data.

The computed water and sediment discharges are plotted in Figures 7.3, 7.4, and 7.5 and show the variation of sediment discharge due to changes in bed material size, slope and temperature for any given water discharge. Figure 7.3 shows that when the bed sediment becomes finer, the sediment discharge increases considerably. The second most important variable affecting sediment discharge is the slope variation (Figure 7.4). Temperature is third in importance (Figure 7.5). The effects of these variables on sediment discharge were studied over approximately the same range of variation for each variable.

Figure 7.6 shows the variation of the sediment discharge due to changes in the depth of flow for any given discharge, computed using Colby's (1964) relations. The values of depth of flow varied from 1.0 to 10.0 ft, the median diameter of the bed sediment is maintained constant equal to 0.030 mm, the water temperature is assumed constant and the concentration of fine sediment is assumed less than 10,000 ppm. The channel width is also maintained constant at 270 ft (82.3 m). In Figure 7.6, the curves for constant depth of flow show a steep slope. This indicates that the capacity of the stream to transport sands increases very fast for a small increase of discharge at constant depth.

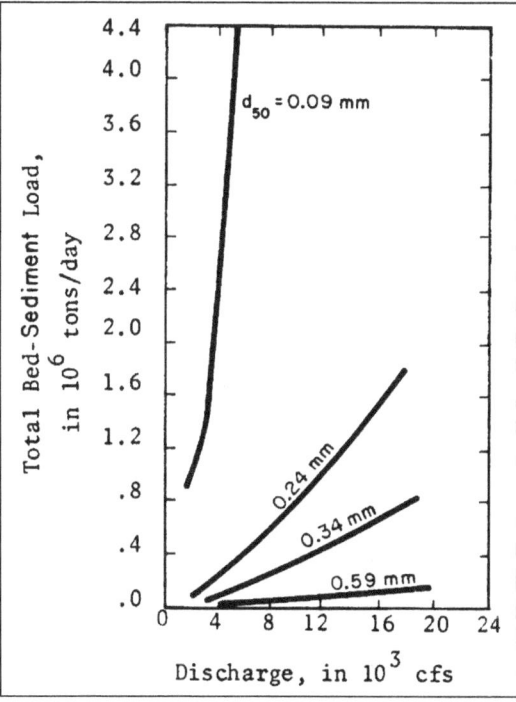

Figure 7.3. Bed-material size effects on bed material transport.

In summary, it is apparent that no single sediment transport equation can encompass all alluvial channel conditions. Computing the bed material sediment transport capacity for a river must be grounded on a firm understanding of the river channel characteristics, sources of sediment and modes of sediment movement. Equations that are well suited for the particular river conditions should be used and, if more than one are well suited, the results should be compared to assess the range of possible outcomes. For a specific river, it is imperative that the results be compared with actual measurements. Reference to HDS 6 (FHWA 2001) is suggested for guidance on selecting an appropriate sediment transport function for given river conditions. HDS 6 (FHWA 2001) and HDS 7 (FHWA 2012a) provide an overview of sediment transport modeling techniques.

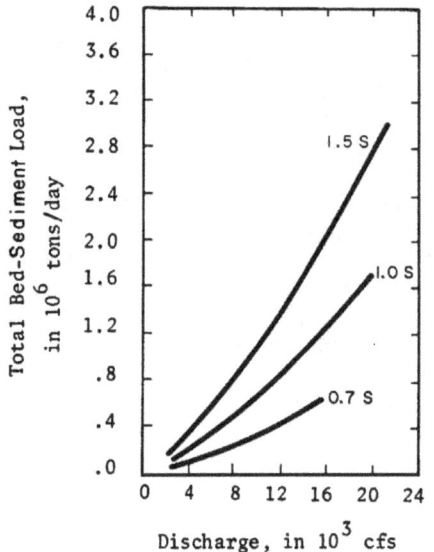

Figure 7.4. Effect of slope on bed material transport.

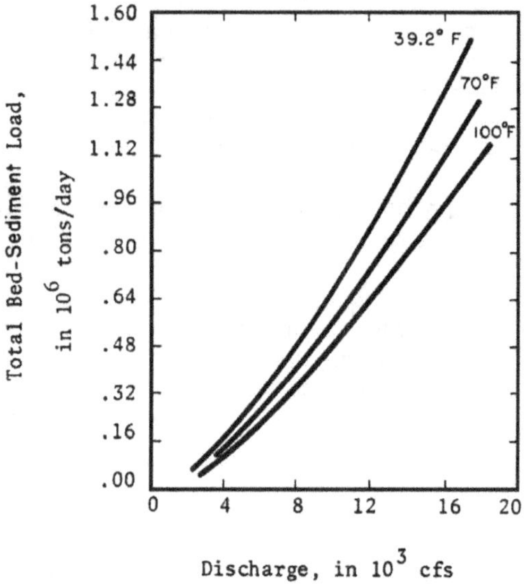

Figure 7.5. Effect of kinematic viscosity (temperature) on bed material transport.

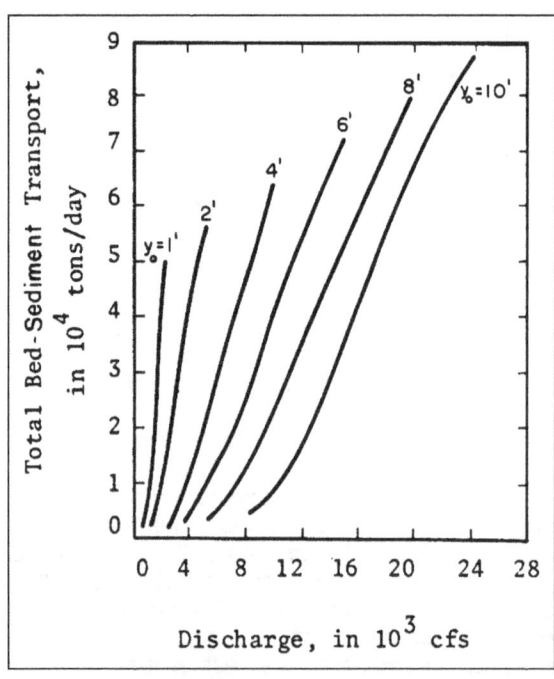

Figure 7.6. Variation of bed material load with depth of flow.

# CHAPTER 8

# CHANNEL STABILITY IN GRAVEL-BED RIVERS

## 8.1 OVERVIEW

The construction and operation of major engineering works on many river systems have had a significant impact on those systems over time. The development and management of water resources, navigation, power generation, land drainage, flood control works, land use changes, and transportation works have resulted in channel dredging and straightening, bank raising and protection, floodwater confinement, vegetation removal, flow modification, and alteration of water quality. The changes are often detrimental to a river system, causing channel instability and/or adversely affecting environmental and ecological characteristics of the channel.

These modifications have a greater impact on rivers flowing in fine alluvial material, which probably explains why much of the early research on flow hydraulics and channel morphology was focused on flow in sand-bed channels. However, gravel-bed rivers have increasingly been affected by these activities resulting in the need for a better understanding of their morphology and associated processes. In recent years, new design and modeling techniques have been developed to predict and minimize the effect of proposed engineering projects on gravel-bed rivers. This chapter provides a general review of the morphology of and processes acting on gravel-bed rivers and compares and contrasts river processes and stability of gravel-bed channels with those in sand-bed channels.

## 8.2 FLUVIAL PROCESSES IN GRAVEL-BED RIVERS

Many of the characteristics of sand-bed channels are found in gravel-bed channels, but gravel-bed channels also have characteristics not common to sand-bed channels. Whereas sand-bed channels are commonly found in flatter alluvial plain areas, gravel-bed channels generally occur in steeper piedmont areas. Gravel-bed channels consist of gravel-sized materials, can have planforms ranging from straight to braided and, where they take the form of a meandering channel, they often have a riffle-pool profile. In general, the processes of bank erosion, lateral migration, and channel instability in gravel-bed channels are similar to those in sand-bed channels. However, flow hydraulics, flow resistance, and sediment transport in gravel-bed channels are distinctly different than those in sand-bed channels. Table 8.1 provides a summary of these differences.

### 8.2.1 Velocity and Flow Resistance in Gravel-Bed Rivers

Velocity, which is strongly related to flow resistance, is one of the most sensitive and variable properties of open-channel flow because of its dependence on a multitude of other factors. Velocity is a vector quantity that has both magnitude and direction. Velocity varies in space and time:

1. *With distance from the streambed* – The velocity usually increases from zero at the bed to the free-stream velocity ($v_s$) at the edge of the boundary layer, the boundary layer being that part of the flow which is retarded by friction at the bed.

2. *Across the stream* – Velocity increases towards the center of a stream as the frictional effect of the channel banks declines, an effect that becomes proportionately less in channels with a high width/depth ratio (>15).

Table 8.1. Generalized Relative Differences of Sand-bed vs. Gravel/Cobble-bed Streams (modified from Bledsoe et al. 2008).

| Parameter | Sand-bed | Gravel/Cobble-bed |
|---|---|---|
| Bed material transport | Continuous | Episodic |
| Functional relationship of sediment transport to velocity | $V^5$ | $V^3$ |
| Armoring | Limited | Significant |
| Bedforms and changes in bed roughness/configuration | Rapidly adjusting for wide range of flow events | Not rapidly adjustable/formed by relatively infrequent events |
| Scour depth | Significant | Limited |
| Variation in scour depth | Rapid | Slow |
| Slope and stream power | Lower | Higher |
| Channel response to changes in hydrology | Rapid | Slower |
| Sensitivity to changed sediment loads | High | Lower |
| Variation in bed material gradation | Small | Large |
| Bankfull shear stress ratio ($\tau/\tau_c$) | ≈ 1 to 10+ | ≈ 0.5 to 1.5 |

3. *Downstream* – Although the slope decreases over long distances along a river, velocity tends to remain relatively constant or increase slightly as the channel becomes hydraulically more efficient and resistance decreases in the downstream direction.
4. *With time* – Under time scales of days, weeks, or months, mean velocity at a stream section responds to fluctuations in discharge. The increase in depth with discharge tends to drown out roughness elements in the bed and thereby produce an increase in velocity.

Thus, velocity is a highly variable quantity in time and space and the character of that variation is important since velocity directly influences the processes of erosion, sediment transport, and sediment deposition.

Flow resistance is one of the most important elements in the interaction between fluid flow and the channel boundary. Several resistance equations have been developed (see HDS 6, Chapter 3) (FHWA 2001). All of the equations assume that resistance approximates that of a steady, uniform flow, but in natural channels with erodible boundaries, the resistance problem becomes much more involved.

Total flow resistance consists of several components (Bathurst 1993):

- *Boundary resistance* resulting from the frictional effect of the bed material itself and the bed forms developed therein
- *Channel resistance* associated with bank irregularities and changes in channel alignment
- *Free surface resistance* stemming from the distortion of the water surface by waves and hydraulic jumps

Boundary resistance is of primary concern with a basic subdivision into 'grain roughness' and 'form roughness' components.

Grain roughness is primarily a function of relative roughness (y/D or R/D), often expressed as:

$$\frac{1}{\sqrt{f}} = c \log\left(a \frac{R}{D_x}\right) \tag{8.1}$$

where:

    c and a   =   Constants
    $D_x$        =   Measure of the size of roughness elements
    f          =   Darcy-Weisbach friction factor (FHWA 2001b)
    R        =   Hydraulic radius

The roughness height of uniform sediment is simply taken as the grain diameter, but with non-uniform sediment the problem is to choose a representative grain diameter. The $D_{84}$ (the diameter at which 84 percent of the sediment is finer) is commonly used since it takes account of the important influence of large particles on flow resistance. The $D_{95}$ and $2D_{65}$ (USDA 2009a) have also been used, but the choice is somewhat arbitrary. Hey (1979) successfully applied a version of Eq. 8.1 to describe flow resistance over riffles in gravel-bed rivers, but the equation becomes less applicable where $R/D_{84} < 4$ and w/d < 15 (Bathurst 1993).

Grain roughness can be the dominant component of resistance where streambeds consist of gravel or cobbles. Equation 8.1 implies that, as depth increases with discharge at a cross section, the effect of grain roughness is drowned out and flow resistance decreases, although possibly at a declining rate with higher discharge.

Form roughness stems from features developed in the bed material and presents a particular problem in that, once grains are set in motion, the shape of the bed can be modified to give a variable form roughness depending on flow conditions. Compared to sand-bed channels, the contribution of form roughness to total flow resistance over coarser beds has received significantly less attention until relatively recently. Bed forms of varying size have recently been identified in this role, from microtopographic pebble clusters to the channel bars characteristic of riffle-pool sequences (Robert 1990). Cluster bed forms, which generally consist of a single obstacle protruding slightly above neighboring grains together with upstream and downstream accumulations of particles, are probably the predominate type of microtopography in gravel-bed rivers (Brayshaw 1985). As yet, the effects on flow resistance produced by the size, shape, and spacing of various bedforms have not been thoroughly investigated. And even less is known about the remaining components of resistance, such as local bank irregularities, channel curvature, and in-channel vegetation.

Neither the flow, form, nor roughness within individual reaches of alluvial channels is uniform. Resistance values are usually obtained by calculation using one of the flow resistance equations provided in Chapter 3 of HDS 6 (FHWA 2001) or by visual comparison with representative reaches rather than by direct measurement, so that computed values tend to include undifferentiated effects of all types of resistance. Attempts to separate total resistance into component parts have achieved limited success and knowledge of the resistance mechanisms in natural stream remains far from complete. With regard to boundary resistance in gravel-bed streams where grain friction might be expected to dominate, form roughness can still be a major contributor (Prestegaard 1983, Griffiths 1989).

The general approach for estimating the resistance to flow in a stream channel is to select a base Manning roughness coefficient value for materials in the channel boundaries assuming a straight, uniform channel, and then to make corrections to the base n value to account for channel irregularities, sinuosity, and other factors which affect the resistance to flow. For gravel and small cobble and boulder-bed channels, analyses of data from many rivers, canals, and flumes shows that channel roughness can be predicted by the following equation (NCHRP 1970):

$$n = K_u D_{50}^{1/6} \tag{8.2}$$

where:

$D_{50}$ = Measured in ft (m)
$K_u$ = 0.0395 (English units)
$K_u$ = 0.0482 (SI units)

Alternately, Limerinos (USGS 1970) developed Equation 8.3 from samples on streams having bed materials ranging in size from small gravel to medium size boulders.

$$n = \frac{K_u R^{1/6}}{1.16 + 2.08 \log\left(\dfrac{R}{D_{84}}\right)} \tag{8.3}$$

where:

R = Hydraulic radius
$D_{84}$ = 84th percentile (percent finer size of bed material measured in ft or m)
$K_u$ = 0.0926 (English units)
$K_u$ = 0.113 (SI units)
y = Flow depth

The flow depth may be substituted for the hydraulic radius in wide channels (w/y > 10).

### 8.2.2 Sediment Characteristics and Armoring

As indicated by the description, gravel-bed channels are composed primarily of gravel. Sediment sizes in gravel-bed channels can generally range from silt and sand to cobbles and small boulders, but the median grain size ($D_{50}$) of the bed material is between 2 and 64 mm (see Chapter 2, Table 2.1). Due to the transport characteristics of gravel-bed channels, the bed surface often is armored or paved and is generally characterized by a unimodal grain size distribution with grain sizes ranging from 16 to 128 mm. The subsurface material usually has a bimodal grain size distribution with one mode being sand size (< 2 mm).

In many gravel-bed rivers that are armored, the median size ($D_{50}$) of the grains at the surface is larger than the median size of the grains in the subsurface (Figure 8.1). Armor (or pavement) is believed to be a fundamental feature of gravel-bed streams as are dunes in sand bed streams (Parker et al. 1982). The armor layer is usually no more than one to two grain diameters thick and acts as a buffer between the flow and moving bedload and the subsurface material, which makes up the bulk of the material temporarily stored in a reach.

Since the particle size distributions of subsurface sediment (which is the sediment found under the streambed surface) and the subarmor sediment (which is the sediment under the armor layer) are generally the same, the term subsurface can be applied to both the subsurface and subarmor sediments.

Although opinions differ as to the formative process, there are three primary processes that are attributed to surface coarsening (USDA 2001):

- Selective scour of fines (erosion pavement)
- Selective deposition of large particles
- Armoring to facilitate equal mobility transport

Selective scour of fines – Selective winnowing of fines from the surface leaving behind a coarse lag deposit on the surface that is about one grain diameter thick. Reasons for surface winnowing include decreased sediment supply and/or increased discharge. Long-term surface coarsening can occur in the absence of sediment supply in places such as the downstream side of log jams, in plunge pools and scour holes, or downstream of dams (erosion pavement).

Selective deposition of large particles – Surface coarsening associated with selective deposition of coarse particles occurs when waning flows are no longer competent to transport the largest particles, which then begin to settle. The supply of finer particles may be low, at least during flows at which they would settle.

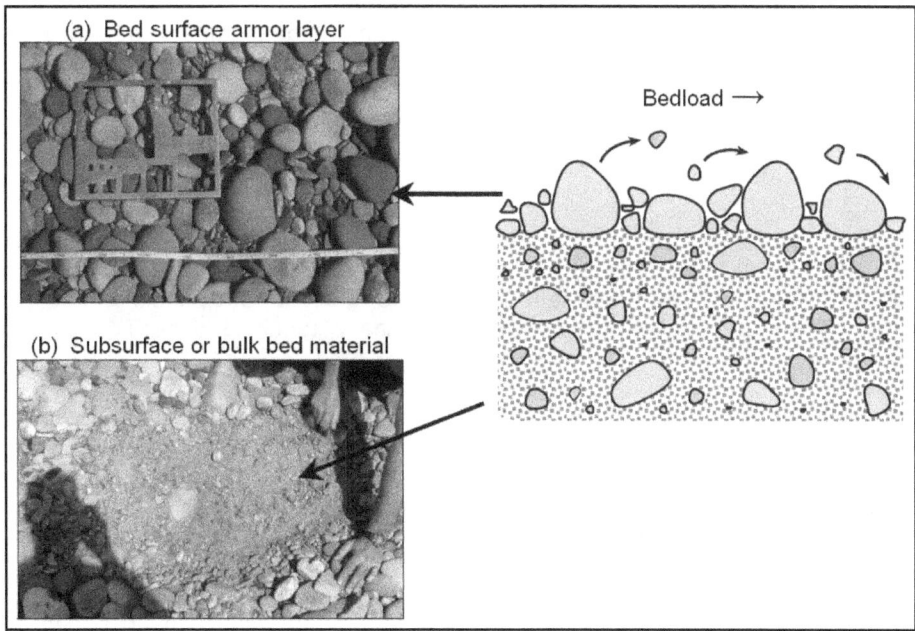

Figure 8.1. Distinction between (a) bed surface armor layer and (b) subsurface sediment common in many gravel-bed rivers and streams (after USDA 2009b).

Armoring to facilitate equal mobility transport – Parker et al. (1982) and Andrews and Parker (1987) suggest that armoring may be a prerequisite for 'equal mobility' transport of coarse and fine particles. If surface and subsurface particle size were the same (i.e., not armored), coarse particles would move less frequently than fine particles. The bedload material then has a finer distribution than the sediment on the channel bed. The often observed similarity between the size distribution of bedload and the subsurface sediment requires that the mobility of coarse particles is increased, while the mobility of small particles is decreased. This mechanism can be facilitated by the presence of a coarse armor layer where coarse particles are exposed to the surface which provides them with an increased chance of transport. Fine particles are hidden below the surface where the probability of their transport is decreased. Therefore, the preferential exposure of larger particles in the armor layer acts to equalize the mobility of coarse and fine particles and eliminates most of the differences in the mobility of those particle sizes.

A surface armor may be less well developed in streams where sediment transport capacity equals the amount of sediment supplied to the reach (e.g., braided streams). The particle-size distributions of the surface and subsurface sediment are relatively similar under these conditions. Where sediment transport capacity is larger than sediment supply, a coarse surface armor becomes prominent. High energy mountain streams usually have high sediment transport capacity, but low sediment supply, which leads to the formation of an erosion pavement that may only be mobilized by the largest floods.

The degree of armoring may be quantified by the ratio of the median grain size of the surface sediment to the median grain size of the subsurface sediment. Therefore, the larger the ratio of surface to subsurface $D_{50}$ the larger the degree of armoring. The ratio approaches a value close to 1 in streams with high sediment supply, whereas streams in which sediment transport capacity exceeds sediment supply, the ratio approaches a factor of approximately 2. The ratio may reach values of 3 or more in high-energy mountain streams or where the sediment supply is cut off (such as below dams) and an immobile, coarse lag deposit forms. Thus, a change in the degree of armoring can be used as an indication of a change in sediment supply or flow regime.

### 8.2.3 Sediment Transport

Sand and gravel-bed systems vary greatly in their transport of sediment and their sensitivity to sediment supply. Whereas sand-bed channels typically have live-beds, which transport sediment continuously even at relatively low flows, gravel-bed channels generally transport the bulk of their bed sediment load more episodically, requiring higher flow events for bed mobility.

Incipient motion is the condition where the hydraulic forces acting on a sediment particle are equal to the forces resisting motion. The particle is at a critical condition where a slight increase in the hydraulic forces will cause the particle to move. The hydraulic forces consist of lift and drag and are usually represented in a simplified form by the shear stress of the flow acting on the particle. Incipient motion conditions can be analyzed using the Shields diagram (FHWA 2001).

The Shields parameter can range from 0.03 to 0.10 for natural sediments based on particle shape, angularity, gradation and imbrication. The use of 0.047 for sand sizes provides reasonable results (Meyer-Peter and Muller 1948, Gessler 1971), but lower values (0.03) are commonly used for gravel and cobble sizes. The Shields relationship can be used to calculate a sediment particle size that will move for a particular hydraulic condition or to calculate the shear stress required to move a particular particle size (see Chapter 6). The average shear stress acting on the channel ($\gamma RS$) includes all the factors contributing to resistance to flow. Only the shear stress acting on the individual particles should be used for this calculation. For coarser grained materials (gravel and larger) the Manning roughness coefficient is a function of grain size and flow depth. The shear stress can be computed from:

$$\tau_o = \frac{\rho V^2}{\left[5.75 \log\left(\frac{12.27 R}{k_s}\right)\right]} \qquad (8.4)$$

where:

    V = Average channel velocity
    R = Hydraulic radius
    $\rho$ = Density of water
    $k_s$ = Grain roughness usually taken as 3.5 $D_{84}$ for gravel and coarser bed material

Equation 8.4 is essentially the Shields relationship with the Limerinos equation (Equation 8.3) substituted for Manning roughness coefficient (see also Equation 6.16). In the Limerinos equation, the grain roughness is equivalent to 3.5 times $D_{84}$, although for poorly graded material grain roughness can be as low as 1.0 to 2.0 times $D_{84}$. The hydraulic depth (channel area divided by topwidth) can be substituted for hydraulic radius, R, in Eq. 8.4 when the width-depth ratio exceeds 10.

With regard to the dynamics of movement, particles roll, slide, or saltate along the bed in a shallow zone only a few grain diameters thick once the entrainment threshold has been exceeded ($\tau_o > \tau_{cr}$). This occurs at a critical shear stress given by:

$$\tau_{cr} = Kg(\rho_s - \rho)D \tag{8.5}$$

where:

    $\rho_S$ = Sediment density
    $\rho$ = Water density
    D = Grain size
    g = Gravity
    K = Constant (on hydraulically rough beds)

K values range from 0.03 to 0.10 with 0.047 now accepted as a good approximation (Komar 1988) for sand size particles, but lower values (0.03) are commonly used for gravel and cobble sizes. Material transported this way constitutes the bed load. Rolling is the primary mode of transport.

Incipient motion conditions can also be evaluated with Equation 6.14 (see Chapter 6, Section 6.4.2).

As previously indicated, armoring occurs in gravel-bed rivers when the hydraulic forces are sufficient to transport most of the sediment supplied to the channel, but insufficient to mobilize or transport the larger sizes in the channel bed. Under these conditions, the smaller material is transported while the coarse material remains as an armor layer on the channel bed. Armor layers often form in gravel-bed rivers during the recession of floods. These armor layers may be disturbed during the next major flood and re-form during the flood recession.

The stability of an armor layer is related to the armor forming discharge. If that discharge is exceeded, degradation can occur. The incipient motion equation can be used to determine the critical size of material that can resist a particular hydraulic condition. If at least five percent of the material is larger than the critical size ($D_{95}$ or smaller), armoring can occur (see Chapter 6, Equation 6.16 and Figure 6.15).

Andrews and Smith (1992) defined different phases of transport in gravel-bed rivers in terms of dimensionless shear stress:

$$\tau_* = \frac{\tau}{g(\rho_s - \rho)D} \tag{8.6}$$

Phase I or marginal transport begins when fluid forces are just sufficient to rotate gravel-sized particles out of the pockets in which they lie, the particles rolling or bouncing over their downstream neighbors until they settle into new resting places. As the dimensionless shear stress increases, so does the number of particles in motion. Such partial movement occurs over the range of $0.02 < \tau_* < 0.06$. During Phase II transport at higher stresses ($\tau_* > 0.06$) most of the bed is mobile and significant saltation occurs. In many gravel-bed rivers, a substantial portion of the bed material will be carried under marginal transport conditions. Where a well-developed armor layer moderates the supply of sediment to the transport process, Phase I is characterized by the passage of the finer fraction over a basically stable coarser bed (Carling 1987). Substantial bed mobility (Phase II transport) requires that the armor is breached, when significant changes in channel morphology can take place. According to the equal-mobility hypothesis, the particle size distributions of the bed load and bed material should be approximately the same at all flow stages. However, coarser fractions tend to be underrepresented during Phase I transport, notably over armored beds, and the bed load becomes progressively coarser with increasing flow, approaching the bed material in size distribution characteristics during Phase II transport when the bed is more fully mobilized (Komar and Shih 1992, Lisle 1995).

### 8.2.4 Channel Pattern and Channel-Scale Bedforms

Gravel-bed channels can form straight, meandering, and braided channel patterns. As shown in Figure 8.2, all three of these patterns are composed of the same basic morphological unit (i.e., pool-bar unit) that includes a different special arrangement for each pattern (Bridge 1993, Ferguson 1993). As with sand-bed channels, the continuum concept of channel pattern holds with gravel-bed channels as well – that is, each pattern can grade into the other – and the sequence of straight-meandering-braided can be associated with not only the slope-discharge relationship, but also increasing bed load and increasing stream power ($\Omega = \gamma QS$) (Schumm and Kahn 1972, Edgar 1984).

The actions of local sorting mechanisms may play a part in riffles having coarser bed material than adjacent pools suggesting that bed topography and particle size characteristics are interrelated. In some cases, riffles and pools have been differentiated based on sediment structure rather than sediment size, with tight structures dominant in riffles and open structures dominant in pools which may explain the maintenance of elevation contrasts between juxtaposed riffles and pools.

Riffle-pool development is traditionally associated with gravel-bed channels since they show little tendency to form in channels carrying uniform sand or silt. Pool (alternating deeps) and riffle (shallows) development, which forms through a combination of scour and deposition, is characteristic of straight and meandering channels with heterogeneous bed material in the size range of 2-256 mm (Figure 8.3). Pools are primarily associated with meander bends. The gravel deposits that form the intervening riffles are generally lobate in shape and often slope alternately toward first one bank and then the other bank forming a sinuous flow path even in straight channels (see Figure 8.2). The general spacing of successive pools or riffles at a distance of 5 to 7 times the channel width, which is at best an average condition, is a feature of riffle-pool geometry. The degree of riffle-pool development varies with bed material size relative to the transport potential in a given reach, since the ability of a stream to reconfigure its bed is dependent on the mobility of the available material and the frequency of competent flows (Knighton 1981).

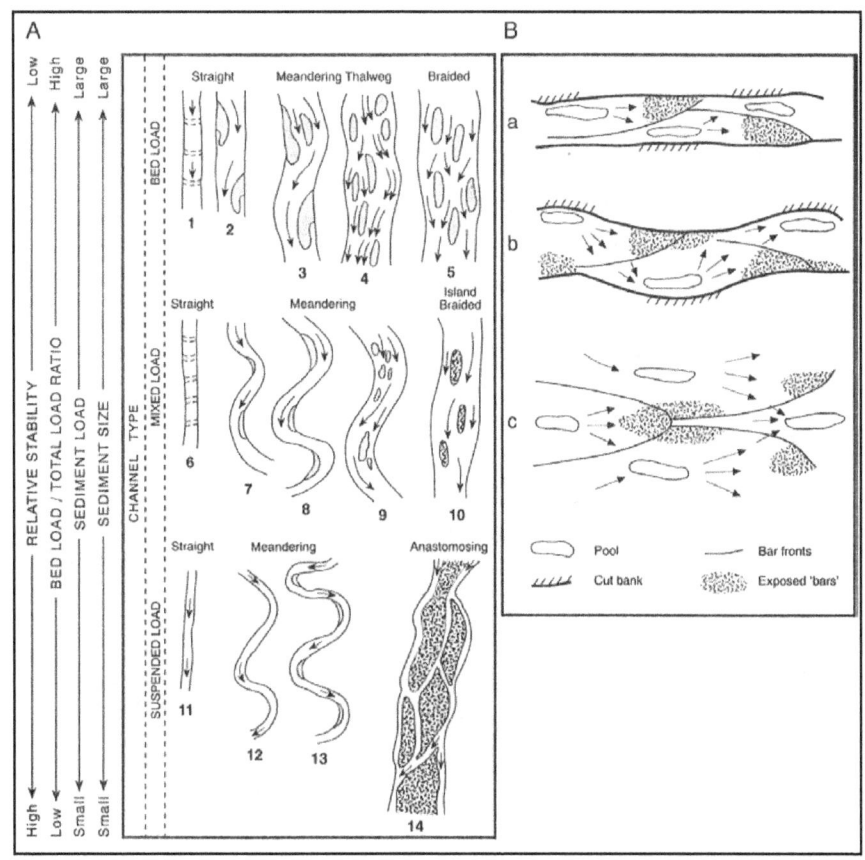

Figure 8.2. (A) Classification of channel pattern (after Schumm 1981) and (B) overlapping pool-bar units in gravel-bed rivers of different channel pattern (after Ferguson 1993; from Knighton 1998).

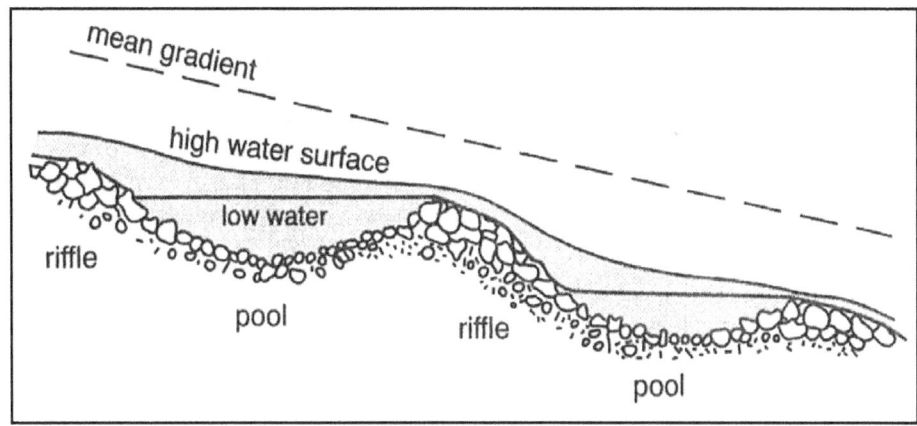

Figure 8.3. Definition sketch for a riffle-pool channel (Knighton 1998).

Riffles and pools appear to have distinctive flow and channel geometries. For example, riffles tend to be wider and shallower at all stages of flow. Because velocity and slope are greater and depth is less over a riffle than in a pool at low flows, the result of these differences in flow geometry produces a convergence or more even distribution of these flow variables along a reach at high flows. Competence, expressed by bed velocity or boundary shear stress ($\gamma RS$), will also tend to become more evenly distributed or may even be reversed resulting in it being higher in pools at those discharges that transport most material in gravel-bed streams (Keller 1971, Lisle 1979, Keller and Florsheim 1993). Given the combination of high-flow transport through pools and low-flow storage on riffles, competence reversal provides a mechanism for the concentration of coarser material in riffles and for the maintenance of the riffle-pool sequence. Even without reversals, competence in pools could still be higher at high flows if their bed sediment has a more open structure (Clifford 1993) and, therefore, a lower entrainment threshold (Sear 1996).

**8.2.5 Bank Erosion**

The way in which a river erodes its banks is the same for both sand-bed channels and gravel-bed channels. The processes responsible for the erosion of material from a bank and the mechanisms of failure resulting from the instability created by those processes must be considered. The following is a brief overview of the processes and mechanisms of bank erosion, which are discussed in greater detail in Appendix B.

Processes of bank erosion fall into two main categories: (a) fluvial entrainment and (b) subaerial/subaqueous weakening and weathering. On non-cohesive banks fluvial processes are the most important, though the reduction of the friction angle by loosening can be very effective in eroding closely packed and imbricated deposits. In the case of composite banks, non-cohesive layers are eroded more quickly than cohesive ones leading to oversteepening where non-cohesive material underlies a cohesive layer, and the formation of a bench where cohesive material underlies a non-cohesive layer.

The operation of processes of erosion and in situ weakening/weathering brings the bank to a state of limiting stability. The mechanism of failure depends on the size, structure, and geometry of the bank and the engineering properties of the bank material. With regard to gravel-bed rivers, rotational failures can occur in composite banks where a non-cohesive layer underlies a cohesive layer – the interface becomes the plane of weakness. The determination of whether the slip is a toe or slope failure is dependent on the location of the interface. Cantilever failures, which are also common on gravel-bed rivers, form in composite banks where a vertical cut in a cohesive layer is stable and undermining is rapid.

Basal endpoint control defines the balance between the rate of supply of debris to the base of the bank by bank failure and surface erosion, and the rate of removal by fluvial entrainment. The state of endpoint control is vital in controlling profile geometry, form, stability, and retreat of banks of all types (see Appendix B, Section B.4).

**8.3 MANAGEMENT OF GRAVEL-BED RIVERS**

In the course of carrying out river engineering works for flood control and channel stabilization, many rivers have been significantly modified. These changes have adversely affected the stability of the engineered and adjacent reaches and, in the process, destroyed or heavily modified the conservational, ecological, and recreational value of the riverine areas. Increasingly there are demands to restore and rehabilitate heavily engineered reaches. This requires the incorporation of less intrusive and more ecologically friendly measures and the adoption of natural solutions to bank and channel stabilization in order to recreate channel features, which are both enduring and in harmony with local flow, sediment transport, and ecological processes. In order to achieve these objectives it is necessary to identify the methods available for predicting the three-dimensional shape of alluvial channels as these are the basis for any natural engineering design or restoration work.

### 8.3.1 Regime Equations

Alluvial rivers possess nine degrees of freedom since they can adjust their average bankfull width (W), depth (d), maximum depth ($d_m$), height ($\Delta$) and wavelength ($\lambda$) of bedforms, slope (S), velocity (V), sinuosity (p), and meander arc length (z) through erosion and deposition. For river reaches that are in regime, implying that they do not systematically change their shape and dimensions over a period of a few years, these can be regarded as dependent variables. In these cases, the sediment load supplied from upstream can be transmitted without net erosion or deposition.

The variables controlling stable river dimensions are the discharge (Q), sediment load ($Q_s$), caliber of the bed material (D), bank material, bank vegetation, and valley slope ($S_v$). Change in any one of these independent variables will eventually result in the development of a new regime channel geometry that is in equilibrium with the changed conditions. When, stable, the channel morphology will be uniquely defined by the new values of the controlling variables.

Under regime conditions most of the controlling variables are effectively constant. The two exceptions are discharge and bed material transport as they can vary considerably through time. To overcome this difficulty, Inglis (1946) suggested that a constant flow, at or about bankfull, produces the same gross shapes and dimensions as the natural sequence of events and could be regarded as the dominant or channel-forming flow. This does not preclude dramatic change during extreme flood events, but rather suggests that the channel morphology is modified by the subsequent events such that over a period of years its long-term average channel dimensions are relatively invariant. There is a considerable body of evidence to suggest that bankfull flow is the dominant or channel forming flow (Ackers and Charlton 1970, Wolman and Leopold (USGS 1957b), Hey 1975, Andrews 1980, Wolman and Miller 1960, Inglis 1946).

For gravel-bed rivers, the 1.5 year flood, based on the annual maximum series, is often regarded as the return period of bankfull flow, although there is considerable scatter around its value (Williams 1978). As this approach disregards all lesser flood events above bed material transport thresholds, it is important to determine return periods using the partial duration series with a threshold discharge set at the initiation of bed material movement (Carling 1988, Hey and Heritage 1988).

A range of empirical regime equations have been developed to predict the geometry of stable alluvial channels. As with all empirical equations, care should be exercised in their use, particularly with regard to their range of application (Hey 1997). This should be restricted to channels that are similar to those used to derive the equations. Ideally, all variables controlling channel shape and dimensions should be used in the derivation of each set of equations and, to ensure their general application, the variability within and between the variables should be maximized. Equally, as the equations are not dimensionless, careful consideration needs to be given to the choice of units.

Regime equations can, for convenience, be classified on the basis of bed material size and divided into mobile- and quasi-fixed-bed channels, with regard to the presence or near-absence of bed material transport, or subdivided on the basis of bank material or vegetation type, which affect the erosional resistance of the banks. For rivers, mobile bed conditions generally prevail, yet relatively few equations have been developed that explicitly allow for bed material transport.

Considerable research has been carried out on the morphology of gravel-bed rivers. Most ignore the effect of bedload transport on channel morphology or assume that its value is zero or trivial (Hey 1997). Kellerhals' (1967) equations actually apply to low or zero load, while Hey and Thorne's (1986) equations were the first to explicitly include sediment load, albeit predicted values.

Hey and Thorne (1986) showed that bedload transport affected both channel width and depth but, as the exponents were so small, its effect on cross-sectional shape could safely be ignored. Equations for width and depth, excluding bedload transport, were derived and these produce practical point estimates for design purposes. These equations, which are given below, define the mean bankfull width, depth, and maximum depth based on data from four cross sections (pool-riffle-pool-riffle) to be obtained over a full meander wavelength.

<u>Hey and Thorne's Equations</u> – The following regime equations (using English units with SI conversions indicated) are from Hey and Thorne (1986):

Database and range of applications:

- Bankfull discharge (Q): 138 - 14,970 ft$^3$/s (3.9 – 424 m$^3$/s)
- Bankfull sediment discharge [as defined by Parker et al.'s (1982) bedload transport equation] ($Q_s$): 0.002 - 251.6 lb/s (0.001 – 114.14 kg/s)
- Median bed material size ($D_{50}$): 0.046 - 0.577 ft (0.014 – 0.176 m)
- Bank material: composite with cohesive fine sand, silt, and clay overlying gravel
- Bedforms: plane bed
- Bank vegetation:  Type I, 0% trees and shrubs
    Type II, 1-5% trees and shrubs
    Type III, 5-50% trees and shrubs
    Type IV, >50% trees and shrubs
- Valley slope ($S_v$): 0.00166 – 0.0219
- Planform: straight and meandering
- Profile: pools and riffles

Equations:

Bankfull width (reach average in feet):

| | |
|---|---|
| $W = 2.39Q^{0.5}$ (ft) Vegetation Type I | (8.7a) |
| $W = 1.84Q^{0.5}$ (ft) Vegetation Type II | (8.7b) |
| $W = 1.51Q^{0.5}$ (ft) Vegetation Type III | (8.7c) |
| $W = 1.29Q^{0.5}$ (ft) Vegetation Type IV | (8.7d) |

The coefficients for Equations 8.7a through 8.7d are 4.33, 3.33, 2.73, and 2.34 for width in meters and discharge in m$^3$/s (SI units).

Bankfull mean depth (reach average):

$$d = 0.22 Q^{0.37} D_{50}^{-0.11} \quad \text{(ft; m) Vegetation Types I-IV} \tag{8.8}$$

This equation is dimensionally consistent (Q in ft³/s and $D_{50}$ in ft, results in d in ft or Q in m³/s and $D_{50}$ in m results in d in m).

Bankfull slope (S):

$$S = 0.153 Q^{-0.43} D_{50}^{-0.09} D_{84}^{0.84} Q_s^{0.10} \quad \text{Vegetation Types I-IV} \tag{8.9}$$

Equation 8.9 is for Q in ft³/s, $D_{50}$ and $D_{84}$ in feet, and $Q_s$ in lb/sec. The coefficient should be 0.087 for SI units (Q in m³/s, $D_{50}$ and $D_{84}$ in meters, and $Q_s$ in kg/s).

Bankfull maximum depth (reach average):

$$d_m = 0.23 Q^{0.36} D_{50}^{-0.56} D_{84}^{0.35} \quad \text{(ft, m) Vegetation Types I-IV} \tag{8.10}$$

The coefficient for Equation 8.10 should be 0.20 for SI units.

Meander arc length (z):

$$z = 6.31 W \quad \text{(ft, m) Vegetation Types I-IV} \tag{8.11}$$

Sinuosity (p):

$$p = S_v / S \quad \text{Vegetation Types I-IV} \tag{8.12}$$

Bankfull riffle width (RW):

$$RW = 1.034 W \quad \text{(ft, m) Vegetation Types I-IV} \tag{8.13}$$

Bankfull riffle depth (Rd):

$$Rd = 0.951 d \quad \text{(ft, m) Vegetation Types I-IV} \tag{8.14}$$

Bankfull riffle maximum depth ($Rd_m$):

$$Rd_m = 0.912 d_m \quad \text{(ft, m) Vegetation Types I-IV} \tag{8.15}$$

Bankfull pool width (PW):

$$PW = 0.966 W \quad \text{(ft, m) Vegetation Types I-IV} \tag{8.16}$$

Bankfull pool depth (Pd):

$$Pd = 1.049 d \quad \text{(ft, m) Vegetation Types I-IV} \tag{8.17}$$

Bankfull pool maximum depth ($Pd_m$):

$$Pd_m = 1.088 d_m \quad \text{(ft, m) Vegetation Types I-IV} \tag{8.18}$$

where:

$D_n$ = Grain diameter of surface bed material (in feet or meters) for which n percent of the sample is finer

Since no direct observations were available regarding bedload transport rates, independent estimates had to be obtained to enable the derivation of mobile gravel-bed regime equations. Before applying these equations it is important to ensure that bedload transport rates can be defined by Parker et al.'s (1982) equation.

The Hey and Thorne regime equations can be applied to predict planform as well as cross-sectional shape of mobile gravel-bed rivers, and pools and riffles can also be incorporated in the design using Equations 8.13 – 8.18. It should be noted that only Hey and Thorne included meandering channels within their database and their equations enable the planform geometry of the channel to be determined. This requires the prediction of sinuosity (Eq. 8.12), given the valley slope and predicted channel slope, and meander arc length (Eq. 8.11).

## 8.3.2 Channel Change and River Response

Analysis of the fluvial morphological system indicates that the channel can adjust in various ways but, because of the number of variables and possibilities involved, channel changes are usually indeterminate and cannot be exactly predicted. Theory predicts the direction of change in variables while case studies show how channels respond to different conditions, but few quantitative models can yet predict the exact style or rate of channel change to be expected. Various causes of channel change, which are common to both sand-bed and gravel-bed channels, have been well documented and a qualitative assessment of some of the potential channel responses have been identified and discussed in Chapter 6 (see also FHWA 2001).

The morphological characteristics of channels are commonly analyzed in two-dimensional slices (i.e., planform, cross-sectional form, and longitudinal form). Most work has been on lateral instability and changes in planform, partly because these are obvious changes and documentation on such changes is most readily available, but arguably also because it is in planform that the greatest adjustment takes place.

Although meander migration is inherent in many streams, changes in characteristics of channel morphology can be influenced by external causes. These may either be natural, such as climate change, or human-induced, such as land-use change and urbanization. There is a basic division between indirect modification (e.g., resulting from basin alteration) and direct, deliberate modification of the channel (e.g., channelization, channel straightening). It is essential to realize that both can result in additional changes to the channel. Case studies of responses to channelization have helped to identify the nature of channel adjustment and to predict the effects of channel disturbances elsewhere.

It is now widely accepted that changes may be complex in spatial and temporal pattern and that sudden changes can occur without a change in external conditions. Assumptions of attainment and stability of equilibrium forms must also be questioned as the evidence grows of continued evolution of patterns over centuries rather than decades. However, morphological changes do occur in response to changes in input of discharge and/or sediment and the expected directions of change were originally set out by Schumm (1969) (Table 8.2). The nature of the channel response depends on the inherent stability, the freedom to adjust and, thus, the sensitivity of different environments and channel reaches. In any engineering analysis it is essential that, prior to work on design, any natural or pre-existing instability is identified. There is also a need to monitor any responses to alterations of the channel.

| Table 8.2. Examples of River Metamorphosis (after Schumm 1969). ||
|---|---|
| Equations | Examples of Change |
| $Q_s^+, Q^{++} \approx S^-, d_{50}^+, D^+, W^+$ | Long-term effect of urbanization. Increased frequency magnitude of discharge. Channel erosion (increasing width and depth). |
| $Q_s^-, Q^+ \approx S^-, d_{50}^+, D^+, W^-$ <br> $Q_s^{--}, Q^+ \approx S^-, d_{50}^+, D^+, W^*$ | Intensification of vegetation cover (through a forestation and improved land management) and in-channel mining reduces sediment supply. Construction of dams shuts off sediment supply. |
| $Q_s^-, Q^+ \approx S^-, d_{50}^+, D^+, W^+$ <br> $Q_s^\circ, Q^+ \approx S^-, d_{50}^+, D^+, W^+$ | Diversion of water into river |
| $Q_s^+, Q^+ \approx S^*, d_{50}^*, D^*, W^+$ <br> $Q_s^-, Q^- \approx S^\pm, d_{50}^*, D^+, W^-$ | Parallel changes of water and sediment discharge with unpredictable changes in slope, depth, and bed material. |
| $Q_s^{++}, Q^+ \approx S^+, d_{50}^-, D^-, W^+$ | Land-use change from forest to crop production. Sediment discharge increasing more rapidly than water discharge. Bed material changes from gravel to sand, wider shallower channels. |
| $Q_s^\circ, Q^- \approx S^+, d_{50}^-, D^-, W^-$ <br> $Q_s^-, Q^{--} \approx S^+, d_{50}^-, D^-, W^-$ | Abstraction of water from river resulting in narrower stream. |
| $Q_s^+, Q^- \approx S^+, d_{50}^-, D^-, W^*$ <br> $Q_s^+, Q^\circ \approx S^+, d_{50}^-, D^-, W^-$ | Increased sediment supply and constant or reduced water discharge (e.g., hydraulic mining activity) |
| $Q_s$ = sediment discharge, $Q$ = water discharge, $D$ = flow depth, $W$ = flow width, $S$ = slope, $d_{50}$ = median diameter of bed material, $\circ$ = no change, $\pm$ = increase/decrease, ++/- - = indicates a change of considerable magnitude, * = unpredictable ||

As shown in Table 8.2, the construction of dams and the mining of sand and gravel from the channel have significant impacts on gravel-bed rivers. Where dams are present, the impacts of the altered flow and sediment transport regime and decreased sediment load can induce downstream-progressing degradation, usually reaching a maximum near the dam and decreasing gradually downriver, accompanied by armoring or paving of the downstream channel bed. The removal of gravel from streams may induce either positive or negative impacts on the stability of stream morphology. For example, removal of sand and gravel during in-channel mining operations can create a sediment deficit at the point of removal, which can be translated downstream. Thus, the downstream channel may degrade and become armored as a result of the decrease in sediment supply. Upstream degradation of the channel can occur because of headward erosion of the channel bed due to a change in base level or creation of an artificial knickpoint.

### 8.3.3 River Stabilization and Training

The boundary characteristics of rivers with gravel beds vary widely. There is a great range in sizes and shapes of bed material in different rivers, while the sizes and grading of bank material often change with elevation and with distance along a bank and between banks of the same river. Before undertaking any river stabilization or training works, it is recommended that a careful survey of the river be conducted to establish the cause of bank erosion and to ensure that an acceptable hydraulic geometry is chosen for the trained or stabilized reach.

There are a variety of methods used to control and train natural river channels. These range from the direct methods of protecting the banks using revetments, through semi-direct methods based on training fences of different types, to the indirect methods including spurs and sills to control the main current, vanes to control secondary currents, and structures to control bed levels and water levels.

There are many types of engineering works that can be constructed to prevent bank erosion. Selection of the best solution is often difficult and the problem is complicated by the need to consider the costs of materials, construction and maintenance of alternative schemes as well as their engineering efficiencies. More recently, bioengineering techniques have been given increasing consideration because of ecological requirements (see FHWA 2009, Chapter 6).

The primary methods for protection against fluvial erosion include bank armoring (e.g., riprap), flow retardation (e.g., fences, jacks, weirs, etc.), and flow deflection (e.g., spurs, dikes, bendway weirs, sills, and vanes). The primary methods for protection against bank failures or slips include methods to reduce seepage pressures, improve drainage, reduce cracking, increase soil strength, and reduce sliding forces. For a detailed description of a variety of stabilization and training methods and required materials, it is recommended that the reader start with HEC-23 "Bridge Scour and Stream Instability Countermeasures" (FHWA 2009) and NCHRP Report 568 "Riprap Design Criteria, Recommended Specifications, and Quality Control" (NCHRP 2006). Where environmental considerations are required in the design, the reader is referred to NCHRP Report 544 "Environmentally Sensitive Channel- and Bank-Protection Measures" (NCHRP 2005).

# CHAPTER 9

# CHANNEL RESTORATION CONCEPTS

## 9.1 INTRODUCTION

Rivers and streams provide essential habitat to countless aquatic plant and animal species, many of which are endangered. Civilization continually encroaches on these waters. Over 80 percent of our riparian ecosystems have been destroyed in some states, and over 50 percent of the nation's wetlands have been destroyed, drained, or filled.

The Clean Water Act was designed to restore and maintain the physical, chemical, and biological integrity of our nation's waters. The Federal Water Pollution Control Act of 1972 gave the act its current form and established a national goal that all waters of the U.S. should be fishable and swimmable. Currently, there is an emphasis on stream restoration in the Clean Water Act Section 404 Regulatory Program. Recent guidance has been issued on mitigation of stream impacts, and in-lieu fee programs for mitigation of regulated discharges are becoming more common. Several states have established stream protection regulations and may require mitigation of impacts. Many states have identified impaired streams and impairment due to highway construction is a common cause.

**The purpose of this manual (HEC-20) is to provide guidelines for identifying stream instability problems at highway stream crossings. Consequently, it is not the intent of this manual to provide definitive guidance on stream restoration design.** However, the qualitative and quantitative techniques for stream stability analysis presented in Chapters 5 and 6 can be used to evaluate the lateral (planform) and vertical (profile) stability of a stream channel and, in some cases, predict the potential for future channel instability. These same techniques can also be applied to the restoration and rehabilitation of degraded stream environments. This chapter provides an introduction to currently available guidelines for channel restoration design, which include suggested design considerations by the U.S. Army Corps of Engineers (USACE) and Natural Channel Design (NCD) procedures.

## 9.2 CHANNEL RESTORATION AND REHABILITATION

Over the last several years, numerous agencies and practitioners have published guidelines for stream corridor restoration and channel rehabilitation design. For example, in 1998, the Natural Resources Conservation Service (NRCS) led an effort resulting in 15 Federal agencies and partners publishing a document (NEH 654) entitled "Stream Corridor Restoration: Principles, Processes and Practices" (FISRWG 1998). This document represents a cooperative effort by the participating agencies to produce a common technical reference on stream corridor restoration. Recognizing that no two stream corridors and no two restoration initiatives are identical, this technical document broadly addresses the elements of restoration that apply in the majority of situations encountered.

As a general goal, the stream corridor restoration manual promotes the use of ecological processes (physical, chemical, and biological) and minimally intrusive solutions to restore self-sustaining stream corridor functions. It provides information necessary to develop and select appropriate alternatives and solutions, and to make informed management decisions regarding valuable stream corridors and their watersheds. In addition, the document recognizes the complexity of most stream restoration work and promotes an integrated approach to restoration. It supports close cooperation among all participants in order to achieve a common set of objectives.

From the perspective of the stream corridor, restoration and rehabilitation are defined as follows:

- Restoration is the process of repairing damage to the diversity and dynamics of ecosystems. Ecological restoration is the process of returning an ecosystem as closely as possible to predisturbance conditions and functions. Implicit in this definition is that ecosystems are naturally dynamic. It is therefore not possible to recreate a system exactly. The restoration process reestablishes the general structure, function, and dynamic, but self-sustaining, behavior of the ecosystem.
- Rehabilitation is making the land useful again after a disturbance. It involves the recovery of ecosystem functions and processes in a degraded habitat. Rehabilitation does not necessarily reestablish the pre-disturbance condition, but does involve establishing geologically and hydrologically stable landscapes that support the natural ecosystem mosaic.

Whether a highway project involves restoration or rehabilitation activities, the complexities of the stream corridor system need to be considered. Previous chapters have emphasized the necessity of a <u>stream system</u> approach to stream stability analyses (see for example, Figure 4.1).

According to Rosgen (1996), implementing a stream restoration project requires answering four basic questions:

- What are the observed problems?
- What caused the problem?
- What stream type should this be?
- What is the probable stable form of the stream type under the present hydrology and sediment regime?

The first step in a channel restoration project is to identify the problems observed in the reach of concern. The stream reconnaissance techniques discussed in Section 5.2 and field reconnaissance check lists of Appendix C support a determination of the nature and extent of the observed problems. A rapid assessment methodology such as that presented in Chapter 5 can help in evaluating how serious the problem is (see also Appendix E).

To determine what caused the problem, a qualitative assessment of important geomorphic factors influencing stream stability (Chapter 2) can provide an initial indication, but generally a more detailed analysis following the Level 1 and Level 2 procedures of Chapter 4 will be required. Understanding land use change in the contributing watershed, and its effects on the delivery (both timing and quantity) of water and sediment to the stream system is critical to identifying the complex interrelationships that are responsible for a stream instability problem (Figures 4.1 and 4.4).

To develop a restoration solution for a degraded stream, it is often useful to review a variety of stream channel classifications based on planform, sediment load, and hydraulic and geomorphic parameters to determine potential stream types consistent with watershed and valley features, and the existing stream system. The classification concepts of Section 5.3 are useful for this purpose. However, a successful restoration project will require developing a stable form for the stream considering the existing hydrologic and sediment regime. Here one must develop a stream that is stable laterally (in planform, see Section 6.2) and stable vertically (in profile, see Section 6.4).

The ultimate test of restoration design is the ability of the reconfigured channel to achieve a state of dynamic equilibrium considering the size and volume of sediment delivered from upstream (see Chapter 7). The sediment continuity concepts of Sections 6.4 and 7.2 can be used for a preliminary evaluation of stream system stability, but a more detailed analysis using water and sediment routing computer models such as the USACE HEC-RAS model (USACE 2010) may be required for large rivers or complex projects. The FHWA HDS 6 manual (FHWA 2001) provides background, concepts and applications of sediment transport technology.

A recent manual, Channel Rehabilitation: Processes, Design, and Implementation, by the U.S. Army Corps of Engineers Engineering Research and Development Center (USACE 1999) recognizes that regardless of the goals of the rehabilitation project, the fundamentals of planning activities should be followed. A typical planning process for channel restoration involves the following general steps:

- Preliminary planning to establish the scope, goals, preliminary objectives, and general approach for restoration
- Baseline assessments and inventories of project location to assess the feasibility of preliminary objectives, to refine the approach to restoration, and to provide for the project design
- Design of restoration projects to reflect objectives and limitations inherent to the project location
- Evaluation of construction to identify, correct, or accommodate for inconsistencies with project design
- Monitoring of parameters important for assessing goals and objectives of restoration

Based on these guidelines, a systematic approach to initiating, planning, analyzing, implementing, and monitoring of channel restoration and rehabilitation projects can be developed.

In addition, the AASHTO Highway Drainage Guidelines (AASHTO 1999a), Volume X, contains detailed guidelines for stream modification and mitigation practices, particularly regarding aquatic habitat and wetland functions. AASHTO's Model Drainage Manual (AASHTO 1999b), Appendix D, suggest a number of strategies to develop channel mitigation geometries when disturbance of a channel is determined to be unavoidable. Three alternatives are suggested to maintain a stream's functional values, including: grade control structures, fish habitat structures, and bendway bank protection. Conceptual sketches for a variety of structures are provided in the Model Drainage Manual.

## 9.3 DESIGN CONSIDERATIONS FOR CHANNEL RESTORATION

Most of the above referenced publications, however, stop short of providing specific guidance tools that cover the full range of treatments, from natural to management to structural, for stream restoration projects. As a result, the NRCS (2007) has developed a detailed handbook (NEH 654) that provides specific "how to" techniques and tools to perform analyses and designs for successful restoration projects.

The U.S. Army Corps of Engineers (USACE) Waterways Experiment Station (WES) has also developed a systematic methodology for hydraulic design of channel restoration projects (USACE 1999). The methodology incorporates both fluvial geomorphologic principles and engineering analysis. It includes use of hydraulic geometry relationships, analytical determination of stable channel dimensions, and a sediment impact assessment. This methodology, which will meet the needs of the highway engineer in many situations, is outlined in this section (from a paper prepared by Copeland and Hall 1998). Reference to the USACE manual, "Channel Rehabilitation: Processes, Design, and Implementation" (USACE 1999), is suggested for more detail.

When the existing channel is stable, the wave length and sinuosity should be maintained in any channel restoration scheme (Copeland and Hall 1998). The USACE methodology is intended for cases where an historically stable channel has been realigned creating instability, or where hydrologic and/or sediment inflow conditions have changed so much that the channel is currently unstable. Stability is defined as the ability to pass the incoming sediment load without significant degradation or aggradation. Bank erosion and bankline migration are natural processes and may continue in a stable channel. When bankline migration is deemed unacceptable, then engineering solutions may be employed to prevent bank erosion. Hydraulic Engineering Circular No. 23 (HEC-23) presents design guidelines for a range of stream instability countermeasures and discusses bioengineering and biotechnical solutions (FHWA 2001).

Rosgen's textbook "Applied River Morphology" (Rosgen 1996) extends his classification system (Figures 5.9 and 5.10) to include concepts and applications for river restoration. He notes that we are now in an unprecedented era of stream restoration, working to "put the kinks back into channelized, over-widened streams." While these restoration efforts are much needed, as with many new programs, restoration efforts run the risk of working counter to natural stability concepts. He concludes that stream classification can assist in river restoration by:

- Enabling more precise estimates of quantitative hydraulic relationships associated with specific stream and valley morphologies.
- Establishing guidelines for selecting stable stream types for a range of dimensions, patterns, and profiles that are in balance with the river's valley slope, valley confinement, depositional materials, streamflow, and sediment regime of the watershed.
- Providing a method for extrapolating hydraulic parameters and developing empirical relationships for use in the resistance equations and hydraulic geometry equations needed for restoration design.
- Developing a series of meander geometry relationships that are uniquely related to stream types and their bankfull dimensions.
- Identifying the stable characteristics for a given stream type by comparing the stable form to its unstable or disequilibrium condition.

In addition, under NCHRP Project 24-19 (NCHRP 2005), McCullah and Gray developed selection criteria, design guidelines, and a compilation of techniques used for environmentally sensitive channel and bank protection measures as well as a rule-based technique selection system presented as an interactive software program entitled "Greenbank." Their hierarchical list and classification of 44 environmentally sensitive channel and bank protection techniques are shown in Table 9.1. McCullah and Gray's 3-level rating system was developed to account for the amount, quality, and reliability of available information.

Table 9.1. Hierarchical List and Classification of Environmentally Sensitive Channel and Bank Protection Techniques (NCHRP 2005).

| Category | Technique | Level |
|---|---|---|
| **River Training** | | |
| Transverse Structures | Spur dikes | I |
| | Vanes | I |
| | Bendway weirs | I |
| | Large woody debris structures | II |
| | Stone weirs | II |
| Longitudinal Structures | Longitudinal stone toe | I |
| | Longitudinal stone toe with spurs | I |
| | Coconut fiber rolls | II |
| | Vegetated gabion basket | I |
| | Live cribwalls | II |
| | Vegetated mechanically stabilized earth | I |
| | Live siltation | II |
| | Live brushlayering | I |
| Channel Planform Measures | Vegetated floodways | II |
| | Meander restoration | II |
| **Bank Armor and Protection** | | |
| Groundcovers | Vegetation alone | II |
| | Live staking | I |
| | Willow posts and poles | II |
| | Live fascines | I |
| | Turf reinforcement mats | II |
| | Erosion control blankets | II |
| | Geocellular containment systems | II |
| Revetments | Rootwad revetments | II |
| | Live brush mattress | I |
| | Vegetated articulated concrete blocks | I |
| | Vegetated riprap | I |
| | Soil and grass covered riprap | II |
| | Vegetated gabion mattress | II |
| | Cobble or gravel armors | II |
| | Trench fill revetment | II |
| **Riparian and Stream Opportunities** | | |
| Top-of-Bank Treatments | Live gully repair | III |
| In-Stream Habitat Improvements | Vanes with J-hooks | I |
| | Cross vanes | I |
| | Boulder clusters | II |
| | Newbury rock riffles | II |
| **Slope Stabilization** | | |
| Drainage Measures | Diversion dike | II |
| | Slope drain | II |
| | Live pole drain | III |
| | Chimney drain | II |
| | Trench drain | II |
| | Drop inlet | II |
| | Fascines with subsurface drain | II |
| Bank Regrading | Slope Flattening | II |
| In-Situ Reinforcement | Stone-fill trenches | II |

*Level I* – Well-established, well documented (good performance and monitoring data available), reliable design criteria based on lab/field studies.

*Level II* – Intermediate, greater uncertainty (used frequently but do not have the level of detail, quality of information, and reliability that characterize Level I); little or inadequate monitoring.

*Level III* – Emerging, promising techniques. Does not have the track record and level of information characterizing Level I or II.

The rule-based Greenbank decision support tool (NCHRP 2005) was developed for use by DOT personnel or consulting engineers. The system does not provide detailed design criteria, but instead offers a list of techniques that match (1) the dominant erosion process and (2) environmental resources of special concern at the site in question.

### 9.3.1 USACE Design Methodology

While the following design methodology continues to be evaluated as part of the U.S. Army Corps of Engineers Flood Damage Reduction and Stream Restoration Research Program, the steps outlined provide a reasonable approach for the highway engineer faced with a channel restoration design requirement.

**Step 1. Determine the design width of the channel.** The design width is related to the idealized "bankfull width" which is the channel topwidth that occurs when the channel-forming (dominant) discharge occurs. Current research by the USACE suggests that the effective discharge is the best representation of the channel forming discharge. The effective discharge is the increment of discharge that transports the most sediment on an annual basis. This discharge may be determined by integrating a sediment transport rating curve with the annual flow-duration curve. Where possible, it is important to attempt to verify this channel-forming discharge with field indicators of bankfull discharge.

Several techniques are available for determining the design width as a function of the channel-forming discharge in stable alluvial streams. In order of preference they are (Copeland and Hall 1998):

1. Develop a width vs. effective discharge relationship for the project stream. This can be accomplished by measuring the average width in stable reaches where the effective discharge can be calculated. These channel reaches may be in the project reach itself or in reference reaches upstream and/or downstream from the project. This is referred to as the analogy method. This technique is inappropriate for streams where the reference reaches are in disequilibrium.

2. Find stable reaches of streams with similar hydrologic, hydraulic, and sediment characteristics in the region and develop a hydraulic geometry relationship for width vs. effective discharge. This technique is also inappropriate for streams where the reference reaches are in disequilibrium.

3. If a reliable width vs. effective discharge relationship cannot be determined from field data, analytical methods discussed in Step 2 may be employed to obtain a range of feasible solutions. If the channel width is constrained due to right-of-way limits, select the required width and be prepared to provide bank protection.

The composition of the bank is very important in the determination of a stable channel width. It has been shown that the percentage of cohesive materials in the bank and the amount of vegetation on the bank significantly affect the stable channel width. General guidance is available in the U.S. Army Engineer Manual EM-1110-2-1418 (USACE 1994a), and Hey and Thorne (1986) (see also Appendix B).

**Step 2. Calculate a stable channel slope and depth.** In sand-bed streams, sediment transport is typically significant and an analytical procedure that considers both sediment transport and bed form roughness is required to determine a stable channel slope and depth. Analytical approaches calculate the design variables of width, slope, and depth from the independent variables of discharge, sediment inflow, and bed-material composition. Three equations are required for a unique solution of the three dependent variables. Flow resistance and sediment transport equations are readily available (see Sections 3.4.4 and 7.5). A hydraulic geometry width predictor can be used as the third equation. Alternatively, the stable-channel analytical method in the U.S. Army Engineer hydraulic design package SAM (USACE 1994b) may be used to determine a depth and slope for the width selected in Step 1 (see also Section 6.4).

**Step 3. Determine a stable channel meander wave length for the planform.** The most reliable hydraulic geometry relationship for meander wave length is wave length vs. width. As with the determination of channel width, preference is given to wave length predictors from stable reaches of the existing stream either in the project reach or in reference reaches. Lacking data from the existing stream, general guidance is available from several literature sources (see Sections 5.5.4 and 6.2.1).

**Step 4. Calculate the channel length for one meander wave length.**

$$\text{meander length} = \frac{\text{wave length} \times \text{valley slope}}{\text{channel slope}} \tag{9.1}$$

**Step 5. Layout a planform using the meander wave length as a guide.** One way to accomplish this task is to cut a string to the appropriate length and lay it out on a map. Another, more analytical approach, is to assume a sine-generated curve for the planform shape as suggested by Langbein and Leopold (USGS 1966a) (see Equation 9.1) and calculate x-y coordinates for the planform. This rather tedious numeric integration can be accomplished using a computer program such as the one in the USACE SAM hydraulic design package (USACE 1994b). The sine-generated curve produces a very uniform meander pattern. A combination of the string layout method and the analytical approach would produce a more natural planform.

Check the design radius of curvature to width ratio, making sure it is within the normal range of 2 to 3 (see Section 6.2.1). If the meander length is too great, or if the required meander belt width is unavailable, grade control may be required to reduce the channel slope. While this bend geometry minimizes resistance to flow, it may maximize the natural channel migration potential. If this migration rate is unacceptable, bankline revetment may be required.

In streams that are essentially straight (sinuosity less than 1.2) riffle and pool spacing may be set as a function of channel width. As an empirical guide, a spacing of 5-7 channel widths can be used, with the lower end for steeper channels and the higher end for flatter channels.

Two times this riffle spacing gives the total channel length through one meander pattern (Section 6.2.1).

**Step 6. Conduct a sediment impact assessment.** Critics argue that at least some level of sediment engineering is required on all stream restoration projects to reduce the risk of undesirable outcomes. Shields et al. (2008) provide guidance on how to prepare and execute a sediment studies plan for a stream restoration project. Copeland et al. (USACE 2001) includes a methodology for assessing hydraulic and sediment transport impacts of alternatives.

The purpose of the sediment impact assessment is to determine the long-term stability of the restored reach in terms of aggradation and/or degradation (Copeland et al. 1999). This can be accomplished using a sediment budget approach for relatively simple projects or by using a numerical model which incorporates solution of the sediment continuity equation for more complex projects (see Section 6.4.3).

With a sediment budget analysis, average annual sediment yield with the design channel is compared to the average annual sediment yield of the existing channel, if the existing channel is stable, or of the upstream supply reach, if the existing channel is unstable. Large differences in calculated sediment yield indicate channel instability. The USACE (USACE 1999) suggests that the most reliable way to determine the long-term effects of changes in a complex mobile-bed channel system is to use a numerical model such as HEC-RAS (USACE 2010), which assumes the streambanks are fixed.

The fact that application of a numerical sediment model requires knowledge of sediment transport and river mechanics should not be a deterrent to its use; that knowledge is required for any responsible design work in a river system. It should be expected that an analysis of system response in a complicated system, such as a mobile-bed river system, will require some engineering effort (Copeland and Hall 1998). Channel restoration design should not be undertaken without reference to the principles of fluvial geomorphology and river engineering hydraulics as presented in this manual (HEC-20), FHWA's HDS 6 (FHWA 2001), and (USACE 2001). Johnson and Niezgoda (2004) argue that an analytical approach, produces cost savings in terms of reduced failure risk, even though the initial costs for design may be higher.

### 9.3.2 Natural Channel Design

River restoration based on the principles of natural channel design is most commonly accomplished by restoring the dimensions, pattern, and profile of a disturbed river system to imitate the natural, stable river (Rosgen 2006a). In general, natural channel design, or NCD, uses engineering, geomorphological, and biological principles to improve the hydrology, habitat, and aesthetics of a stream, considering current and future watershed conditions.

Skidmore et al. (2001) proposed a common terminology and categorization of approaches to natural channel design as:

*Analog Approach* – Adopts templates from historic or adjacent channel characteristics and assumes equilibrium between channel form and sediment and hydrologic inputs.

*Empirical Approach* – Uses equations that relate various channel characteristics derived from regionalized or "universal" data sets, and also assumes equilibrium conditions.

*Analytical Approach* – Makes use of hydraulic models and sediment transport functions to derive equilibrium conditions, and thus is applicable to situations where historic or current channel conditions are not in equilibrium with existing or predicted sediment and hydrologic inputs.

All channel design is based on the premise that "natural" channels tend toward equilibrium between channel form and sediment and hydrologic inputs (USGS 1953). Channel form is dictated by independent variables of discharge, sediment supply, and boundary material characteristics, including vegetation. Dependent variables are those physical characteristics that define channel form (width, depth, slope, and planform), which can be selected using various approaches to channel design. Analog approaches can be conducted without any quantification of independent variables. Empirical approaches require only dominant discharge and, therefore, can be conducted without any quantification or consideration of sediment supply. Analytical design methods require some quantification of independent variables in some instances, and can be used to quantify independent variables in other instances. Shields and Copeland (2006) combined the "analog" and "empirical" category into a single "empirical" category for the purposes of contrasting the two approaches used in NCD.

Since the late 1990s, the Rosgen classification system (see Chapter 5 Section 5.3) and its associated methods have become synonymous with natural channel design of stream restoration projects. The term "natural channel design" has been adopted by Rosgen (2006a) and others advocating and using the Rosgen system (Hey 2006). According to Malakoff (2004), Rosgen's Natural Channel Design approach is the most influential approach in river restoration and probably the most controversial (Nagle 2007, Lave 2009, Lave et al. 2010).

Rosgen (2006a) indicates that NCD uses a geomorphic approach that incorporates a combination of analog, empirical, and analytical methods for assessment and design. He notes that since all rivers within a wide range of valley types do not exhibit similar morphological, sedimentological, hydraulic, or biological characteristics, it is necessary to group rivers of similar characteristics into discrete stream types. These characteristics are obtained from stable "reference reach" (Rosgen 1998) locations by discrete valley types, and then converted to dimensionless ratios for extrapolation to disturbed stream reaches of various sizes, while hydraulic, sedimentologic, and morphologic relations are obtained for both the reference and impaired conditions.

The Rosgen Natural Channel Design method is summarized in Chapter 11 of the USDA Natural Resources Conservation Service's "Stream Restoration Design" handbook (NRCS 2007). Rosgen's geomorphic approach to natural channel design consists of eight phases as shown in the flow chart in Figure 9.1. The flow chart defines the complex steps associated with this method, including detailed, quantitative assessments of the causes of river instability; field measurements required to quantify hydraulic and sedimentological relations; and designs that implement analog, empirical, and analytical methods. The eight phases of Rosgen's (2006a) geomorphic channel design method are as follows:

*Phase I*: Restoration Goal/Objectives – Define specific restoration objectives associated with physical, biological, and/or chemical process.

*Phase II*: Regional and Local Relations – Develop regional and localized specific information on geomorphologic characterization, hydrology, and hydraulics.

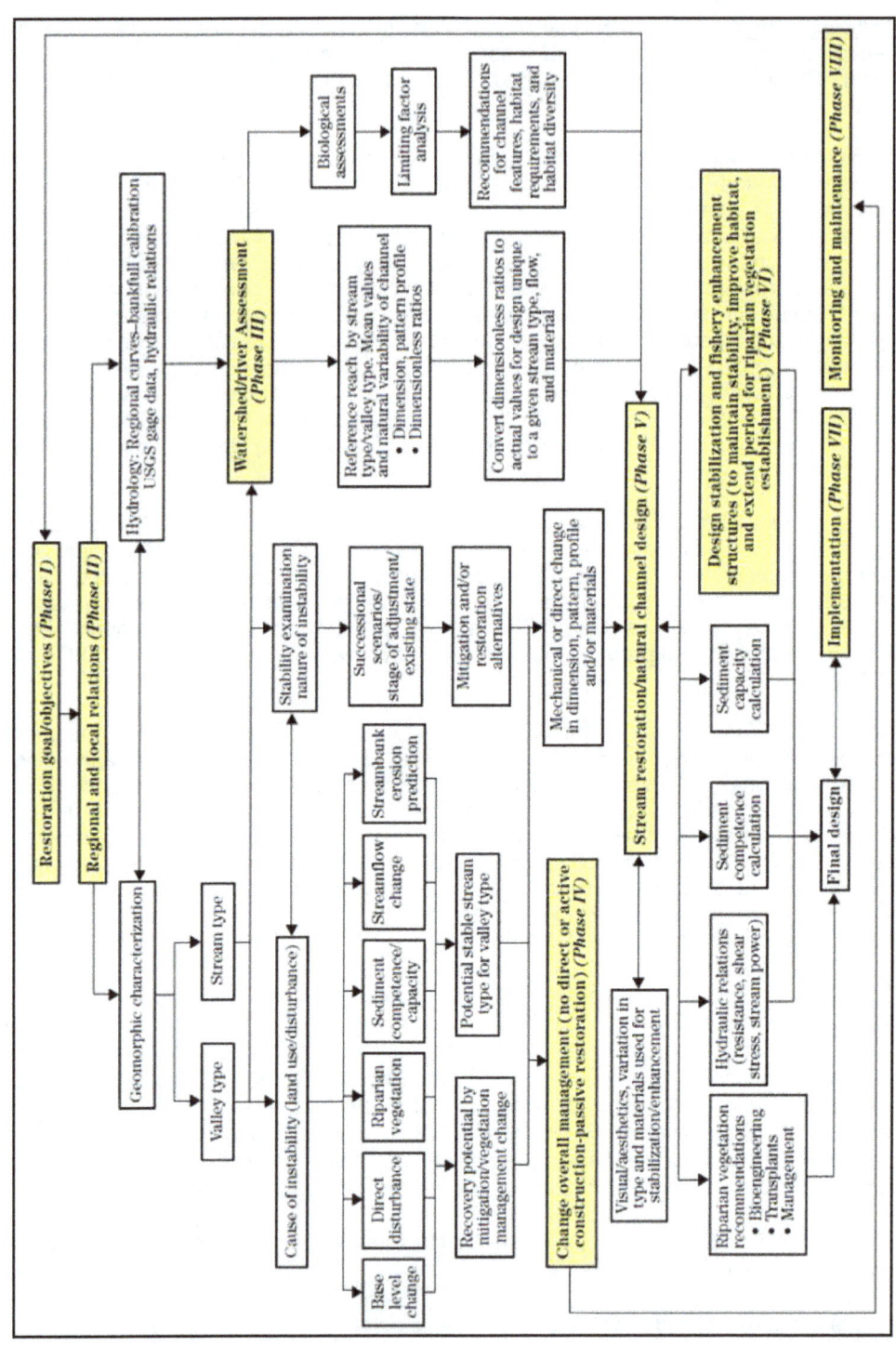

Figure 9.1. Flow chart depicting the sequence of implementation of Rosgen's (2006a) eight sequence phases associated with natural channel design using a geomorphic approach.

*Phase III*: Watershed/River Assessment – Conduct a watershed/river assessment to determine river potential, current state, and the nature, magnitude, direction, duration, and consequences of change.

*Phase IV*: Change Overall Management (Passive Restoration) – Consider passive restoration recommendations based on land use change prior to considering mechanical restoration.

*Phase V*: Stream Restoration/Natural Channel Design – Initiate natural channel design with subsequent analytical testing of hydraulic and sediment transport (competence and capacity) relations.

*Phase VI*: Design Stabilization and Fisheries Enhancement Structures – Select and design stabilization/enhancement/vegetative establishment measures and materials to maintain dimension, pattern, and profile to meet stated objectives.

*Phase VII*: Implementation – Implement the proposed design and stabilization measures involving layout, water quality control, and construction staging.

*Phase VIII*: Monitoring and Maintenance Plan – Design a plan for effectiveness, validation, and implementation monitoring to ensure stated objectives are met, prediction methods are appropriate, and construction is implemented as designed.

However, critics (Miller and Skidmore 2001; Skidmore et al. 2001; Committee on Applied Fluvial Geomorphology 2004; Simon et al. 2005, 2007; Shields and Copeland 2006) indicate that problems are encountered with using Rosgen's classification system, specifically with identifying bankfull dimensions, particularly in incising channels and with the mixing of bed and bank sediment into a single population for use in hydraulic and sediment transport analyses. They also indicate that its use for engineering design and restoration may be flawed by ignoring some processes governed by force and resistance and the imbalance between sediment supply and transporting power in unstable systems (Simon et al. 2007). Rosgen (2008) counters that the stream classification does not substitute for a detailed stability analysis, which is provided under his Level III and IV analyses and is also conducted as part of the "Watershed Assessment for River Stability and Sediment Supply" (WARSSS) methodology developed by Rosgen for the U.S. Environmental Protection Agency (Rosgen 2006b, USEPA 2006).

Simon et al. (2008) note that the primary criticism of the Rosgen Classification, NCD, and WARSSS methodology is that the use of this form-based approach is not and cannot be used to predict stable morphologies in currently unstable alluvial systems. Additionally, they note that even though the Rosgen approach has been updated to included supplementary features (Rosgen 2006a, 2008), the methodology still fails to account for fundamental elements of adjusting natural streams because of fundamental flaws in the overall NCD and WARSSS approaches. They provide examples of how the supplementary features (such as sediment supply and streambank contributions) in Rosgen's NCD (Rosgen 2006a, 2007) and WARSSS (Rosgen 2006b) approaches do not stand up to analytical scrutiny, field testing, and validation.

Although there remains continued debate and criticism of the applicability and efficacy of the Rosgen methodology for natural channel design (Lave 2009, Lave et al. 2010), Nagle (2007) notes that some critics of NCD do not seem to grasp the need to deal immediately with pressing problems in severely degraded streams. Nagle (2007) suggests that NCD can work adequately at least over the short- to medium-term (5-10 years), but the long-term (>10 years) viability of projects is unknown. Often, available information on the varied performance of older NCD projects is anecdotal, lacking in the necessary documentation to evaluate outcomes and likely future conditions.

### 9.3.3 Restoration Project Uncertainties

For any restoration project, there are inherent uncertainties that must be accounted for. These uncertainties include: lack of scientific and other information; limitations of analytical methods and tools; complexities of river systems; and needs to make value-laden judgments at all stages of river restoration problem identification, analysis, and solution implementation (Lemons and Victor 2008). The prediction of river response is accompanied by high levels of uncertainty and "success" tends to be poorly defined in restoration projects (Palmer et al. 2005). Darby and Sear (2008) have attempted to provide a rational theoretical analysis of the uncertain basis of restoration and practical guidance on managing the implications of uncertainty.

### 9.3.4 In-Stream Flow Control Structures

Various structural measures that are intended to deflect stream currents, induce sediment deposition, induce scour, or in some other way alter the flow and sediment regimes of a stream are commonly referred to as river training. In-stream structures can be classified under two fundamental categories, sills and deflectors (Radspinner et al. 2010). Sills are low structures that span the entire channel and deflectors are structures that extend from one bank into the channel without reaching the other bank (Shields 1983). Figure 9.2 shows a typical in-stream deflector and sill.

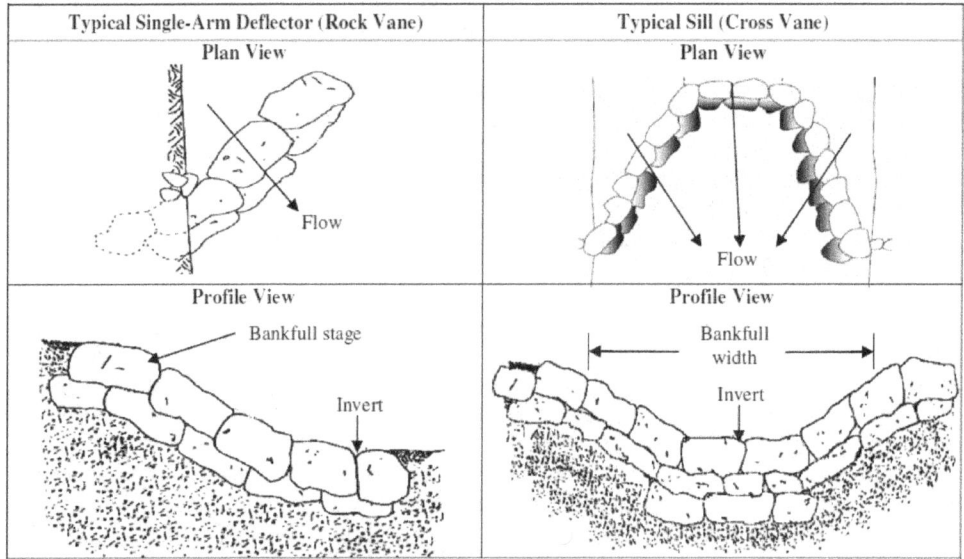

Figure 9.2. Typical in-stream deflector and sill (Radspinner et al. 2010).

Because of the potential cost effectiveness and ecological benefits, the use of in-stream flow training structures in channel restoration and stabilization has become increasingly popular. Proper structure design and placement are necessary to avoid unintended channel aggradation, localized bed scour, and bank erosion, all of which can induce significant damage to the stream and nearby property and accelerate the adverse effects they were initially installed to prevent.

Stream restoration and river training measures currently rely heavily on an analog method that emphasizes a prescribed design approach rather than the application of physically-based hydraulic engineering principles to attain performance-based criteria (Simon et al. 2007; Slate et al. 2007). Radspinner et al. (2010) note that the lack of engineering standards for stream restoration and river training techniques is underscored in the design and installation of the wide variety of shallow, in-stream, low-flow structures in small- to medium-sized rivers.

As part of NCHRP Project No. 24-33, Radspinner et al. (2010) performed a survey of a wide variety of practicing engineers and scientists in the United States, including transportation and natural resource departments from a majority of states as well as federal agency employees and practicing engineers to document and synthesize recent experiences with in-stream structures for river training and erosion control in small- to medium-sized rivers. The goal of the survey was to determine the most popular structures currently in use and the most common field conditions in which they are installed. The in-stream structures documented as being the most commonly used include cross vanes, J-hook vanes, rock vanes, W weirs, submerged vanes, stream barbs, bendway weirs, spurs, and constructed riffles. Descriptions of these types of structures can be found in Rosgen (2001) and NRCS's NEH 654 (NRCS 2007). Radspinner et al. (2010) noted that there was a consensus among survey respondents that these structures can provide a more cost-effective alternative than traditional channel stabilization measures including riprap revetment, dredging, and concrete.

However, results from Radspinner et al.'s (2010) case studies and discussions with practitioners highlight widespread ambiguity in the design, construction, and maintenance of these structures. They note that: "The lack of a comprehensive theory addressing in-stream structure design guidelines, in spite of intense scour-related research for well over 100 years, can be attributed to the unsteady three-dimensional character of the flow in the vicinity of structures, the significant flow and stream variability from site to site, and the previously mentioned complex and poorly understood interaction between flow and the bed." It should be emphasized that these structures are temporary in nature and can all be undermined by active incision/degradation and, even in stable channels, can only be expected to withstand a flood up to some critical value.

A qualitative evaluation of the performance and failure mechanisms of rock weirs at 127 sites by Mooney et al. (USBR 2007) underscores the need for a greater understanding of sediment transport, geomorphology, and physical processes encountered by these types of structures. They found that 74% of the 127 structures evaluated could be considered to be in partial or complete failure and hypothesized that the primary failure mechanism was growth of the scour pool and geotechnical slumping of the footer rocks. In a follow up study, Holburn et al. (USBR 2009) conducted a quantitative investigation of the same rock weir sites and identified several discernable relationships between structure parameters and degree of failure. The most notable include the relationships with recurrence intervals of high flows, structure throat width, planform angles, and scour offset from the structure, while structure spacing, planform location, and scour depth were found to be important variables that relate to structure performance.

## 9.4 MANAGING ROADWAY IMPACTS ON STREAM ECOSYSTEMS

Planning for transportation routes, has traditionally sought to find the shortest route with the least cost. Constructability and construction cost have been a large deciding factor. More recently, the potential impact of roadways, bridges, and other hydraulic structures on stream systems has become a part of the evaluation process. Understanding these potential impacts is a key component of roadway planning and maintenance of existing infrastructure.

Potential impacts of specific roadway features such as culverts, bridges, lateral fills, channel encroachments, and floodplain encroachments include channel erosion/incision, degradation, aggradation, loss of natural floodplain, reduction of instream habitat, reduction or loss of aquatic mobility (blockages/barriers), and riparian degradation. Therefore, transportation engineers and designers involved with stream restoration or stabilization work around bridges and other infrastructure will need to recognize and evaluate these impacts, including stream classification, reach delineations and assessments, analytical approaches, and mitigation opportunities.

Restoration should re-establish functional stream habitats and optimize stream stability and productivity, as well as consider the value of adjacent riparian communities. At the same time, restorations must minimize the risk of damage to economic resources and infrastructure, and human life and property. The tools and guidelines as presented in this manual (HEC-20) can be used to assist in identifying and mitigating the impacts of highway planning, construction, operation, and maintenance on streams and their associated ecosystems. Highways are often built in stream valleys and may require that the stream channel be modified or relocated. Because the impacts to streams from highways are varied and complex, appropriate stream channel design and restoration can benefit stream ecosystems as well as minimize the risk to critical highway infrastructure and other economic resources.

In addition, the FHWA National Highway Institute (NHI) offers a three-day course, entitled "Managing Road Impacts on Stream Ecosystems," that introduces participants to the basic concepts related to the impacts that roadways have on streams and stream ecosystems (NHI 2008). The course is structured to first address the ecological and physical characteristics of stream ecosystems, discuss the impacts that roadways can have on those ecosystems, and then turn to tools that the practitioner can use to help avoid and mitigate those effects. Through the use of Case Examples, discussion, and other application techniques, the participants are afforded an opportunity to use critical thinking to identify solutions and preventative measures related to the impacts of roads on streams and their riparian communities. The key points covered in the course are:

- Components of Stream Assessment/Restoration Plan
    - Introduction to elements of a good plan
    - State and Federal regulatory compliance requirements
    - Interdisciplinary expertise and collaboration

- Physical Aspects of Stream Systems
    - Natural channel dynamics
    - Impacts to streams affecting form and function

- Ecological Aspects of Stream Systems
    - Ecological and water quality components
    - Effects to ecological functions from changes to river systems

- Assessing the Impacts of Roadways on Rivers/Streams
  - Typical roadway and transportation corridor impacts
  - Implications from a physical and biological perspective

- Opportunities and Constraints for Stream Restoration
  - Approaches and techniques for more sensitive planning
  - Large-scale natural resource components

- Performance Measures and Monitoring
  - Physical monitoring techniques
  - Instream biological methods

- Risk Assessment
  - Stream stabilization techniques and risk for implementation
  - Pre- and post-construction risks

(page intentionally left blank)

# CHAPTER 10

AASHTO Task Force on Hydrology and Hydraulics, 1982, Highway Subcommittee on Design, "Hydraulic Analyses for the Location and Design of Bridges," <u>Volume VII, Highway Drainage Guidelines</u>, American Association of State Highway and Transportation Officials, 444 North Capitol Street, N.W., Suite 225, Washington, D.C.

Abad, J.D. and M.H. Garcia, 2006, Discussion of "Efficient Algorithm for Computing Einstein Integrals by Junke Guo and Pierre Y. Julien," Journal of Hydraulic Engineering, Vol. 130, No. 12, pp. 1198-1201, 2004), ASCE, 132(3), 332-334.

Ackers, P. and F.G. Charlton, 1970, "Meander Geometry with Varying Flows," Journal of Hydrology, Vol. 11, pp. 230-252.

American Association of State Highway and Transportation Officials, 1999a, "Highway Drainage Guidelines," published by AASHTO, Washington, D.C.

American Association of State Highway and Transportation Officials, 1999b, "Model Drainage Manual," published by AASHTO, Washington, D.C.

American Society of Civil Engineers, 2008, "Sedimentation Engineering Processes, Measurements, Modeling, and Practice," M.H. Garcia (ed.), ASCE Manuals and Reports on Engineering Practice No. 110, Reston, VA.

Andrews, E.D., 1980, "Effective and Bankfull Discharges of Streams in Yampa Basin," Colorado and Wyoming, Journal of Hydrology, Vol. 46, pp. 311-330.

Andrews, E.D. and G. Parker, 1987, "Formation of a Coarse Surface Layer as the Response to Gravel Mobility," in: C.R. Thorne, J.C. Bathurst and R.D. Hey (Eds.), Sediment Transport in Gravel-Bed Rivers, John Wiley and Sons, New York, NY, pp. 269-300.

Andrews, E.D. and J.D. Smith, 1992, "A Theoretical Model for Calculating Marginal Bed Load Transport Rates of Gravel," in: P. Billi, R.D. Hey, C.R. Thorne, P. Tacconi (Eds.), Dynamics of Gravel-Bed Rivers, Wiley, Chichester, UK, pp. 41-52.

Annandale, G.W. 1994, "Guidelines for the Hydraulic Design and Maintenance of River Crossings," Volume VI: Risk Analysis of River Crossing Failure. Department of Transport, Report No. TRH 25:1994, Pretoria, Republic of South Africa.

Annandale, G.W. 1999, "Risk Analysis of River Bridge Failure," In Stream Stability and Scour at Highway Bridges. Ed. E.V. Richardson and P.F. Lagasse, ASCE, Reston, VA, 1003-1012.

ASCE Task Committee on Hydraulics, Bank Mechanics, and Modeling of River Width Adjustment 1998, "River width adjustment I: processes and mechanisms," Journal of Hydraulic Engineering, 1249, 881-902.

Bathurst, J.C., 1993, "Flow Resistance Through the Channel Network," in: K. Bevin and M.J. Kirkby (Eds.), Channel Network Hydrology, Wiley, Chichester, UK, pp. 69-98.

Bathurst, J.C., C.R. Thorne, and R.D. Hey, 1979, "Secondary flow and shear stress at bends," American Society of Civil Engineers, Journal of Hydraulics (HY10), Vol. 105, pp. 1277-1295.

Beeson, C.E. and P.F. Doyle, 1995, "Comparison of Bank Erosion at Vegetated and Non-Vegetated Channel Bends," American Water Resources Association, Water Resources Bulletin, Vol. 31, No. 6.

Bledsoe, B., Hawley, R., and Stein, E.D., 2008, "Stream Channel Classification and Mapping Systems: Implications for Assessing Susceptibility to Hydromodification Effects in Southern California," Southern California Coastal Waters Research Project, Technical Report 562, 38 p.

Brayshaw, A.C., 1985, "Bed Microtopography and Entrainment Thresholds in Gravel-Bed Rivers," Bulletin of the Geological Society of America, Vol. 96, pp. 218-223.

Bridge, J.S., 1993, "The Interaction Between Channel Geometry, Water Flow, Sediment Transport and Deposition in Braided Rivers," in: J.L. Best and C.S. Bristow (Eds.), Braided Rivers, Special Publication of the Geological Society of London, Vol. 75, pp. 13-71.

Bull, W.B., 1968, "Alluvial fans," Journal of Geology Education, Vol. 16.

Carling, P.A., 1987, "Bed Stability in Gravel Streams, with Reference to Stream Regulation and Ecology," in: K.S. Richards (Ed.), River Channels: Environment and Process, Blackwell Press, Oxford, UK, pp. 321-347.

Carling, P., 1988, "The Concept of Dominant Discharge Applied to Two Gravel-Bed Streams in Relation to Channel Stability Thresholds," Earth Surface Processes and Landforms, Vol. 13, pp. 355-367.

Clifford, N.J., 1993, "Differential Bed Sedimentology and the Maintenance of Riffle-Pool Sequences," Catena, Vol. 20, pp. 447-468.

Colby, B.R., 1964, "Practical Computations of Bed-Material Discharge," ASCE Hydr. Div., Jour., Vol. 90, No. HY2.

Committee on Applied Fluvial Geomorphology, 2004, Position statement on "Applied Fluvial Geomorphology," Committee on Applied Fluvial Geomorphology, Quaternary Geology and Geomorphology Division, Geological Society of America.

Copeland, R.R. and B.R. Hall, 1998, "Channel Restoration Hydraulic Design Procedure," In ASCE Proceedings for 1998 Wetlands Engineering and River Restoration Conference, Engineering Approaches to Ecosystem Restoration, Denver, CO.

Copeland, R.R., C.R. Thorne, and P.J. Soar, 1999, "Continuity of Sediment in Channel Restoration Design," ASCE, Proceedings of the International Water Resource Engineering Conference, Seattle, WA.

Cowan, W.L., 1956, "Estimating Hydraulic Roughness Coefficients," Agricultural Engineering, Vol. 37, pp. 473-475.

Darby, S. and D. Sear, 2008, "River Restoration - Managing the Uncertainty in Restoring Physical Habitat," John Wiley & Sons Ltd., Chichester, UK.

Diehl, T.H. and B.A. Bryan, 1993, "Supply of large woody debris in a stream channel," In: H.W. Shen, S.T. Su, and F. Wen (eds), Hydraulic Engineering '93, Proceedings of the 1993 National Conference on Hydraulic Engineering, ASCE, New York, p. 1055-1060.

Dietrich, W.E. and J.D. Smith, 1983, "Influence of the point bar on flow through curved channels," Water Resources Research, Vol. 19, pp. 1173-1192.

Dietrich, W.E., 1987, "Mechanics of flow and sediment transport in river bends," In: K.S. Richards (ed), River channels: environment and process, Blackwell, Oxford, pp. 179-227.

Edgar, D.E., 1984, "The Role of Geomorphic Thresholds in Determining Alluvial Channel Morphology," in: C.M. Elliott (Ed.), River Meandering, ASCE Proceedings of the Conference Rivers '83, New Orleans, LA, pp. 44-54.

Einstein, H.A., 1950, "The Bed Load Function for Sediment Transportation in Open Channel Flows," U.S. Dept. of Agriculture, Soil Conservation Serv., Tech. Bull. 1026.

Einstein, H.A. and H.W. Shen, 1964, "A study of meandering in straight alluvial channels," Journal of Geophysical Research, Vol. 69, pp. 5239-5247.

Faustini, J.M., R. Kaufmann, and T. Herlihy, 2009, "Downstream Variation in Bankfull Width of Wadeable Streams Across the Conterminous United States," Geomorphology, Vol. 108, Elsevier, pp. 292-311.

Federal Highway Administration, 1978a, "Countermeasures for Hydraulic Problems at Bridges, Vol. 1, Analysis and Assessment," Report No. FHWA/RD-78-162, Federal Highway Administration, Washington, D.C. (Brice, J.C. and J.C. Blodgett).

Federal Highway Administration, 1978b, "Countermeasures for Hydraulic Problems at Bridges," Vol. 2, Case Histories for Sites 1-283," Report No. FHWA/RD-78-163, Federal Highway Administration, Washington, D.C. (Brice, J.C. and J.C. Blodgett).

Federal Highway Administration, 1978c, "Hydraulics of Bridge Waterways," Hydraulic Design Series No. 1, U.S. Department of Transportation, FHWA (Bradley, J.N.).

Federal Highway Administration, 1980a, "Stream Channel Degradation and Aggradation: Analysis of Impacts to Highway Crossings," FHWA/RD-80-159, Federal Highway Administration, Washington, D.C. (Brown, S.A., R.S. McQuivey, and T.N. Keefer).

Federal Highway Administration, 1980b, "Interim Report - Stream Channel Degradation and Aggradation: Causes and Consequences to Highways," FHWA/RD-80/038, Federal Highway Administration, Washington, D.C. (Keefer, T.N., R.S. McQuivey, and D.B. Simons).

Federal Highway Administration, 1981, "Methods for Assessment of Stream-Related Hazards to Highways and Bridges," FHWA/RD-80/160, Federal Highway Administration, Washington, D.C. (Shen, H.W., S.A. Schumm, J.D. Nelson, D.O. Doehring, and M.M. Skinner).

Federal Highway Administration, 1982, "Stream channel stability assessment," Federal Highway Administration, Offices of Research and Development, Report No. FHWA/RD-82/021, 42 pp (Brice, J.C.).

Federal Highway Administration, 1985, "Streambank Stabilization Measures for Highway Stream Crossings--Executive Summary," FHWA/RD-84/099, Federal Highway Administration, Washington, D.C. (Brown, S.A.).

Federal Highway Administration, 1989a, "Laboratory Studies of the Effect of Footings and Pile Groups on Bridge Pier Scour," Proceeding of 1989 Bridge Scour Symposium, FHWA, Washington, D.C. (Jones, J.S.).

Federal Highway Administration, 1989b, "Evaluation, Modeling, and Mapping of Potential Bridge Scour, West Tennessee," Proceedings of the National Bridge Scour Symposium, Federal Highway Administration Report FHWA-RD-90-035, 112-129 (Simon, A., G.S. Outlaw, and R. Thomas).

Federal Highway Administration, 1994a, "Advanced Technology for Soil Slope Stability," Volume 1, Slope Stability Manual, FHWA-SA-94-005, U.S. Department of Transportation, Washington, D.C.

Federal Highway Administration, 1994b, "Advanced Technology for Soil Slope Stability," Volume 2, Sample Problems and Case Histories, FHWA-SA-94-006, U.S. Department of Transportation, Washington, D.C.

Federal Highway Administration, 1995, "Recording and Coding Guide for the Structure Inventory and Appraisal of the Nation's Bridges, Report No. FHWA-PD-96-001, U.S. Department of Transportation, Washington, D.C.

Federal Highway Administration, 1997a, "Potential Drift Accumulation at Bridges," Federal Highway Administration Report No. FHWA-RD-97-028, Washington, D.C. (Diehl, T.H.).

Federal Highway Administration, 1997b, "Introduction to Highway Hydraulics," Hydraulic Design Series No. 4, Federal Highway Administration, Washington, D.C. (Schall, J.D. and E.V. Richardson).

Federal Highway Administration, 2001, "River Engineering for Highway Encroachments, Highways in the River Environment," Report FHWA NHI 01-004, Federal Highway Administration, Hydraulic Design Series No. 6, Washington, D.C. (Richardson, E.V., D.B. Simons, and P.F. Lagasse).

Federal Highway Administration, 2002, "Hydrologic Design of Highways," Hydraulic Design Series No. 2, Federal Highway Administration, Washington, D.C. (McCuen, R.H., P.A. Johnson, and R.M. Ragan).

Federal Highway Administration, 2006, "Assessing Stream Channel Stability at Bridges in Physiographic Regions," Report No. FHWA-HRT-05-072, Federal Highway Administration, Washington, D.C. (Johnson, P.A).

Federal Highway Administration, 2009, "Bridge Scour and Stream Instability Countermeasures - Experience, Selection, and Design Guidelines," Third Edition, Report FHWA NHI 09-111, (Volume 1) and FHWA NHI 09-112 (Volume 2), Federal Highway Administration, Hydraulic Engineering Circular No. 23, U.S. Department of Transportation, Washington, D.C. (Lagasse, P.F., L.W. Zevenbergen, L.A. Arneson, P.E. Clopper, J.E. Pagán-Ortiz, J.D. Schall, and L.G. Girard).

Federal Highway Administration, 2012a, "Hydraulic Design of Safe Bridges," Report FHWA-HIF-12-018 Hydraulic Design Series No. 7, Washington, D.C. (Zevenbergen, L.W., L.A. Arneson, J.H. Hunt, and A.C. Miller).

Federal Highway Administration, 2012b, "Evaluating Scour at Bridges," Fifth Edition, Report FHWA-HIF-12-003, Hydraulic Engineering Circular No. 18, U.S. Department of Transportation, Washington, D.C. (Arneson, L.A., L.W. Zevenbergen, P.F. Lagasse, and P.E. Clopper).

Federal Interagency Stream Restoration Working Group (FISRWG), 1998, "Stream Corridor Restoration - Principles, Processes, and Practices," National Technical Information Service, Order No. PB98-158348INQ, Washington, D.C.

Ferguson, R.I., 1993, "Understanding Braided Processes in Gravel-Bed Rivers: Progress and Unsolved Problems," in: J.L. Best and C.S. Bristow (Eds.), Braided Rivers, Special Publication of the Geological Society of London, Vol. 75, pp. 73-87.

Gellis, A., R. Hereford, S.A. Schumm, and B.R. Hayes, 1991, "Channel evolution and hydrologic variations in the Colorado River basin: Factors influencing sediment and salt loads," Journal of Hydrology, Vol. 124, p. 317-344.

Gessler, J., 1971, "Beginning and Ceasing of Sediment Motion," River Mechanics, Chapter 7, edited by H. W. Shen, Fort Collins, CO, 22 pp.

Griffiths, G.A., 1989, "Form Resistance in Gravel Channels with Mobile Beds," ASCE Journal of Hydraulic Engineering, Vol. 115, pp. 340-355.

Harvey, M.D. and C.C. Watson, 1986, "Fluvial processes and morphological thresholds in incised channel restoration," Water Resources Bulletin, 223, 359-368.

Hey, R.D., 1975, "Design Discharge for Natural Channels," in: R.D. Hey and T.D. Davies (Eds.), Science, Technology and Environmental Management, Saxon House, UK, pp. 73-88.

Hey, R.D., 1979, "Flow Resistance in Gravel-Bed Rivers," ASCE Journal of the Hydraulics Division, Vol. 105, No. HY4, pp. 365-379.

Hey, R.D., 1997, "Stable River Morphology," in: C.R. Thorne, R.D. Hey, and M.D. Newson (Eds.), Applied Fluvial Geomorphology for River Engineering and Management, John Wiley & Sons, Chichester, UK , pp. 223-236.

Hey, R.D., 2006, "Fluvial Geomorphological Methodology for Natural Channel Design," Journal of the American Water Resources Association, Vol. 42, Issue 2.

Hey, R.D. and C.R. Thorne, 1986, "Stable Channels with Mobile Gravel Beds," Journal of Hydraulic Engineering, American Society of Civil Engineers, Vol. 112, No. 8, pp. 671-689.

Hey, R.D. and G.L. Heritage, 1988, "Dominant Discharge in Alluvial Channels," Proceedings of an International Conference on Fluvial Hydraulics, Budapest, Hungary, pp. 143-148.

Hickin, E.J. and G.C. Nanson, 1975, "The character of channel migration on the Beaton River, northeast British Columbia, Canada," Geological Society of America Bulletin, Vol. 86, pp. 487-494.

Hickin, E.J. and G.C. Nanson, 1984, "Lateral migration rates of river bends," American Society of Civil Engineers, Journal of Hydraulic Engineering, Vol. 110, pp. 1557-1567.

Hooke, J.M., 1977, "The distribution and nature of changes in river channel patterns: The example of Devon," in K.J. Gregory (ed) River channel changes: Wiley, Chichester, p. 265-280.

Inglis, C.C., 1946, "Meanders and Their Bearing on River Training," Maritime Paper No. 7, Institution of Engineers, London.

Johnson, P.A., 2005, "Preliminary assessment and rating of stream channel stability," Journal of Hydraulic Engineering, 13110, 845-852.

Johnson, P.A., G.L. Gleason, and R.D. Hey, 1999, "Rapid assessment of channel stability in vicinity of road crossing," ASCE Journal of Hydraulic Engineering, Vol. 125, No. 6, p. 645-651.

Johnson, P.A. and S.L. Niezgoda, 2004, "Risk-based method for selecting bridge scour countermeasures," Journal of Hydraulic Engineering, Vol. 130, No. 2.

Julien, P.Y., 2010, "Erosion and Sedimentation," second edition, Cambridge University Press. Cambridge, UK.

Julien, P.Y. and G.J. Klaassen, 1995, "Sand-Dune Geometry of Large Rivers During Floods," American Society of Civil Engineers, Journal of Hydraulic Engineering, Vol. 121, No. 9.

Julien, P.Y. and J. Wargadalam, 1995, "Alluvial Channel Geometry: Theory and applications," Journal of Hydraulic Engineering, Vol. 121, No. 4, April.

Karim, F., 1999, "Bed-Form Geometry in Sand-Bed Flows," Journal of Hydraulic Engineering, Vol. 125, No. 12, December.

Keller, E.A., 1971, "Areal Sorting of Bed Load Material: The Hypothesis of Velocity Reversal," Bulletin of the Geological Society of America, Vol. 82, pp. 753-756.

Keller, E.A. and J.L. Florsheim, 1993, "Velocity-Reversal Hypothesis: A Model Approach, Earth Surface Processes and Landforms," Vol. 18, pp. 733-740.

Kellerhals, R., 1967, "Stable Channels with Gravel-Paved Beds," ASCE Journal of the Waterways and Harbors Division, Vol. 93 (WW1), pp. 63-83.

Kennedy, R.C., 1895, "The Prevention of Silting in Irrigation Canals," Min. Proc. Instn. Civil Engr., Vol. CXIX.

Knighton, A.D., 1981, "Local Variations of Cross-Sectional Form in a Small Gravel-Bed Stream," Journal of Hydrology (New Zealand), Vol. 20, pp. 131-146.

Knighton, D., 1998, "Fluvial Forms and Processes – A New Perspective," Arnold, London.

Komar, P.D., 1988, "Sediment Transport by Floods," in: V.R. Baker, R.C. Kochel, and P.C. Patton (Eds.), Flood Geomorphology, Wiley-Interscience, New York, NY, pp. 97-111.

Komar, P.D. and S.M. Shih, 1992, "Equal Mobility Versus Changing Bedload Grain Sizes in Gravel-Bed Streams," in: P. Billi, R.D. Hey, C.R. Thorne, P. Tacconi (Eds.), Dynamics of Gravel-Bed Rivers, Wiley, Chichester, UK, pp. 73-93.

Lacy, G., 1930, "Stable Channels in Alluvium," Proc. Inst. Civil Engrs., 229 p.

Lane, E.W., 1955, "The Importance of Fluvial Geomorphology in Hydraulic Engineering," ASCE Proceeding, Vol. 81, No. 745, pp. 1-17.

Lave, R., 2009, "The Controversy Over Natural Channel Design: Substantive Explanations and Potential Avenues for Resolution," Journal of the American Water Resources Association, Vol. 45, Issue 6.

Lave, R., M. Doyle, and M. Robertson, 2010, "Privatizing stream restoration in the US," Social Studies of Science, Vol. 40, Issue 5.

Lee, Jong-Seok and P.Y. Julien, 2006, "Donwstream Hydraulic Geometry of Alluvial Channels," ASCE Journal of Hydraulic Engineering, Vol. 32, No. 12, December, pp. 1347-1352.

Leliavsky, S., 1955, "An Introduction to Fluvial Hydraulics," Constable & Company Ltd., London, England, 257 p.

Lemons, J. and R. Victor, 2008, "Uncertainty in River Restoration," In: Darby, S. and D. Sear (Eds.), "River Restoration - Managing the Uncertainty in Restoring Physical Habitat," John Wiley & Sons Ltd., Chichester, UK.

Leopold, L.B. and M.G. Wolman, 1960, "River Meanders," Geological Society of America Bulletin Vol. 71.

Lisle, T., 1979, "A Sorting Mechanism for a Riffle-Pool Sequence," Bulletin of the Geological Society of America, Vol. 90, Part 2, PP. 1142-1157.

Lisle, T.E., 1995, "Particle Size Variations Between Bed Load and Bed Material in Natural Gravel Bed Channels," Water Resources Research, Vol. 31, pp. 1107-1118.

Lohnes, R. and R.L. Handy, 1968, "Slope angles in friable loess," Journal of Geology, Vol. 76, pp. 247-258.

Malakoff, D., 2004, "The River Doctor," Science, Vol. 305, Issue 5686.

Markham, A.J. and C.R. Thorne, 1992, "Geomorphology of gravel-bed river bends," In: P. Billi, R.D. Hey, C.R. Thorne, and P. Tacconi (eds), Dynamics of gravel-bed rivers, Wiley, Chichester, pp. 433-450.

Meyer-Peter, E. and R. Muller, 1948, "Formulas for Bed Load Transport, Proceedings of Third Meeting of the IAHR, Stockholm, pp. 39-64.

Miller, D.E. and P.B. Skidmore, 2001, "Natural Channel Design: How Does Rosgen Classification-Based Design Compare with Other Methods?," Proceedings of the ASCE 2001 Wetlands Engineering and River Restoration Conference, 27-31 August 2001, Reno, Nevada.

Montgomery, D.R. and J.M. Buffington, 1997, "Channel-reach morphology in mountain drainage basins," Geological Society of America Bulletin, Vol. 109, No. 5, p. 596-611.

Montgomery, D.R. and L.H. MacDonald, 2002, "Diagnostic approach to stream channel assessment and monitoring," Journal of the American Water Resources Association, 381, 1-16.

Myers, T.J. and S. Swanson, 1992, "Variation of stream stability with stream type and livestock bank damage in Northern Nevada," Water Resources Bulletin, 284, 743-754.

Myers, T.J. and S. Swanson, 1996, "Temporal and geomorphic variations of stream stability and morphology: Mahogany Creek, Nevada," Water Resources Bulletin, 322, 253-265.

Nagle, G., 2007, "Evaluating 'natural channel design' stream projects," Invited Commentary, Hydrological Processes, Vol. 21, Issue 18.

National Cooperative Highway Research Program, 1970, "Tentative Design Procedure for Riprap-Lined Channels," Highway Research Board, NCHRP Report Number 108, National Academies of Science, Washington, D.C. (Anderson, A.G., A.S. Paintal, and J.T. Davenport).

National Cooperative Highway Research Program, 2004a, "Handbook for Predicting Stream Meander Migration," NCHRP Report 533, Transportation Research Board, National Academies of Science, Washington, D.C. (Lagasse, P.F., Spitz, W.J., Zevenbergen, L.W., and D.W. Zachmann).

National Cooperative Highway Research Program, 2004b, "Methodology for Predicting Channel Migration," Final Report for NCHRP Report Project 24-16, NCHRP Web Document 67, Transportation Research Board, National Academies of Science, Washington, D.C. (Lagasse, P.F., Zevenbergen, L.W., Spitz, W.J., and C.R. Thorne).

National Cooperative Highway Research Program, 2004c, "Pier and Contraction Scour in Cohesive Soils," NCHRP Report 516, Transportation Research Board, National Academies of Science, Washington, D.C. (Briaud, J.L., H.C. Chen, Y. Li, P. Nurtjahyo, and J. Wang).

National Cooperative Highway Research Program, 2005, "Environmentally Sensitive Channel- and Bank-Protection Measures," NCHRP Report 544, Transportation Research Board, National Academies of Science, Washington, D.C., 50 p. (McCullah, J. and D. Gray).

National Cooperative Highway Research Program, 2006, "Riprap Design Criteria, Recommended Specifications, and Quality Control," NCHRP Report 568, Transportation Research Board, National Academies of Science, Washington, D.C., 148 p. (Lagasse, P.F., P.E. Clopper, L.W. Zevenbergen, and J.F. Ruff).

National Cooperative Highway Research Program, 2010, "Effects of Debris on Bridge Pier Scour," NCHRP Report 653, Transportation Research Board, National Academies of Science, Washington, D.C., 115 p. (Lagasse, P.F., P.E. Clopper, L.W. Zevenbergen, W.J. Spitz, and L.G. Girard).

National Cooperative Highway Research Program, 2011, "Scour at Bridge Foundations on Rock," Final Report, NCHRP Project 24-29 (Keaton, J.R., S.K. Mishra, and P.E. Clopper).

National Highway Institute (NHI), 2008, "Managing Road Impacts on Stream Ecosystems: An Interdisciplinary Approach," NHI Course No. 142048A Instructors Guide, National Highway Institute, Federal Highway Administration, Publication No. FHWA-NHI-09-014, Washington, D.C.

National Transportation Safety Board, 1988, "Collapse of the New York Thruway (I-90) Bridge Over Schoharie Creek, near Amsterdam, New York, April 5, 1987," Highway Accident Report No. NTSB/HAR-88/02, Washington, D.C.

National Transportation Safety Board, 1990, "Collapse of the Northbound U.S. Route 51 Bridge Spans over the Hatchie River near Covington, TN, April 1, 1989," Highway Accident Report No. NTSB/HAR-90/01, Washington, D.C.

Natural Resources Conservation Service (NRCS), 2007, "Stream Restoration Design," National Engineering Handbook Part 654 (NEH 654), U.S. Department of Agriculture, Natural Resources Conservation Service, Washington, D.C.

Neill, C.R., (Editor), 1972, "Guide to Bridge Hydraulics," First Edition, prepared by Project Committee on Bridge Hydraulics, Roads and Transportation Association of Canada, University of Toronto Press.

Neill, C.R., (Editor), 2004, "Guide to Bridge Hydraulics," Second Edition, prepared by Project Committee on Bridge Hydraulics, Roads and Transportation Association of Canada (TAC), University of Toronto Press.

Niezgoda, S.L. and P.A. Johnson, 2005, "Improving the urban stream restoration effort: identifying critical form and processes relationships," Environmental Management, 355, 579-592.

Pacific Southwest Interagency Committee (PSIAC), 1968, "Report on Factors Affecting Sediment Yield in the Pacific Southwest Areas," Water Management Subcommittee Sediment Task Force.

Palmer, M.A., E.S. Bernhardt, J.D. Allan, P.S. Lake, G. Alexander, S. Brooks, J. Carr, S. Clayton, C.N. Dahm, J. Follstad-Shah, D.L. Galat, S.G. Loss, P. Goodwin, D.D. Hart, B. Hassett, R. Jenkinson, G.M. Kondolf, R. Lave, J.L. Meyer, T.K. O'Donnell, L. Pagano, and E. Sudduth, 2005, "Standards for ecologically successful river restoration," Journal of Applied Ecology, Vol. 42, No. 2.

Parker, G., P.C. Klingeman, and D.L. McLean,, 1982, "Bedload and Size Distribution in Paved Gravel-Bed Streams," ASCE Journal of Hydraulic Engineering, Vol. 108, pp. 544-571.

Prestegaard, K.L., 1983, "Bar Resistance in Gravel Bed Streams at Bankfull Stage," Water Resources Research, Vol. 19, pp. 472-476.

Radspinner, R.R., P. Diplas, A.F. Lightbody, and F. Sotiropoulos, 2010, "River Training and Ecological Enhancement Potential Using In-Stream Structures," American Society of Civil Engineers, Journal of Hydraulic Engineering, Vol. 136, No. 12.

Rhodes, J. and Trent, R., 1993, "Economics of Floods, Scour, and Bridge Failures, Hydraulic Engineering, ASCE, Proceeding of the 1993 National Conference, v.1, pp. 928-933.

Richards, K.S., 1982, "Rivers: Form and Process in Alluvial Channels," Methuen, London.

Robert, A., 1990, "Boundary Roughness in Coarse-Grained Channels," Progress in Physical Geography, Vol. 14, pp. 42-70.

Rosgen, D.L., 1994, "A classification of natural rivers," Catena, Vol. 22, p. 169-199.

Rosgen, D.L., 1996, "Applied River Morphology," Wildland Hydrology Books, Pagosa Springs, Colorado.

Rosgen, D.L., 1998, "The Reference Reach - A Blueprint for Natural Channel Design," Proceedings of the ASCE 1998 Wetlands Engineering and River Restoration Conference, 22-27 March 1998, Denver, Colorado

Rosgen, D.L., 2001, "The cross vane, W-weir and J-hook structures: Their description, design and application for stream stabilization and river restoration," Proceedings of the ASCE 2001 Wetlands Engineering and River Restoration Conference, 27-31 August 2001, Reno, Nevada.

Rosgen, D.L., 2006a, "River Restoration Using A Geomorphic Approach For Natural Channel Design," Proceedings of the $8^{th}$ Federal Interagency Sedimentation Conference, 2-6 April 2006, Reno, Nevada.

Rosgen, D.L., 2006b, "Watershed Assessment of River Stability & Sediment Supply (WARSSS)," Wildland Hydrology Books, Fort Collins, Colorado.

Rosgen, D.L., 2007, "Chapter 11 – Rosgen Geomorphic Channel Design," USDA Natural Resources Conservation Service, National Engineering Handbook Part 654 (NEH 654), Washington, D.C.

Rosgen, D.L., 2008, Discussion – "Critical Evaluation of How the Rosgen Classification and Associated 'Natural Channel Design' Methods Fail to Integrate and Quantify Fluvial Processes and Channel Response," by Simon, A., Doyle, M., Kondolf, M., Shields, F.D., Jr., Rhoads, B., Grant, G., Fitzpatrick, F., Juracek, K., McPhillips, M. and J. MacBroom, Journal of the American Water Resources Association, Vol. 44, Issue 3.

Rouse, H., 1937, "Modern Conceptions of the Mechanics of Fluid Turbulence," ASCE Trans., Vol. 102, Reston, VA.

Santos-Cayado, J., 1972, "State Determination for High Discharges," Ph.D. Dissertation, Department of Civil Engineering, Colorado State University, Fort Collins, CO.

Schumm, S.A., 1969, "River Metamorphosis," ASCE Journal of the Hydraulics Division, Vol. 95, pp. 255-273.

Schumm, S.A., 1971, "Fluvial Geomorphology: Channel Adjustment and River Metamorphosis," In River Mechanics, Vol. 1, H.W. Shen (ed), pp. 5.1-5.22.

Schumm, S.A., 1977, "The Fluvial System," John Wiley & Sons, New York, 338 pp.

Schumm, S.A., 1981, "Evolution and response of the fluvial system," Sedimentologic Implications: SEPM Special Publication 31, p. 19-29.

Schumm, S.A. and H.R. Kahn, 1972, "Experimental study of channel patterns," Geological Society of America Bulletin, Vol. 83, p. 1755-1770.

Schumm, S.A. and P.F. Lagasse, 1998, "Alluvial Fan Dynamics - Hazards to Highways," ASCE Water Resources Engineering 98, Proceedings of the International Water Resource Engineering Conference, Memphis, TN.

Schumm, S.A., M.D. Harvey, and C.C. Watson, 1984, "Incised channels: Morphology, dynamics and control," Water Resources Publications, Littleton, CO.

Sear, D.A., 1996, "Sediment Transport in Pool-Riffle Sequences," Earth Surface Processes and Landforms, Vol. 20, pp. 629-647.

Shen, H.W. and S. Komura, 1968, "Meandering tendencies in straight alluvial channels," American Society of Civil Engineers, Journal of Hydraulics (HY4), Vol. 94, pp. 997-1016.

Shields, F.D., Jr. 1983, "Design of habitat structures for open channels," Journal of Water Resources Planning and Management, Vol. 109, No. 4.

Shields, F.D., Jr. 1996, "Hydraulic and Hydrologic Stability in River Channel Restoration: Guiding Principles for Sustainable Projects," Eds. A. Brookes and F.D. Shields, Jr., Wiley, NY, Chapter 2.

Shields, F.D., Jr. and R.R. Copeland, 2006, "Empirical and Analytical Approaches for Stream Channel Design," Proceedings of the 8th Federal Interagency Conference, 2-6 April 2006, Reno, Nevada.

Shields, F.D., Jr., R.R. Copeland, P.C. Klingeman, M.W. Doyle, and A. Simon, 2008, "Stream Restoration," In Garcia, M.H., (Ed.), "Sedimentation Engineering – Processes, Measurements, Modeling, and Practice," ASCE Manuals and Reports on Engineering Practice No. 110, Reston, Virginia.

Shields, I.A., 1936, "Application of Similarity Principles and Turbulence Research to Bed-Load Movement," a translation from the German by W.P. Ott and J.C. van Vchelin, U. S. Soil Consdrv. Service Coop. Lab., California Inst. of Tech., Pasadena, CA.

Simon, A., 1989, "A model of channel response in disturbed alluvial channels," Earth Surface Processes and Landforms, Vol. 14, p. 11-26.

Simon, A. and P.W. Downs, 1995, "An interdisciplinary approach to evaluation of potential instability in alluvial channels," Geomorphology, Vol. 12, p. 215-232.

Simon, A., M. Doyle, M. Kondolf, F.D. Shields, Jr., B. Rhoads, G. Grant, F. Fitzpatrick, K. Juracek, M. McPhillips, and J. MacBroom, 2005, "Response to discussion How Well do the Rosgen Classification and Associated 'Natural Channel Design' Methods Integrate and Quantify Fluvial Processes and Channel?," Proceedings of the ASCE 2005 World Water and Environmental Resources Congress, 15-19 May 2005, Anchorage, Alaska.

Simon, A., M. Doyle, M. Kondolf, F.D. Shields, Jr., B. Rhoads, and M. McPhillips, 2007, "Critical Evaluation of How the Rosgen Classification and Associated 'Natural Channel Design' Methods Fail to Integrate and Quantify Fluvial Processes and Channel Response," Journal of the American Water Resources Association, Vol. 43, Issue 5.

Simon, A., M. Doyle, M. Kondolf, F.D. Shields, Jr., B. Rhoads, and M. McPhillips, 2008, Reply to Discussion - "Critical Evaluation of How the Rosgen Classification and Associated 'Natural Channel Design' Methods Fail to Integrate and Quantify Fluvial Processes and Channel Response," by Dave Rosgen, Journal of the American Water Resources Association, Vol. 44, Issue 3.

Simons, D.B. and F. Senturk, 1992, "Sediment Transport Technology," Water Resources Pub., Littleton, CO.

Skidmore, P.B., F.D. Shields, Jr., M.W. Doyle, and D.E. Miller, 2001, "A Categorization of Approaches to Natural Channel Design," Proceedings of the 2001 ASCE Wetlands Engineering and River Restoration Conference, 27-31 August 2001, Reno, Nevada.

Slate, L.O., Shields, F.D., Schwartz, J.S., Carpenter, D.D. and G.E. Freeman, 2007, "Engineering Design Standards and Liability for Stream Channel Restoration," American Society of Civil Engineers, Journal of Hydraulic Engineering, Vol. 133, No.10.

Strahler, A.N., 1965, "Physical Geography, Wiley, New York, 442 p.

Thompson, A., 1986, "Secondary flows and the pool-riffle unit: a case study of the processes of meander development," Earth Surface Processes and Landforms, Vol. 11, pp. 631-641.

Thorne, C.R., 1981, "Field measurements of rates of bank erosion and bank material strength," In: Erosion and Sediment Transport Measurement, IAHS Publication No. 133, pp. 503-512.

Thorne, C.R., 1992, "Bend scour and bank erosion on the meandering Red River, Louisiana," In: P.A. Carling and G.E. Petts (eds), Lowland Floodplain Rivers - Geomorphological Perspectives, Wiley, Chichester, pp. 95-115.

Thorne, C.R., 1997, "Channel types and morphological classification," Chapter 7 in: C.R. Thorne, R.D. Hey, and M.D. Newsom (eds), Applied Fluvial Geomorphology for River Engineering and Management, John Wiley & Sons, Chichester.

Thorne, C.R., 1998, "Stream Reconnaissance Handbook," John Wiley & Sons, Chichester.

Thorne, C.R., R.G. Allen, and A. Simon, 1996, "Geomorphological river channel reconnaissance for river analysis, engineering, and management," Transactions of the Institute of British Geographer, NS 21, 469-483.

U.S. Army Corps of Engineers, 1944, "Geological investigation of the alluvial valley of the lower Mississippi River," U.S. Army Corps of Engineers, Mississippi River Commission, Vicksburg, Mississippi (Fisk, H.N.).

U.S. Army Corps of Engineers, 1947, "Fine grained alluvial deposits and their effects on the Mississippi River activity," U.S. Army Corps of Engineers Waterways Experiment Station, Vicksburg, Mississippi (Fisk, H.N.).

U.S. Army Corps of Engineers, 1957, "A Study of the Shape of Channels Formed by Natural Streams Flowing in Erodible Material," M.R.D. Sediment Series No. 9, U.S. Army Engineers Division, Missouri River, Corps of Engineers, Omaha, NE (Lane, E.W.).

U.S. Army Corps of Engineers, 1994a, "Channel Stability Assessment for Flood Control Projects, EM 1110-2-1418, Washington, D.C.

U.S. Army Corps of Engineers, 1994b, "Application of Channel Stability Methods - Case Studies," TR HL-94-11, USACE Waterways Experiment Station, Vicksburg, MS (Copeland, R.R.).

U.S. Army Corps of Engineers, 1999, "Channel Rehabilitation: Processes, Design, and Implementation (Draft)," U.S. Army Engineer, Engineering Research and Development Center, Vicksburg, MS (Watson, C.C., D.S. Biedenharn, and S.H. Scott).

U.S. Army Corps of Engineers, 2001, "Hydraulic Design of Stream Restoration Projects," Coastal and Hydraulics Laboratory Report No. ERDC/CHL TR-01-28, U.S. Army Corps of Engineers, Washington, D.C. (Copeland, R.R., McComas, D.N., Thorne, C.R., Soar, P.J., Jonas, M.M., and Fripp, J.B.)

U.S. Army Corps of Engineers, 2002, "SAM Hydraulic Design Package for Channels," Coastal and Hydraulics Laboratory, U.S. Army Engineer Research and Development Center, Vicksburg, MS (Thomas, M.A., Copeland, E.R., and McComas, D.N.)

U.S. Army Corps of Engineers, 2010, "River Analysis System," HEC-RAS, User's Manual Version 4.1, Hydrologic Engineering Center, Davis, CA.

U.S. Army Research Office, 1975, "Air photo interpretation of the form and behavior of alluvial rivers," Final report to the U.S. Army Research Office (Brice, J.C.).

U.S. Bureau of Reclamation, 1984, "Computing Degradation and Local Scour," Technical Guideline for Bureau of Reclamation (E.L. Pemberton and J.M. Lara).

U.S. Bureau of Reclamation, 1988, "1986 Lake Powell Survey," U.S. Bureau of Reclamation Report REC-ERC-88-6, Denver, CO (Ferrari, R.L.).

U.S. Bureau of Reclamation, 2007, "Qualitative Evaluation of Rock Weir Field Performance and Failure Mechanisms," U.S. Department of the Interior, Bureau of Reclamation, Technical Service Center, Sedimentation and River Hydraulics Group, Denver, CO (Mooney, D.M., C.L. Holmquist-Johnson, and E. Holburn).

U.S. Bureau of Reclamation, 2009, "Quantitative Investigation of Field Performance of Rock Weirs," U.S. Department of the Interior, Bureau of Reclamation, Technical Service Center, Sedimentation and River Hydraulics Group, SRH-2009-46, Denver, CO (Holburn, E., D. Varyu, and K. Russell).

U.S. Department of Agriculture, 1978, "Predicting Rainfall Erosion Losses," Agricultural Handbook 537, Science and Education Administration, USDA (Wischmeier, W.H. and D.D. Smith).

U.S. Department of Agriculture, 2001, "Sampling Surface and Subsurface Particle-Size Distributions in Wadable Gravel- and Cobble-Bed Streams for Analyses in Sediment Transport, Hydraulics, and Streambed Monitoring," USDA Forest Service, Rocky Mountain Research Station, General Technical Report RMRS-GTR-74, 428 p. (Bunte, K. and S.R. Abt).

U.S. Department of Agriculture, 2009a, "Sediment Transport Primer Estimating Bed-Material Transport in Gravel-Bed Rivers," USDA Forest Service, Rocky Mountain Research Station, General Technical Report RMRS-GTR-226, 78 p. (Wilcock, P., J. Pitlick, and C. Yantao).

U.S. Department of Agriculture, 2009b, "Manual for Computing Bed Load Transport Using BAGS (Bedload Assessment for Gravel-bed Streams) Software," USDA Forest Service, Rocky Mountain Research Station, General Technical Report RMRS-GTR-223, 45 p. (Pitlick, J., Y. Cui, and P. Wilcock).

U.S. Department of Transportation, Federal Highway Administration, 1988, "Scour at Bridges," Technical Advisory T5140.20, updated by Technical Advisory T5140.23, October 28, 1991, "Evaluating Scour at Bridges," U.S. Department of Transportation, Washington, D.C.

U.S. Department of Transportation, Federal Highway Administration, 2001, "Revision of Coding Guide, Item 113 - Scour Critical Bridges," Memorandum, HIBT-30, April 27, Washington, D.C.

U.S. Department of Transportation, Federal Highway Administration, 2003, "Compliance with National Bridge Inspection Standards; Plan of Action for Scour Critical Bridges," Memorandum, HIBT-20, July 24, Washington, D.C.

U.S. Department of Transportation, Federal Highway Administration, 2004, "National Bridge Inspection Standards," Federal Register, Volume 69, No. 239, 23CFR Part 650, FHWA Docket No. FHWA-2001-8954, Final Rule, December 14, 2004, effective January 13, 2005, Washington, D.C.

USDA Forest Service, 1978, "Stream Reach Inventory and Channel Stability Evaluation," Unpublished Report, U.S. Department of Agriculture, Forest Service, Northern Region, 25 pp. (Pfankuch, D.J.).

U.S. Environmental Protection Agency (USEPA), 2006, "WARSSS - Watershed Assessment of River Stability & Sediment Supply," EPA, Washington, D.C.

U.S. Geological Survey, 1953, "The Hydraulic Geometry of Stream Channels and Some Physiographic Implications," USGS Professional Paper 252, Reston, VA, 57 p. (Leopold, L.B. and T. Maddock, Jr.).

U.S. Geological Survey, 1957a, "River channel patterns – braided, meandering, and straight," U.S. Geological Survey Professional Paper 282-B, pp. 39-85 (Leopold, L.B. and M.G. Wolman).

U.S. Geological Survey, 1957b, "River Flood Plains: Some Observations on Their Formation," U.S. Geological Survey Professional Paper 282-C, pp. 87-107 (Wolman. M.G. and L.B. Leopold).

U.S. Geological Survey, 1960, "Some aspects of the shape of river meanders," U.S. Geological Survey Professional Paper 282-E (Bagnold, R.A.).

U.S. Geological Survey, 1964, "Aspects of Flow Resistance and Sediment Transport, Rio Grande near Bernalillo, New Mexico," USGS Water Supply Paper 1498-4, 41 p. (Nordin, C.F.).

U.S. Geological Survey, 1966a, "River meanders – theory of minimum variance," U.S. Geological Survey Professional Paper 422-H (Langbein, W.B. and L.B. Leopold).

U.S. Geological Survey, 1966b, "Channel and Hillslope Processes in a Semiarid Area, New Mexico," U.S. Geological Survey Professional Paper 352-G, U.S. Geological Survey, Washington, D.C. (Leopold, L.B., W.W. Emmett, and R.M. Myrick).

U.S. Geological Survey, 1966c, "An approach to the sediment transport problem from general physics," U.S. Geological Survey Professional Paper 422-I (Bagnold, R.A.).

U.S. Geological Survey, 1967, "Roughness Characteristics of Natural Channels," U.S. Geological Survey Water Supply Paper 1849, Washington, D.C. (Barnes, H.H.).

U.S. Geological Survey, 1968, "River adjustment to altered hydrologic regime – Murrumbidgee River and paleochannels, Australia," U.S. Geological Survey Professional Paper 598 (Schumm, S.A.).

U.S. Geological Survey, 1970, "Determination of the Manning's Coefficient for Measured Bed Roughness in Natural Channels," U.S. Geological Survey Water Supply Paper 1891-B (Limerinos, J.T.).

U.S. Geological Survey, 1982, "Measurement and Computation of Streamflow: Volume 2." Computation of Discharge, U.S. Geological Survey Water Supply Paper 2175, Washington, D.C. (Rantz, S.E. et al.).

U.S. Geological Survey, 1984, "Guide for Selecting Manning's Roughness Coefficients for Natural Channels and Flood Plains," U.S. Geological Survey Water Supply Paper 2339, Washington, D.C. (Arcement, G.K. and V.R. Schneider).

U.S. Geological Survey, 1992, "Geomorphic and Vegetative Recovery Processes along Modified Stream Channels of West Tennessee," USGS Open File Report 91-502, Nashville, Tennessee (Simon, A., and Hupp, C.R.).

U.S. Geological Survey, 1995, "Potential-Scour Assessments and Estimates of Maximum Scour at Selected Bridges in Iowa," Water Resources Investigations Report 95-4051, prepared in cooperation with the Iowa Highway Research Board and the Highway Division of the Iowa Department of Transportation, Washington, D.C.

U.S. Geological Survey, 1997, "Resistance, Sediment Transport, and Bedform Geometry Relationships in Sand-Bed Channels," in: Proceedings of the U.S. Geological Survey (USGS) Wediment Workshop, February 4-7 (Bennett, J.P.).

Vermont Water Quality Division 2001, "Provisional Hydraulic Geometry Curves," Vermont Department of Environmental Conservation, Waterbury, VT.

Water Resources Council, Hydrology Committee, 1981, "Guidelines for Determining Flood Frequency," <u>Bulletin 17B</u>, U.S. Water Resources Council, Washington, D.C.

Williams, G.P., 1978, "Bankfull Discharges of Rivers," Water Resources Research, Vol. 14, pp.1141-1154.

Wolman, M.G. and J.P. Miller, 1960, "Magnitude and Frequency of Forces in Geomorphic Processes," Journal of Geology, Vol. 68, pp. 54-74.

Woodward, S.M., 1920, "Hydraulics of the Miami Flood Control Project," Technical Reports, Pt. VII, Miami Conservancy District, Dayton, OH.

Yalin, M.S., 1971, "On the formation of dunes and meanders," Proceedings of the $14^{th}$ International Congress of the International Association for Hydraulic Research, Vol. 3, Paper C13, pp. 1-8.

Yalin, M.S., 1992, "River Mechanics," Pergamon, Oxford.

Yang, C.T., 1996, "Sediment Transport: Theory and Practice," B.J. Clark and J.M. Morris (eds), McGraw-Hill Companies, Inc.

Yang, C.T., 2003, "Sediment Transport: Theory and Practice," Krieger Publishing Company.

# APPENDIX A

# METRIC SYSTEM, CONVERSION FACTORS, AND WATER PROPERTIES

(page intentionally left blank)

# APPENDIX A

## Metric System, Conversion Factors, and Water Properties

The following information is summarized from the Federal Highway Administration, National Highway Institute (NHI) Course No. 12301, "Metric (SI) Training for Highway Agencies." For additional information, refer to the Participant Notebook for NHI Course No. 12301.

In SI there are seven base units, many derived units and two supplemental units (Table A.1). Base units uniquely describe a property requiring measurement. One of the most common units in civil engineering is length, with a base unit of meters in SI. Decimal multiples of meter include the kilometer (1000m), the centimeter (1m/100) and the millimeter (1 m/1000). The second base unit relevant to highway applications is the kilogram, a measure of mass which is the inertial of an object. There is a subtle difference between mass and weight. In SI, mass is a base unit, while weight is a derived quantity related to mass and the acceleration of gravity, sometimes referred to as the force of gravity. In SI the unit of mass is the kilogram and the unit of weight/force is the newton. Table A.2 illustrates the relationship of mass and weight. The unit of time is the same in SI as in the English system (seconds). The measurement of temperature is Centigrade. The following equation converts Fahrenheit temperatures to Centigrade, $°C = 5/9 (°F - 32)$.

Derived units are formed by combining base units to express other characteristics. Common derived units in highway drainage engineering include area, volume, velocity, and density. Some derived units have special names (Table A.3).

Table A.4 provides useful conversion factors from English to SI units. The symbols used in this table for metric units, including the use of upper and lower case (e.g., kilometer is "km" and a newton is "N") are the standards that should be followed. Table A.5 provides the standard SI prefixes and their definitions.

Table A.6 provides physical properties of water at atmospheric pressure in SI system of units. Table A.7 gives the sediment grade scale and Table A.8 gives some common equivalent hydraulic units.

| Table A.1. Overview of SI Units. ||||
|---|---|---|---|
| | Base Units | Units | Symbol |
| Base units | length | meter | m |
| | mass | kilogram | kg |
| | time | second | s |
| | temperature* | kelvin | K |
| | electrical current | ampere | A |
| | luminous intensity | candela | cd |
| | amount of material | mole | mol |
| Supplementary units | angles in the plane | radian | rad |
| | solid angles | steradian | sr |
| *Use degrees Celsius (°C), which has a more common usage than kelvin. ||||

| Table A.2. Relationship of Mass and Weight. ||||
|---|---|---|---|
| System | Mass | Weight or Force of Gravity | Force |
| English | slug, pound-mass | pound, pound-force | pound, pound-force |
| metric | kilogram | newton | newton |

| Table A.3. Derived Units With Special Names. | | | |
|---|---|---|---|
| Quantity | Name | Symbol | Expression |
| Frequency | hertz | Hz | $s^{-1}$ |
| Force | newton | N | kg • m/s$^2$ |
| Pressure, stress | pascal | Pa | N/m$^2$ |
| Energy, work, quantity of heat | joule | J | N • m |
| Power, radiant flux | watt | W | J/s |
| Electric charge, quantity | coulomb | C | A • s |
| Electric potential | volt | V | W/A |
| Capacitance | farad | F | C/V |
| Electric resistance | ohm | Ω | V/A |
| Electric conductance | siemens | S | A/V |
| Magnetic flux | weber | Wb | V • s |
| Magnetic flux density | tesla | T | Wb/m$^2$ |
| Inductance | henry | H | Wb/A |
| Luminous flux | lumen | lm | cd • sr |
| Illuminance | lux | lx | lm/m$^2$ |

Table A.4. Useful Conversion Factors.

| Quantity | From English Units | To Metric Units | Multiply by * |
|---|---|---|---|
| Length | mile | km | 1.609 |
| | yard | m | 0.9144 |
| | foot | m | 0.3048 |
| | inch | mm | 25.40 |
| Area | square mile | $km^2$ | 2.590 |
| | acre | $m^2$ | 4047 |
| | acre | hectare | 0.4047 |
| | square yard | $m^2$ | 0.8361 |
| | square foot | $m^2$ | 0.09290 |
| | square inch | $mm^2$ | 645.2 |
| Volume | acre foot | $m^3$ | 1233 |
| | cubic yard | $m^3$ | 0.7646 |
| | cubic foot | $m^3$ | 0.02832 |
| | cubic foot | L (1000 $cm^3$) | 28.32 |
| | 100 board feet | $m^3$ | 0.2360 |
| | gallon | L (1000 $cm^3$) | 3.785 |
| | cubic inch | $cm^3$ | 16.39 |
| Mass | lb | kg | 0.4536 |
| | kip (1000 lb) | metric ton (1000 kg) | 0.4536 |
| Mass/unit length | plf | kg/m | 1.488 |
| Mass/unit area | psf | $kg/m^2$ | 4.882 |
| Mass density | pcf | $kg/m^3$ | 16.02 |
| Force | lb | N | 4.448 |
| | kip | kN | 4.448 |
| Force/unit length | plf | N/m | 14.59 |
| | klf | kN/m | 14.59 |
| Pressure, stress, modulus of elasticity | psf | Pa | 47.88 |
| | ksf | kPa | 47.88 |
| | psi | kPa | 6.895 |
| | ksi | MPa | 6.895 |
| Bending moment, torque | ft-lb | N·m | 1.356 |
| | ft-kip | kN·m | 1.356 |
| Moment of mass | lb·ft | m | 0.1383 |
| Moment of inertia | lb·$ft^2$ | kg·$m^2$ | 0.04214 |
| Second moment of area | $in^4$ | $mm^4$ | 416200 |
| Section modulus | $in^3$ | $mm^3$ | 16390 |
| Power | ton (refrig) | kW | 3.517 |
| | Btu/s | kW | 1.054 |
| | hp (electric) | W | 745.7 |
| | Btu/h | W | 0.2931 |

*4 significant figures; underline denotes exact conversion

| Table A.4. Useful Conversion Factors (continued). | | | |
|---|---|---|---|
| Quantity | From English Units | To Metric Units | Multiply by * |
| Volume rate of flow | ft$^3$/s | m$^3$/s | 0.02832 |
|  | cfm | m$^3$/s | 0.0004719 |
|  | cfm | L/s | 0.4719 |
|  | mgd | m$^3$/s | 0.0438 |
| Velocity, speed | ft/s | m/s | <u>0.3048</u> |
| Acceleration | f/s$^2$ | m/s$^2$ | <u>0.3048</u> |
| Momentum | lb · ft/sec | kg · m/s | 0.1383 |
| Angular momentum | lb · ft$^2$/s | kg · m$^2$/s | 0.04214 |
| Plane angle | degree | rad | 0.01745 |
|  | degree | mrad | 17.45 |
| *4 significant figures; underline denotes exact conversion | | | |

| Table A.5. Prefixes. | | | | | |
|---|---|---|---|---|---|
| Submultiple Name | Submultiple Factor | Submultiple Symbol | Multiple Name | Multiple Factor | Multiple Symbol |
| deci | $10^{-1}$ | d | deka | $10^1$ | da |
| centi | $10^{-2}$ | c | hecto | $10^2$ | h |
| milli | $10^{-3}$ | m | kilo | $10^3$ | k |
| micro | $10^{-6}$ | μ | mega | $10^6$ | M |
| nano | $10^{-9}$ | n | giga | $10^9$ | G |
| pica | $10^{-12}$ | p | tera | $10^{12}$ | T |
| femto | $10^{-15}$ | f | peta | $10^{15}$ | P |
| atto | $10^{-18}$ | a | exa | $10^{18}$ | E |
| zepto | $10^{-21}$ | z | zetta | $10^{21}$ | Z |
| yocto | $10^{-24}$ | y | yotto | $10^{24}$ | Y |

| Table A.6. Physical Properties of Water at Atmospheric Pressure in SI Units. | | | | | | | | |
|---|---|---|---|---|---|---|---|---|
| Temperature | Temperature | Density | Specific weight | Dynamic Viscosity | Kinematic Viscosity | Vapor Pressure | Surface Tension[1] | Bulk Modulus |
| Centigrade | Fahrenheit | kg/m$^3$ | N/m$^3$ | N·s/m$^2$ | m$^2$/s | N/m$^2$ abs. | N/m | GN/m$^2$ |
| 0 | 32 | 1,000 | 9,810 | 1.79 x 10$^{-3}$ | 1.79 x 10$^{-6}$ | 611 | 0.0756 | 1.99 |
| 5 | 41 | 1,000 | 9,810 | 1.51 x 10$^{-3}$ | 1.51 x 10$^{-6}$ | 872 | 0.0749 | 2.05 |
| 10 | 50 | 1,000 | 9,810 | 1.31 x 10$^{-3}$ | 1.31 x 10$^{-6}$ | 1,230 | 0.0742 | 2.11 |
| 15 | 59 | 999 | 9,800 | 1.14 x 10$^{-3}$ | 1.14 x 10$^{-6}$ | 1,700 | 0.0735 | 2.16 |
| 20 | 68 | 996 | 9,790 | 1.00 x 10$^{-3}$ | 1.00 x 10$^{-6}$ | 2,340 | 0.0728 | 2.20 |
| 25 | 77 | 997 | 9,781 | 8.91 x 10$^{-4}$ | 8.94 x 10$^{-7}$ | 3,170 | 0.0720 | 2.23 |
| 30 | 86 | 996 | 9,771 | 7.97 x 10$^{-4}$ | 8.00 x 10$^{-7}$ | 4,250 | 0.0712 | 2.25 |
| 35 | 95 | 994 | 9,751 | 7.20 x 10$^{-4}$ | 7.24 x 10$^{-7}$ | 5,630 | 0.0704 | 2.27 |
| 40 | 104 | 992 | 9,732 | 8.53 x 10$^{-4}$ | 6.58 x 10$^{-7}$ | 7,380 | 0.0696 | 2.28 |
| 50 | 122 | 988 | 9,693 | 5.47 x 10$^{-4}$ | 5.53 x 10$^{-7}$ | 12,300 | 0.0679 | |
| 60 | 140 | 983 | 9,843 | 4.68 x 10$^{-4}$ | 4.74 x 10$^{-7}$ | 20,000 | 0.0662 | |
| 70 | 158 | 978 | 9,694 | 4.04 x 10$^{-4}$ | 4.13 x 10$^{-7}$ | 31,200 | 0.0644 | |
| 80 | 176 | 972 | 9,535 | 3.54 x 10$^{-4}$ | 3.64 x 10$^{-7}$ | 47,400 | 0.0626 | |
| 90 | 194 | 965 | 9,467 | 3.15 x 10$^{-4}$ | 3.26 x 10$^{-7}$ | 70,100 | 0.0607 | |
| 100 | 212 | 958 | 9,398 | 2.82 x 10$^{-4}$ | 2.94 x 10$^{-7}$ | 101,300 | 0.0589 | |

[1]Surface tension of water in contact with air

| Table A.7. Physical Properties of Water at Atmospheric Pressure in English Units. | | | | | | | | |
|---|---|---|---|---|---|---|---|---|
| Temperature | Temperature | Density | Specific Weight | Dynamic Viscosity | Kinematic Viscosity | Vapor Pressure | Surface Tension[1] | Bulk Modulus |
| Fahrenheit | Centigrade | Slugs/ft$^3$ | Weight lb/ft$^3$ | lb-sec/ft$^2$ x 10$^{-4}$ | ft$^2$/sec x 10$^{-5}$ | lb/in$^2$ | lb/ft | lb/in$^2$ |
| 32 | 0 | 1.940 | 62.416 | 0.374 | 1.93 | 0.09 | 0.00518 | 287,000 |
| 39.2 | 4.0 | 1.940 | 62.424 | | | | | |
| 40 | 4.4 | 1.940 | 62.423 | 0.323 | 1.67 | 0.12 | 0.00514 | 296,000 |
| 50 | 10.0 | 1.940 | 62.408 | 0.273 | 1.41 | 0.18 | 0.00508 | 305,000 |
| 60 | 15.6 | 1.939 | 62.366 | 0.235 | 1.21 | 0.26 | 0.00504 | 313,000 |
| 70 | 21.1 | 1.936 | 62.300 | 0.205 | 1.06 | 0.36 | 0.00497 | 319,000 |
| 80 | 26.7 | 1.934 | 62.217 | 0.180 | 0.929 | 0.51 | 0.00492 | 325,000 |
| 90 | 32.2 | 1.931 | 62.118 | 0.160 | 0.828 | 0.70 | 0.00486 | 329,000 |
| 100 | 37.8 | 1.927 | 61.998 | 0.143 | 0.741 | 0.95 | 0.00479 | 331,000 |
| 120 | 48.9 | 1.918 | 61.719 | 0.117 | 0.610 | 1.69 | 0.00466 | 332,000 |
| 140 | 60.0 | 1.908 | 61.386 | 0.0979 | 0.513 | 2.89 | | |
| 160 | 71.1 | 1.896 | 61.006 | 0.0835 | 0.440 | 4.74 | | |
| 180 | 82.2 | 1.883 | 60.586 | 0.0726 | 0.385 | 7.51 | | |
| 200 | 93.3 | 1.869 | 60.135 | 0.0637 | 0.341 | 11.52 | | |
| 212 | 100 | 1.847 | 59.843 | 0.0593 | 0.319 | 14.70 | | |

[1]Surface tension of water in contact with air

| Table A.8. Sediment Particles Grade Scale. ||||||| 
| Size |||| Approximate Sieve Mesh Openings Per Inch || Class |
| Millimeters | Millimeters | Microns | Inches | Tyler | U.S. Standard | Name |
|---|---|---|---|---|---|---|
| 4000-2000 | | | 160-80 | | | Very large boulders |
| 2000-1000 | | | 80-40 | | | Large boulders |
| 1000-500 | | | 40-20 | | | Medium boulders |
| 500-250 | | | 20-10 | | | Small boulders |
| 250-130 | | | 10-5 | | | Large cobbles |
| 130-64 | | | 5-2.5 | | | Small cobbles |
| 64-32 | | | 2.5-1.3 | | | Very coarse gravel |
| 32-16 | | | 1.3-0.6 | | | Coarse gravel |
| 16-8 | | | 0.6-0.3 | 2.5 | | Medium gravel |
| 8-4 | | | 0.3-0.16 | 5 | 5 | Fine gravel |
| 4-2 | | | 0.16-0.08 | 9 | 10 | Very fine gravel |
| 2-1 | 2.00-1.00 | 2000-1000 | | 16 | 18 | Very coarse sand |
| 1-1/2 | 1.00-0.50 | 1000-500 | | 32 | 35 | Coarse sand |
| 1/2-1/4 | 0.50-0.25 | 500-250 | | 60 | 60 | Medium sand |
| 1/4-1/8 | 0.25-0.125 | 250-125 | | 115 | 120 | Fine sand |
| 1/8-1/16 | 0.125-0.062 | 125-62 | | 250 | 230 | Very fine sand |
| 1/16-1/32 | 0062-0031 | 62-31 | | | | Coarse silt |
| 1/32-1/64 | 0.031-0.016 | 31-16 | | | | Medium silt |
| 1/64-1/128 | 0.016-0.008 | 16-8 | | | | Fine silt |
| 1/128-1/256 | 0.008-0.004 | 8-4 | | | | Very fine silt |
| 1/256-1/512 | 0.004-0.0020 | 4-2 | | | | Coarse clay |
| 1/512-1/1024 | 0.0020-0.0010 | 2-1 | | | | Medium clay |
| 1/1024-1/2048 | 0.0010-0.0005 | 1-0.5 | | | | Fine clay |
| 1/2048-1/4096 | 0.0005-0.0002 | 0.5-0.24 | | | | Very fine clay |

| Table A.9. Common Equivalent Hydraulic Units. | | | | | | | | |
|---|---|---|---|---|---|---|---|---|
| Volume | | | | | | | | |
| Unit | cubic inch | liter | U.S. gallon | cubic foot | cubic yard | cubic meter | acre-foot | sec-foot-day |
| liter | 61.02 | 1 | 0.264 2 | 0.035 31 | 0.001 31 | 0.001 | 810.6 E-9 | 408.7 E-9 |
| U.S. gallon | 231 | 3.785 | 1 | 0.133 7 | 0.004 95 | 0.003 79 | 3.068 E-6 | 1.547 E-6 |
| cubic foot | 1,728 | 28.32 | 7.481 | 1 | 0.037 04 | 0.028 32 | 22.96 E-6 | 11.57 E-6 |
| cubic yard | 46,660 | 764.6 | 202 | 27 | 1 | 0.746 60 | 619.8 E-6 | 312.5 E-6 |
| meter$^3$ | 61,020 | 1,000 | 264.2 | 35.31 | 1.308 | 1 | 810.6 E-6 | 408.7 E-6 |
| acre-foot | 75.27 E+6 | 1,233,000 | 325,900 | 43,560 | 1,613 | 1,233 | 1 | 0.5042 |
| sec-foot-day | 149.3 E+6 | 2,447,000 | 646,400 | 86,400 | 3,200 | 2,447 | 1.983 | 1 |
| Discharge (Flow Rate, Volume/Time) | | | | | | | | |
| Unit | | gallon/min | liter/sec | acre-foot/day | foot$^3$/sec | million gal/day | meter$^3$/sec | |
| gallon/minute | | 1 | 0.063 09 | 0.004 419 | 0.002 228 | 0.001 440 | 63.09 E-06 | |
| liter/second | | 15.85 | 1 | 0.070 05 | 0.035 31 | 0.022 82 | 0.001 | |
| acre-foot/day | | 226.3 | 14.28 | 1 | 0.504 2 | 0.325 9 | 0.014 28 | |
| feet$^3$/second | | 448.8 | 28.32 | 1.983 | 1 | 0.646 3 | 0.028 32 | |
| million gal/day | | 694.4 | 43.81 | 3.068 | 1.547 | 1 | 0.043 82 | |
| meter$^3$/second | | 15,850 | 1,000 | 70.04 | 35.31 | 22.82 | 1 | |

# APPENDIX B

# BANK EROSION AND FAILURE MECHANISMS

(page intentionally left blank)

# APPENDIX B

## Bank Erosion and Failure Mechanisms

### B.1 FACTORS INFLUENCING BANK RETREAT

The erosion, instability, and/or retreat of a stream bank is dependent on the processes responsible for the erosion of material from the bank and the mechanisms of failure resulting from the instability created by those processes. Bank retreat is often a combination of these processes and mechanisms varying at seasonal and sub-seasonal timescales. Bank retreat processes may be grouped into three categories: weakening and weathering processes, direct fluvial entrainment, and mass failure. The general factors which influence the various bank retreat processes and mechanisms are shown in Table B.1. The impact of these processes on bank retreat is dependent on site characteristics, especially near-bank hydraulic conditions, bank height, and the geotechnical properties of the bank material. As indicated in Chapter 2, the resistance of a stream bank to erosion and failure is closely related to several characteristics of the bank material, which can be broadly classified as noncohesive, cohesive, or composite.

| Table B.1. Factors Influencing Bank Retreat Processes and Mechanisms (after Lawler et al. 1997). ||
|---|---|
| Subareal Processes | Microclimate, especially temperature |
|  | Bank composition, especially silt/clay percentage |
| Fluvial Processes | Stream power |
|  | Shear stress |
|  | Secondary currents |
|  | Local slope |
|  | Bend morphology |
|  | Bank composition |
|  | Vegetation |
|  | Bank moisture content |
| Mass Failure | Bank Height |
|  | Bank angle |
|  | Bank composition |
|  | Bank moisture content or pore water pressure/tension |

### B.2 PROCESSES OF WEAKENING AND WEATHERING

The processes of weakening and weathering reduce the strength of intact bank material and decrease bank stability. Mass wasting of bank materials is related to these processes, which in turn are associated directly with soil moisture conditions (Thorne 1982, Hagerty et al. 1983). The processes, which depend on both climatic conditions and on the properties of the bank, fall into two groups: those operating within the bank to reduce its strength, and those acting on the bank surface to loosen and detach particles or aggregates.

#### B.2.1 Strength Reduction

The effective strength of poorly drained banks can be reduced by positive pore-water pressure. The most critical condition occurs during heavy or prolonged precipitation, snowmelt runoff, or rapid drawdown after a high flow stage. Positive pore water pressures in a bank act to reduce friction and effective cohesion, which can lead to liquefaction (a complete loss of strength and flow-type failure) in extreme cases. Even if no significant pore water pressures develop, the stability of a saturated bank will be reduced due to the increase in unit weight that results from saturation.

Cycles of wetting and drying cause shrinkage and swelling of the soil material, which leads to the development of micro-failure planes, desiccation cracks, and downslope soil creep. Freezing and thawing of water in pores, cracks, or fissures can break soil units apart and weaken bank material by reducing granular interlocking and, hence, the friction angle, and by destroying any cohesion. A similar effect can be created by the relaxation of normal load and lateral earth pressure due to lateral stream cutting or overburden removal. Movement of water through the bank can lead to leaching of clay particles by solution or suspension and softening of the bank material, thereby causing a reduction in bank material cohesion.

### B.2.2 Surface Erosion

Overland flow occurs when bank materials become fully saturated or when the rate of precipitation locally exceeds the infiltration capacity of the soil mass. In turn, this can lead to surface erosion of the bank through the processes of sheet erosion, rilling, and gullying. Similar types of erosion can occur as a result of return flow from flooded overbank and floodplain areas. In addition, the importance of these surface erosion processes is largely dependent on the vegetative cover of the bank. The presence of dense riparian vegetation can reduce surface erosion rates by several orders of magnitude when compared to non-vegetated banks (Thorne 1982). However, the introduction of trees into the channel from bank failures can cause local scour and significant additional local bank erosion.

### B.3. FLUVIAL ENTRAINMENT

Bank retreat is produced by fluvial entrainment in two ways. First, sediment may be directly entrained from the bank (by detaching and/or moving grains or aggregates) and transported downstream. Second, flow may scour the channel bed at the base of the bank (increasing bank angle and height) and induce the gravitational failure of the bank. This type of failure mechanism is probably of greatest importance when the banks are located on the concave (outside) margin of a bend where scour depths during a flood may range from 1.75 to 2 times the depth of flow in sand-bed streams (Wolman and Leopold 1957).

Shear stress along the bed and banks as generated by flow in the channel is directly proportional to the velocity gradient close to the channel boundary. In order for the boundary material to remain in equilibrium it must supply an internally derived, equal and opposite shear strength. If the velocity gradient becomes steeper, a point is eventually reached where the internal shear strength (the resistance to motion of the boundary material) equals the fluid shear stress. Any subsequent increase in the fluid shear stress results in entrainment of the boundary material.

### B.3.1 Noncohesive Material

Individual grains from noncohesive materials are entrained by pivoting, rolling, or sliding. The stability of the surface grain can be assessed by resolving the forces which act on the grain into those that tend to cause motion and those that tend to resist it (Lane 1955). However, Parker has shown that this approach is inappropriate in gravel-bed streams that are transporting significant amounts of bedload (Parker 1979). The Task Committee on Sedimentation (ASCE 1966) developed a relationship that was dependent on defining the critical boundary shear stress, but, as Thorne indicates, this method has limited usefulness because of the stochastic nature of the distribution and fluctuation of shear stress and particle size distributions (Thorne 1982).

### B.3.2 Cohesive Material

The mechanics of fluvial entrainment of cohesive bank material are even less well understood. Tractive stress approaches have been attempted, but these suffer from the fact that little consideration has been given to the nature of the soil unit which is entrained or to the mechanism of failure at the time of entrainment (Thorne 1982). Further, delineation of materials as cohesive on the basis of grain-size distributions may be misleading because most fine-grained cohesive materials form very strongly bonded aggregates, which are composed of clay, silt and sand. In fact, many fine-grained aggregate particles can behave as low density sand and gravel particles (Barden 1972, Nanson et al. 1986). Thus, fluvial erosion of cohesive soil often occurs through entrainment of aggregates rather than discrete particles.

### B.4. BASAL ENDPOINT CONTROL

Material is delivered to the basal area of a bank by mechanical bank failures and erosion. The removal of this material from the basal area depends almost entirely on fluvial entrainment and downstream transport (Figure B.1). The amount of basal accumulation of bank material depends on the relative rates of supply by bank failures and erosion and removal by fluvial entrainment. Where the flow is able to remove all the sediment supplied to the basal area and scour of the basal area continues, bank erosion will also continue. In contrast, where the rate of supply exceeds the rate of removal, bank stability will be increased with respect to gravity failures because loading and buttressing the base of the slope effectively reduces the bank angle and height. Neill (1984) has argued that the bedload transport rate must set an upper limit to local erosion rates over a period of time, and Nanson and Hickin (1986) support this view. Carson and Kirkby (1972) characterize the balance between basal supply and removal in terms of three states of basal endpoint control, as follows:

(a) <u>Impeded Removal</u>. If bank failures supply material to the base at a higher rate than it is removed, then basal accumulation results, thus decreasing the bank angle and vertical height and increasing bank stability.

(b) <u>Unimpeded Removal</u>. Bank failures and erosion supply material to the base at the same rate that it is removed resulting in bank recession by parallel retreat, the rate being controlled by the degree of fluvial activity at the base of the bank. Slope angle and basal elevation remain relatively unchanged.

(c) <u>Excess Basal Capacity</u>. Basal scour is greater than the rate of supply of material. This causes bed scour and basal lowering which increases the bank height and angle and promotes bank failure.

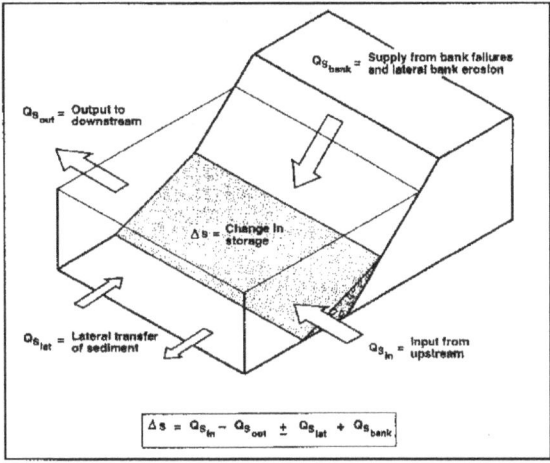

Figure B.1. Schematic representation of sediment fluxes to and from river bank basal zones (Thorne and Osman 1988).

## B.5 MECHANICS OF BANK FAILURE

The mechanics of bank failure, which result from the operation of the processes of erosion as outlined above are closely related to the size, geometry, and stratigraphy of the banks and to the geotechnical properties of the bank material. Based on the stratigraphy and physical properties, banks can be classified as noncohesive, cohesive, and composite as described in Chapter 2. Hey et al. (1991) compiled a useful summary of bank failure modes and characteristics which is shown in Figure B.2.

### B.5.1 Noncohesive Banks

The shear strength (s) of noncohesive banks can be described by the modified Coulomb equation with no cohesion which accounts for the normal stress on the bank ($\sigma$), pore water pressure of the bank ($\mu$), and the apparent angle of internal friction of the bank material ($\phi'$) (Terzaghi 1943). The effect of pore water pressure on the shear strength of noncohesive banks is dependent on whether the banks are drained, undrained, or submerged.

Under drained conditions, pore water pressure is not a factor and, therefore, the stability of the bank becomes dependent only on the bank angle and the angle of internal friction (Taylor 1948). Failure occurs by dislodgment of individual grains from the surface of the bank or by shallow slip along a plane or slightly curved surface (Figure B.2a). Deep-seated failures are rare in noncohesive banks because the shear strength increases with depth more rapidly than shear stress (Selby 1982). Weakening and weathering processes can act to decrease packing densities and granular interlocking, thus reducing the friction angle to less than the slope angle, thereby resulting in failure. The friction angles for loosely-packed, noncohesive materials range from 20 to 35 degrees. Erosion of the lower part of a noncohesive bank or the bed adjacent to the bank can cause bank oversteepening, which results in slip failures higher up the bank (Figure B.2b).

Under undrained conditions the shear strength of noncohesive banks is significantly affected by pore water pressure. A positive pore water pressure, which may occur during rapid drawdown, results in a limiting slope angle that is smaller than the friction angle. If the bank is partially saturated the pore water pressure is negative allowing the bank angle to exceed the friction angle. The noncohesive materials can behave like weakly cohesive soils under this condition due to the capillary effects in partly filled pores (Thorne 1982). This condition disappears if the material is completely dry or fully saturated.

Undrained noncohesive banks fail in a manner similar to drained ones with the added effects of positive pore water pressure. Shallow slips and individual grain detachment are the common modes of failure. Piping in the lower bank, caused by high seepage pressures, can cause failure higher up the bank due to oversteepening. In addition, fully saturated, loosely packed, cohesionless materials may fail by liquifaction.

### B.5.2 Cohesive Banks

The shear strength for cohesive banks in the modified Coulomb equation increases by adding the cohesion (c') of the bank material. Unlike noncohesive banks where stability is independent of bank height, both the bank angle and height determine the stability of cohesive banks. Failure mechanisms in cohesive materials fall into three categories: rotational slip, shallow slip, and plane slip. Although shallow slips do occur in cohesive material (Thorne 1982), failure generally occurs by deep-seated slip because the strength of cohesive materials increases at a lesser rate with depth than does shear stress (Terzaghi 1943).

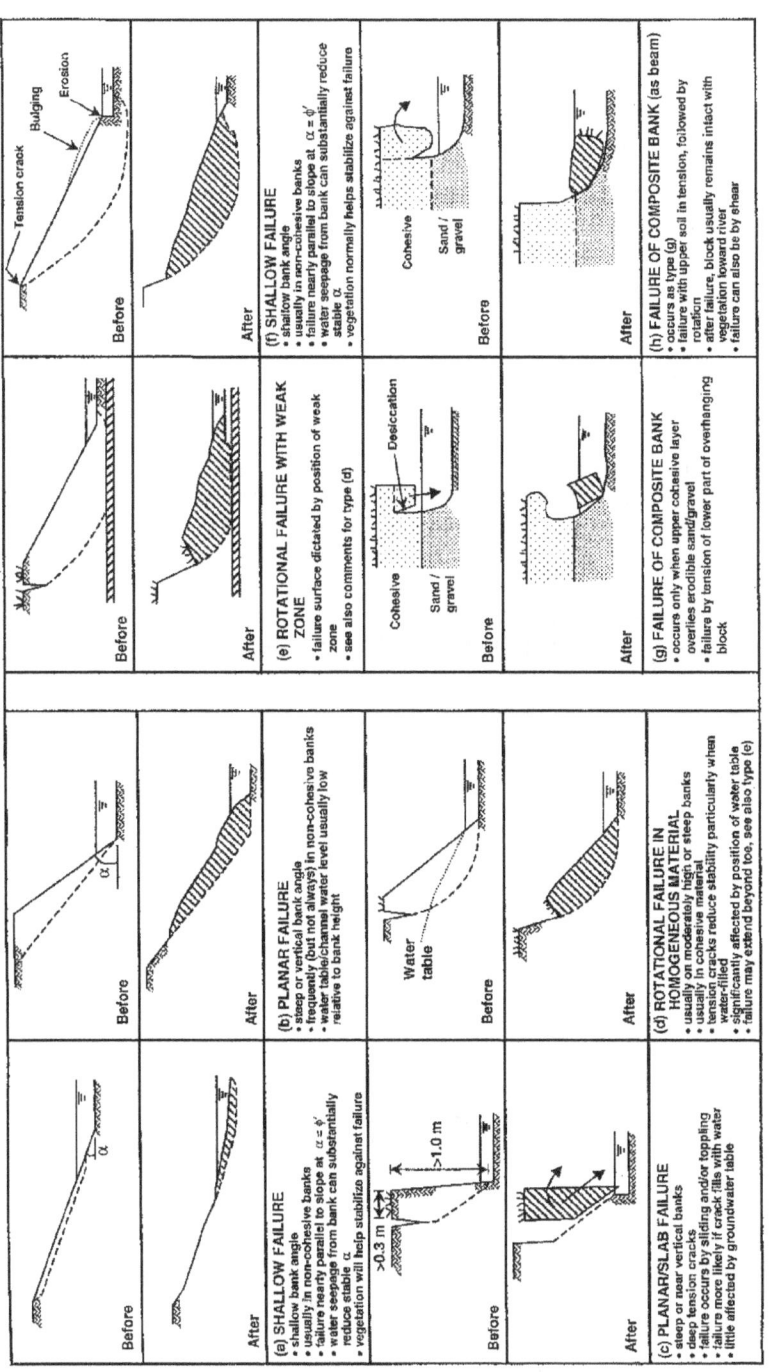

Figure B.2. Models and characteristics of bank failure (Hey et al. 1991).

The stability of a cohesive bank can be evaluated by considering the ratio of disturbing and restoring forces acting on the most critical failure surface to produce a factor of safety. This approach requires that the shape of the failure surface be known. For low, steep banks, the most simple and reasonable approach is to use the Culmann method, which assumes a planar surface passing through the toe of the bank producing a planar or slab type failure (Figure B.2c). The Culmann formula for this type of failure assumes a planar shear surface along which slab or wedge failure occurs and is based on total, rather than effective, stress principles.

In addition, the effects of cracks or fissures in the soil must be accounted for when analyzing the stability of banks. Cracks may be inherent in the soil fabric, or they may develop to relieve tension stress at the top of a steep slope.

Rotational Slip Failure. This type of failure can be further characterized as a base, toe, or slope failure depending on where the failure arc intersects the ground surface (Figure B.3, see also Figure B.2d). The ratio of restoring to disturbing moments about the center of the failure arc defines the factor of safety. The simplified solution for the factor of safety per unit length along the bank assumes that interslice forces act horizontally and accounts for pore water pressure (Bishop 1955). Since there is no simple way to locate the critical slip circle, a number of possible locations must be evaluated, which requires iterative calculations. Therefore, stability charts that predict the worst case have been developed (Bishop and Morgenstern 1960, Morgenstern 1963, Ponce 1978). However, failure surfaces are seldom circular and undrained conditions may be critical, both of which limit the applicability of these charts for natural river banks (Thorne and Tovey 1979).

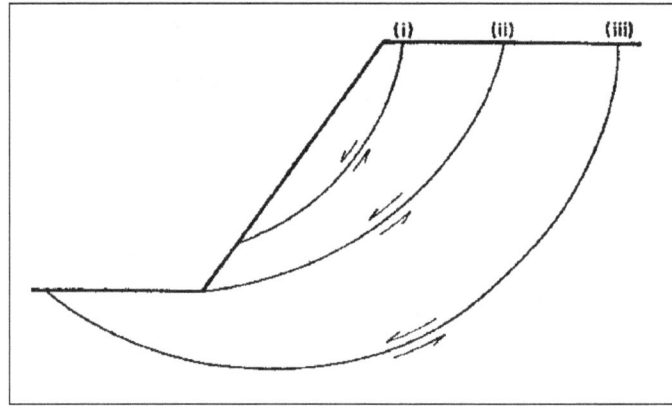

Figure B.3. Rotational slip failures in a cohesive bank: (i) slope failure, (ii) toe failure, and (iii) base failure (Thorne 1982).

Well established procedures developed in geotechnical engineering may be used to analyze rotational slips (Schofield and Wroth 1968, Fredlund 1987). Research indicates that rotational slips mainly occur in cohesive banks with angles less than about 60 degrees. Osman (1986) and Thorne (1988) attained reasonable results using a slope stability program developed to assess bank stability with respect to rotational slip for a variety of undercut and oversteepened banks.

Shallow Slip Failure. Shallow slips occur frequently, but have less impact on a river bank than deep seated rotational failures (Thorne 1982). Shallow slip failure takes place along an almost planar surface parallel to the bank surface. Theoretical analysis of shallow slips by the method of slices suggests that these should be confined to noncohesive materials, but shallow slips in cohesive material do occur naturally. The discrepancy can be explained by the presence of tensile stress in the soil due to lateral stream cutting which causes fissures in the soil. This leads to the movement of water through the soil causing softening, leaching, and possible piping, all of which reduce the effective cohesion, and makes the cohesive soil behave like a noncohesive one.

Plane Slip Failure. In low, steep banks the most critical failure surface is almost planar, passes through the toe of the bank, and produces a slab or block of soil that slides downward and outwards followed by toppling forward into the channel (Figure B.2c). Plane slip failure is the most common type failure for eroding river banks. As slope angle decreases and height increases, plane slip becomes much less likely. The Culmann analysis, which is based on the total stress principles and assumes a planar shear surface along which slab or wedge failure occurs, is used to analyze this type of failure (Figure B.4a). The critical bank height for the plane slip type of failure is proportional to 4 times the bank material cohesion (c), inversely proportional to the unit weight of the bank material ($\gamma$), and is related to the bank slope angle ($\theta$), and the bank material friction angle ($\phi$).

As the bank angle decreases, the assumption of a planar shear surface rapidly becomes invalid since deep-seated failures of high banks with low slope angles are usually curved as a result of changes with depth in the orientation of the principal stresses in the soil. Plane slips become less likely and the Culmann analysis seriously overestimates bank stability as slope angle decreases and bank height increases.

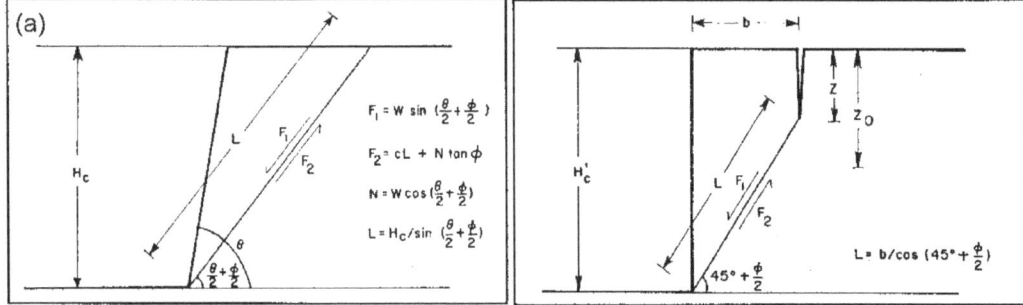

Figure B.4. Culmann analysis for (a) plane slip failure and (b) plane slip failure modified to account for tension cracking (Thorne 1982).

In many cases river banks are very steep and almost vertical. In the case of a vertical bank, the bank slope angle is removed from the Culmann analysis. In addition, the Culmann analysis described above does not account for possible tension cracking. Therefore, for a vertical bank with a tension crack that may extend about one-half the bank height in soils of negligible tensile strength, Thorne has modified the Culmann analysis such that the depth of the tensile stress of the bank ($z_o$) is subtracted from the original critical bank height (Figure B.4b) (Thorne 1982). The depth of the tensile stress may be calculated from the Mohr diagram (see any standard civil engineering reference (Landberg 1992).

Lawler (1992) constructed a series of Culmann-type bank stability curves that can be used to predict the critical bank height required to produce wedge or slab failures for a given range of saturated bulk unit weights ($\gamma$), cohesions (c), and friction angles ($\phi$). In high banks, the presence of a tension crack does not significantly change the failure surface geometry since $z_o$ is only a few percent of the bank height. Therefore, the potential for cracking in high banks can be accounted for by simply reducing the length of the failure surface by that portion within the tensile zone.

### B.5.3 Composite Banks

Composite banks are composed of cohesive and noncohesive materials stratified into discrete and discontinuous layers. In alluvial materials, the interfingering of cohesive and noncohesive materials can be related to lateral migration of channels and the resulting juxtaposition of channel and non-channel depositional environments. However, fluvial entrainment of the failed or eroded basal material is vital to the process. Thus, the rate of retreat of composite banks in the medium to long term is dependent on the stability of the lower bank and toe zone. The individual erosion processes and failure mechanics operating on a bank composed of a single type of material are combined to reflect the multiplicity of the bank material types in composite banks (Thorne 1982). Failure mechanisms include rotational slip, plane slip, and cantilever slip.

Rotational Slip Failure. Where a cohesive layer underlies a noncohesive layer (e.g., gravel) at depth in a high composite bank (Figure B.5a, see also Figure B.2e), fluvial erosion of the lower bank can result in oversteepening and failure of the cohesive upper bank. The likelihood of rotational failure increases with increasing thickness of the cohesive layer. The critical slip surface is classified as a toe or slope failure depending on the height of the contact surface in the bank (Figure B.5a).

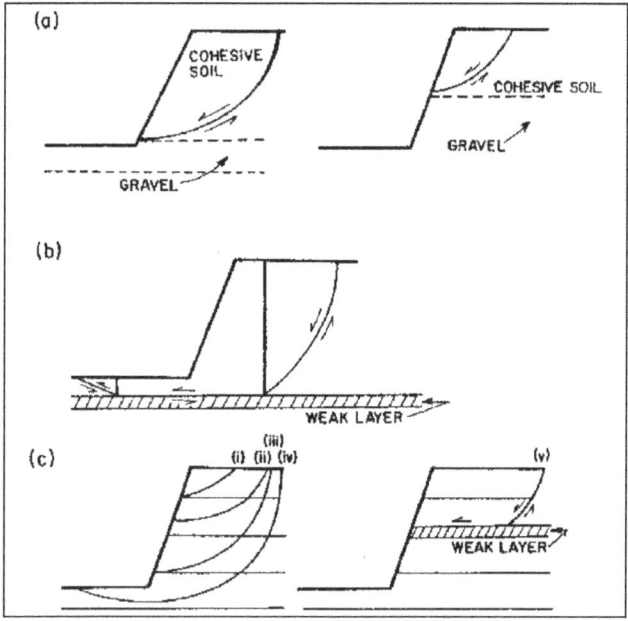

Figure B.5. (a) Rotational slip failure of the upper cohesive unit of a high composite bank. Toe or slope failure is determined by location of the contact between the soil and gravel. (b) Composite failure surface relative to a weak substratum. (c) Slip failures in a multilayered bank: (i)-(iv) possible failure surface locations within or between layers depending on soil properties and bank geometries; (v) critical influence of a weak layer (Thorne 1982).

If there is excess pore water pressure or softening of the base of the cohesive layer, the contact surface between the cohesive and noncohesive units can become a plane of weakness. The critical failure surface takes on a composite form if this weak layer is present (Figure B.5b). Bank stability is estimated based on a comparison of the forces causing and resisting movement of the central failure block away from the bank. The most critical surface must be located through a number of trial calculations because the points of intersection of the composite failure surface with the plane of weakness are unknown (Terzaghi and Peck 1948). Calculation of the location of the critical failure surface in a multilayered bank (Figure B.5c) also requires a number of iterations since failure may occur within one layer or between layers. Where one or more weak layers are present, the longest part of the failure surface will probably be located in the weakest layer (Terzaghi and Peck 1948). Although Morgenstern and Price (1965) and Sarma (1979) have developed improved stability analyses dealing with composite banks, their analyses have not been evaluated with regard to field data (Thorne 1982).

Plane Slip Failure. Plane slips and slab failures occur on low banks in general and can be expected on high banks with thin cohesive layers. A thin cohesive layer underlain by a noncohesive layer (e.g., sand or gravel) is often well drained, so pore water pressure can be ignored and desiccation cracking may occur. When a cohesive layer underlies a noncohesive layer, the decrease in permeability may produce a plane of strong seepage pressure, which can lead to piping or liquefaction of the noncohesive material resulting in oversteepening and failure higher up the bank.

Cantilever Failure. Cantilever failures occur when cohesive material overlies a noncohesive layer, which is removed by fluvial erosion, resulting in an overhanging block of cohesive material (Figure B.6). They generally occur on low banks and are most common in settings where the river is transporting a significant gravel load because floodplain stratigraphy is usually composed of a fining-upward sequence. An increase in the width of the overhang or cantilever by further undercutting, weakening by wetting, or cracking eventually exceeds the equilibrium state of the block and it fails. Failure occurs by shear, beam, or tensile failure and is dependent on block geometry (Thorne and Tovey 1981). Tension and desiccation cracks are of considerable importance and must be accounted for. Once cantilever failure of the block occurs, its removal is dependent on fluvial entrainment.

## B.6 ESTIMATING CRITICAL BANK HEIGHT

As previously indicated, the stability of the bank with respect to mass failure is dependent on soil properties and bank geometry. Bed lowering and lateral erosion are the two most common processes that act to steepen the bank and cause bank instability. For estimating critical bank height for steep, cohesive banks, a simple slope stability analysis can be developed. Reference is suggested to the analysis approach derived by Osman and Thorne (1988) to predict bank stability response to lateral erosion and bed degradation.

Thorne and Osman (1988) also developed a modeling technique to study the effects of channel widening and bank-sediment contribution on flow energy, stream power, and the rate and extent of bed lowering during degradation, and the influence of outer bank stability on bed and failure using a critical shear stress concept to account for lateral erosion and a slope stability criterion for mass failure. Again, a review by the reader is recommended prior to evaluating lateral erosion and bank instability problems in detail for a given site.

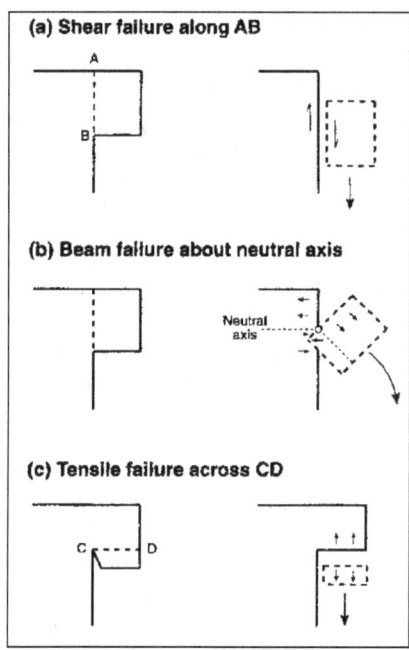

Figure B.6. Mechanisms of cantilever failure on composite banks (Thorne and Tovey 1979).

## B.7 REFERENCES

Barden, L., 1972, "The influence of structure on deformation and failure in clay soil," Géotechnique, Vol. 22, No. 1, p. 159-163.

Bishop, A.W., 1955, "The use of the slip circle in the stability analysis of slopes," Géotechnique, Vol. 5, p. 7-17.

Bishop, A.W. and N.R. Morgenstern, 1960, "Stability coefficients for earthslopes," Géotechnique, Vol. 10, No. 4, p. 129-150.

Carson, M.A. and M.J. Kirkby, 1972, "Hillslope Form and Process," Cambridge University Press, Cambridge.

Fredlund, D.G., 1987, "Slope stability analysis incorporating the effect of soil suction," In: M.G. Anderson, and K.S. Richards, (eds), Slope Stability, Wiley, Chichester, p. 113-144.

Hagerty, D.J., M. Sharifounnasab, and M.F. Spoor, 1983, "River bank erosion: A case study," Bulletin American Association Engineering Geology, Vol. XX, No. 4, p. 411-437.

Hey, R.D., G.L. Heritage, N.K. Tovey, R.R. Boar, N. Grant, and R.K. Turner, 1991, "Streambank Protection in England and Wales," R&D Note 22, National Rivers Authority, London.

Howard, A.D., 1967, "Drainage analysis in geological interpretation: a summation," Bulletin of the American Association of Petroleum Geologists, Vol. 56, p. 275-370.

Landberg, M.R., 1992, "Civil Engineering Reference Manual," Sixth Edition, Professional Publications, Inc., Belmont, CA.

Lane, E.W., 1955, "Design of stable channels," Transactions of the American Society of Civil Engineers, Vol. 120, p. 1-34.

Lawler, D.M., 1992, "Process dominance in bank erosion systems," In: P. Carling and G.E. Petts (eds), Lowland Floodplain Rivers: Geomorphological Perspectives, Wiley, Chichester, p. 117-143.

Lawler, D.M., C.R. Thorne, and J.M. Hooke, 1997, "Bank Erosion and Instability," Chapter 6 in: C.R. Thorne, R.D. Hey, and M.D. Newsom (eds), Applied Fluvial Geomorphology for River Engineering and Management, John Wiley & Sons, Chichester.

Morgenstern, N.R., 1963, "Stability charts for earth slopes during rapid drawdown," Géotechnique, Vol. 13, No. 2, p. 121-131.

Morgenstern, N.R. and V.E. Price, 1965, "The analysis of the stability of general slip surfaces," Géotechnique, Vol. 15, p. 79-93.

Nanson, G.C. and E.J. Hickin, 1986, "A statistical analysis of bank erosion and channel migration in western Canada," Geological Society of America Bulletin, Vol. 97, p. 497-504.

Nanson, G.C., B.R. Rust, and G. Taylor, 1986, "Coexistent mud braids and anastomosing channels in an arid-zone river, Cooper Creek, Central Australia," Geology, Vol. 14, p. 175-178.

Neil, C.R., 1984, "Bank erosion versus bedload transport in a gravel river," In: C.M. Elliot, (ed), River Meandering, Proceedings of the Conference Rivers '83, American Society of Civil Engineers, New York, p. 204-211.

Osman, A.M., 1986, "Channel width response to changes in flow Hydraulics and Sediment Load," unpublished Ph.D. Thesis, Colorado State University, Fort Collins, CO, USA.

Osman, A.M. and C.R. Thorne, 1988, "Riverbank Stability Analysis I: Theory, Proceedings of the American Society of Civil Engineers, Journal of Hydraulic Engineering, Vol. 114, No. 2, New York.

Parker, G., 1979, "Hydraulic geometry of active gravel rivers," Journal of the Hydraulics Division, American Society of Civil Engineers, Vol. 105, No. HY9, p. 1185-1201.

Ponce, V.M., 1978, "Generalized stability analysis of channel banks," Journal of the Irrigation and Drainage Division, American Society of Civil Engineers, Vol. 104, No. IR4, p.343-350.

Sarma, S.K., 1979, "Stability analysis of embankments and slopes," Journal of the Geotechnical Engineering Division, American Society of Civil Engineers, Vol. 105, No. GT12, p. 1511-1524.

Schofield, A.N. and P. Wroth, 1968, "Critical State Soil Mechanics," McGraw-Hill.

Selby, M.J., 1982, "Hillslope Materials and Processes," Oxford University Press, Oxford.

Task Committee on Sedimentation, American Society of Civil Engineers, 1966, "Sediment transport mechanics: initiation of motion," Journal of the Hydraulics Division, Vol. 92, No. HY2, p. 291-314.

Taylor, D.W., 1948, "Fundamentals of Soil Mechanics," John Wiley & Sons, NY.

Terzaghi, K., 1943, "Theoretical Soil Mechanics," John Wiley & Sons, NY.

Terzaghi, K. and R.B. Peck, 1948, "Soil Mechanics and Engineering Practice," John Wiley & Sons, NY.

Thorne, C.R., 1982, "Processes and mechanisms of river bank erosion," In: R.D. Hey, J.C. Bathurst, and C.R. Thorne, (eds), Gravel-bed Rivers, John Wiley & Sons, Chichester, p. 227-259.

Thorne, C.R., 1988, "Analysis of bank stability in the DEC watersheds, Mississippi," Final Technical Report to the U.S. Army European Research Office, London, under contract No. DAJA45-87-C-0021, Queen Mary College, University of London.

Thorne, C.R., 1998, "Stream Reconnaissance Handbook," John Wiley & Sons, Chichester.

Thorne, C.R. and N.K. Tovey, 1979, "Discussion on 'Generalized stability analysis of channel banks' by V.M. Ponce, Journal of the Irrigation and Drainage Division," American Society of Civil Engineers, Vol. 105, No. IR4, p. 436-437.

Thorne, C.R. and N.K. Tovey, 1981, "Stability of composite river banks," Earth Surface Processes and Landforms, Vol. 6, p. 469-484.

Thorne, C.R. and A.M. Osman, 1988a, "The influence of bank stability on regime geometry of natural channels," In: W.R. White (ed), International Conference on River Regime, Hydraulics Research Limited, John Wiley & Sons, Chichester, p. 135-147.

Thorne, C.R. and A.M. Osman, 1988b, "Riverbank stability analysis. II: Applications," Proceedings of the American Society of Civil Engineers, Journal of Hydraulic Engineering, Vol. 114, No. 2, New York.

Varnes, D.J., 1958, "Landslide types and processes," in E.B. Eckel (ed), Landslides and Engineering Practice, Special Report 29, NAS-NRC Publication 544, Highway Research Board, Washington, D.C., p. 20-47.

Wolman, M.G. and L.B. Leopold, 1957, "River floodplains, some observations on their formation," U.S. Geological Survey Professional Paper 282C, p. 87-107.

# APPENDIX C

# STREAM RECONNAISSANCE RECORD SHEETS

(page intentionally left blank)

## C.1 OVERVIEW

Thorne (1998) developed a comprehensive handbook that can be used to document stream channel and watershed conditions. Thorne's handbook includes stream reconnaissance record sheets and guidelines for a detailed geomorphological stream reconnaissance. This appendix presents a modified version of the stream reconnaissance sheets that can be used in documenting the stability and conditions of a stream system. The stream reconnaissance record sheets consist of four sections. Section 4 can be used for each bank (left and right) and should be properly identified. The four sections are:

Section 1 – Scope and Purpose  
Section 2 – Region and Valley Description  
Section 3 – Channel Description  
Section 4 – (Left & Right) Bank Survey

Although the sheets appear complex, they were designed to produce a comprehensive record of the morphology of the stream and its surroundings, and be applicable to a wide range of river types and sizes in diverse settings. With this in mind, one should resist the temptation to omit filling out parts of the sheets for the purposes of expediency or because of perceived irrelevance, since the data may be used for other applications in the future. However, this does not preclude the customization of the sheets to a particular region, basin, or river through the removal of extraneous material rather than the omission of entire topics or sections.

## C.2 SECTION 1 – SCOPE AND PURPOSE

This section is used to document the purpose of the study, the nature of any morphology related problems to be addressed, basic logistical information, and the limits of the study area. The reconnaissance trip should have a clearly defined purpose, aims, and objectives. Omission of any sections or topics should be recorded and justified in this section.

## C.3 SECTION 2 – REGION AND VALLEY DESCRIPTION (PARTS 1-5).

This section is used: first, to describe the surrounding landscape, establish the nature of the river basin, and define the relationship between the channel and its valley; and, second, to identify any lateral or vertical channel instability problems relative to the valley in terms of trend, severity, and extent. The geological setting, sedimentary characteristics, and land use practices in the basin, valley, and floodplain surrounding the stream channel are documented. Much of this information can be completed in the office through the use of topographic, geologic, and land use maps together with aerial photography or satellite images. In large basins, or where access is limited, an overflight may be necessary. Detailed descriptions of many of the geomorphic features discussed in the following paragraphs can be found in Chapter 2.

**In Part 1 (Figure C.1)**, the objective is to characterize the landscape surrounding the river valley relative to terrain (topography), drainage pattern, surface geology, rock type, land use, and vegetation. The *terrain*, or topography, defines the amount of energy available to do geomorphic work and the responsiveness of the watershed to stability problems. The planform or *drainage pattern* of the drainage network, which is generally indicative of the underlying geology and topography, is also identified. The eight common types of patterns are shown in Figure C.2. *Surface geology* and underlying *rock type* directly and indirectly determine erosion resistance and strongly affect sediment yield and delivery. *Land use* strongly influences the runoff hydrograph and has an impact on sediment yields. *Vegetation* also has a significant influence on basin hydrology, affecting both runoff and sediment yield.

# STREAM RECONNAISSANCE RECORD SHEETS

(Modified from Thorne, 1998)

## SECTION 1 - SCOPE AND PURPOSE

**Brief Problem Statement:**

**Purpose of Stream Reconnaissance:**

**Logistics of Reconnaissance Trip:**

| RIVER: | LOCATION: | DATE: |
| --- | --- | --- |
| | | From:         To: |
| PROJECT: | STUDY REACH: | |
| SHEET COMPLETED BY: | | |
| RIVER STAGE: | TIME START: | TIME FINISH: |

**General Notes and Comments on Reconnaissance Trip:**

| PART 1 AREA AROUND RIVER VALLEY | | | | | |
|---|---|---|---|---|---|
| Terrain | Drainage Pattern | Surface Geology | Rock Type | Land Use | Vegetation |
| Mountains | Dendritic | Bed rock | Sedimentary | Natural | Temperate forest |
| Uplands | Parallel | Weathered soils | Metamorphic | Managed | Boreal forest |
| Hills | Trellis | Glacial Moraine | Igneous | Cultivated | Woodland |
| Plains | Rectangular | Glacio/Fluvial | None | Urban | Savanna |
| Lowlands | Radial | Fluvial | | Suburban | Grassland |
| | Annular | Lake Deposits | Specific Rock Types (if known) | | Desert scrub |
| | Multi-Basin | Wind blown (loess) | | | Extreme Desert |
| | Contorted | | | | Tundra or Alpine |
| | | | | | Agricultural land |

Figure C.1. Reconnaissance sheet - area around river valley.

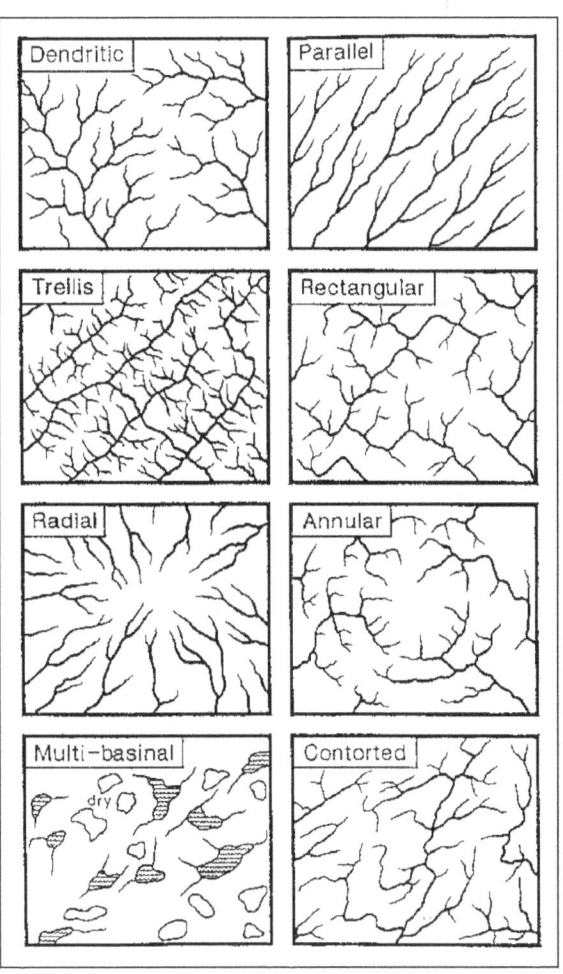

Figure C.2. Basic drainage patterns (from Howard 1967).

In **Part 2 (Figure C.3)**, the characteristics of the river valley and valley sides are described. The *location of the river* is defined relative to a valley or some other physiographic setting (the rest of this section may remain uncompleted if the river is not confined between valley walls). *Valley shape* is noted relative to being symmetrical or asymmetrical. The *height* and *side slope angle* of the valley sides define the potential to drive large-scale channel instability through the input of debris or sediment to the fluvial system. *Valley side failures*, their frequency, and *location* may be indicative of large scale, lateral geomorphic activity and possible valley widening, and defines how sediment is delivered to the system. Valley side mass failure mechanisms are shown in Figure C.4.

| PART 2: RIVER VALLEY AND VALLEY SIDES | | | | | |
|---|---|---|---|---|---|
| Location of River | Height | Side Slope Angle | Valley Shape | Valley Side Failures | Failure Locations |
| In Valley | < 5 m | < 5 degrees | Symmetrical | None | None |
| On Alluvial Fan | 5 - 10 m | 5-10 degrees | Asymmetrical | Occasional | Away from river |
| On Alluvial Plain | 10 - 30 m | 10-20 degrees | | Frequent | Along river |
| IN a Delta | 30 - 60 m | 20-50 degrees | | | (Undercut) |
| In Old Lake Bed | 60 - 100 m | > 50 degrees | | | |
| | > 100 m | | **Failure Type** | | (see Figure 4.4) |

Figure C.3. Reconnaissance sheet - river valley and valley sides.

In **Part 3 (Figure C.5)**, the valley floor and floodplain (see Chapter 2, Figure 2.6) characteristics are documented. *Valley floor type* and *width* define the susceptibility of the stream to destabilization by valley side slope failures relative to runoff and erosion processes operating on the valley side slopes. *Flow resistance* records Manning "n" values of flow resistance for the left and right overbank areas, which can be used in hydraulic analysis of flow and sediment movement during flooding (see Chapter 3). *Surface geology* defines the composition and, consequently, the resistance to erosion, of the surficial materials making up the valley floor or floodplain. *Land use*, especially urban or industrial development of the floodplain or valley floor, has a considerable effect on the stream system because of the impacts of channelization, bank protection, and flood control works. *Vegetation* is important in terms of floodplain hydrology, overbank hydraulics, and sediment dynamics. Vegetation is also important relative to the potential for drift (vegetative debris), accumulations along the stream and on bridges (see Chapter 5). In turn, the *riparian buffer strip* and its *width* have a significant influence on natural stream processes, channel stability, and environmental conditions.

| PART 3: FLOODPLAIN (VALLEY FLOOR) | | | | | |
|---|---|---|---|---|---|
| Valley Floor Type | Valley Floor Data | Surface Geology | Land Use | Vegetation | Riparian Buffer Strip |
| None | None | Bed rock | Natural | None | None |
| Indefinite | 1 river width | Glacial Moraine | Managed | Unimproved Grass | Indefinite |
| Fragmentary | 1-5 river widths | Glacio/Fluvial | Cultivated | Improved Pasture | Fragmentary |
| Continuous | 5-10 river widths | Fluvial: Alluvium | Urban | Orchards | Continuous |
| | > 10 river widths | Fluvial: Backswamp | Suburban | Arable Crops | **Strip Width** |
| | **Flow Resistance*** | Lake Deposits | Industrial | Shrubs | None |
| Left Overbank Manning n value ___ | | Wind Blown (loess) | | Deciduous Forest | < 1 river width |
| Rght Overbank Manning n value ___ | (*note: n value for channel is recorded in Part 6) | | | Coniferous Forest | 1-5 river widths |
| | | | | Mixed Forest | > 5 river widths |

Figure C.5. Reconnaissance sheet - floodplain (valley floor).

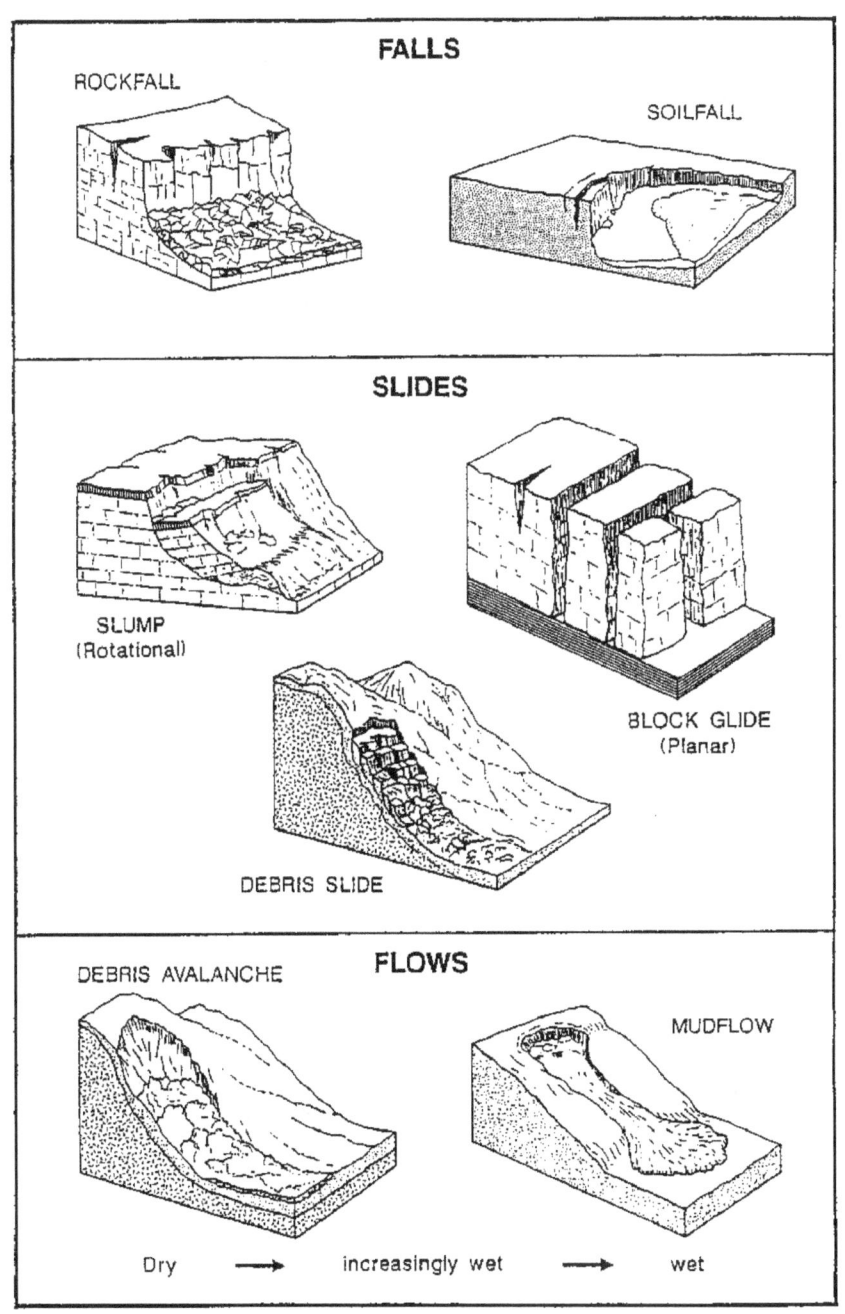

Figure C.4. Typical valley side mass failure mechanisms (after Varnes 1958).

**In Part 4 (Figure C.6)**, the present relationship between the stream and its valley is established relative to being aggraded, adjusted, or incised. *Terraces*, which form a stepped appearance in the valley cross-section profile (Figure C.7), are indicative of past vertical instability. The nature and magnitude of past vertical instability is indicated by the number of terraces which demonstrates the potential for dynamic vertical adjustment of the system through floodplain cut and fill sequences. The presence or absence of *trash lines* in the overbank areas, the vertical locations, and their recurrence relative to frequent versus infrequent events help define whether the channel is aggradational, degradational, or stable.

| PART 4: VERTICAL RELATION OF CHANNEL TO VALLEY | | | | | |
|---|---|---|---|---|---|
| Terraces | Overbank Deposits | Levees | Levee Data | Levee Description | Trash Lines |
| None | None | None | Height (m) | None | Absent |
| Indefinate | Silt | Natural | Side Slope (o) | Indefinite | Present |
| Fragmentary | Fine sand | Constructed | **Levee Condition** | Fragmentary | Height above |
| Continuous | Medium sand | **Instability Status** | None | Continuous | floodplain (m) |
| Number of Terraces | Coarse sand | Stable | Intact | Left Bank | |
| | Gravel | Degrading | Local Failures | Right Bank | |
| | Boulders | Aggrading | Frequent failures | Both Banks | |

Figure C.6. Reconnaissance sheet - vertical relation of channel to valley.

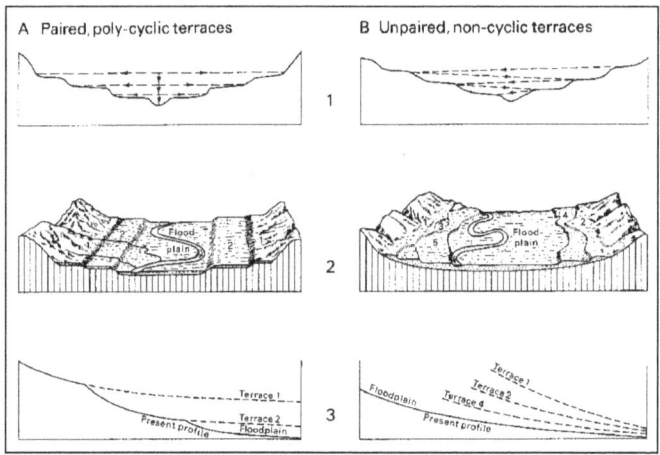

Figure C.7. Valley cross-sections, plan view, and profiles of (A) paired, polycyclic and (B) unpaired, non-cyclic river terraces (Chorley et al. 1984).

Although vertical floodplain accretion in a dynamically stable system is normal, heavy *overbank deposits* (sedimentation) may be indicative of an aggrading system. *Levees* and *levee descriptions* deal with natural and artificial levees that are present along a channel as a result of overbank flooding. Natural levees, which form by overbank sedimentation during flooding, may become prominent if the stream is carrying a heavy sediment load and has a high frequency of overbank flooding. Artificial levees are constructed in areas where flooding is unacceptable. *Levee data* and *levee condition* record the height, side slope angle, and stability of existing levees. The instability status defines whether vertical instability in the system is ongoing or has ceased.

**In Part 5 (Figure C.8)**, the present relationship between the stream and its valley is established relative to lateral stability or channel migration, and fluvial landforms indicative of lateral instability. *Planform* represents the geometry of the channel as viewed from above and can be classified based on a simple classification scheme shown in Figure C.9 (see Chapter 2 and Chapter 5 for additional discussion). It should be noted that in reality there is a continuum of river patterns. *Planform data* as shown in Figure C.10 is recorded on the characteristic dimensions of a typical meander. *Lateral activity* and resultant *floodplain features* as shown in Figure C.11 are recorded for the current type of planform evolution.

| PART 5 LATERAL RELATION OF CHANNEL TO VALLEY | | | | | | | |
|---|---|---|---|---|---|---|---|
| Planform | | Planform Data | | Lateral Activity | | Floodplain Features | |
| Straight | | Bend Radius | | None | | None | |
| Sinuous | | Meander belt width | | Meander progression | | Meander scars | |
| Irregular | | Wavelength | | Increasing amplitude | | Scroll bars+sloughs | |
| Regular meanders | | Meander Sinuosity | | Progression+cutoffs | | Oxbow lakes | |
| Irregular meanders | | **Location in Valley** | | Irregular erosion | | Irregular terrain | |
| Tortuous meanders | | Left | | Avulsion | | Abandoned channel | |
| Braided | | Middle | | Brading | | Braided deposits | |
| Anastomosed | | Right | | | | | |

Figure C.8. Reconnaissance sheet - lateral relation of channel to valley.

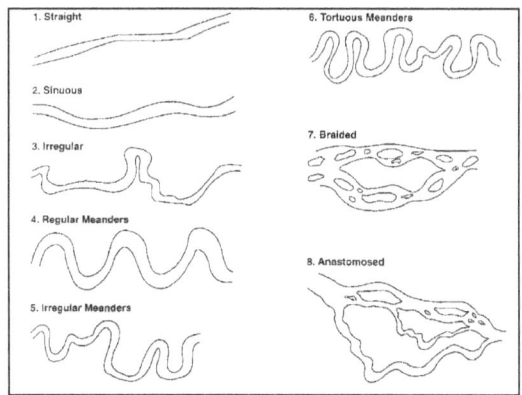

Figure C.9. Guide to classification of river channel planform pattern (Thorne 1998).

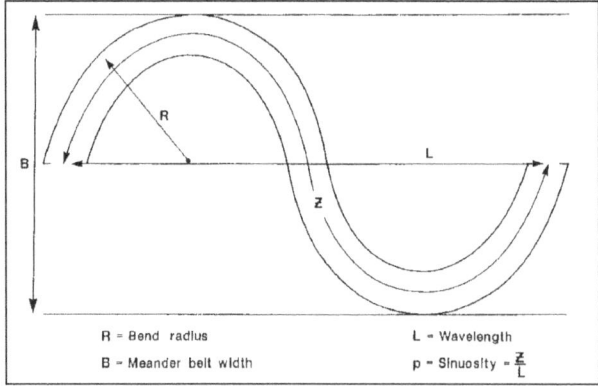

Figure C.10. Definition of meander planform parameters (Thorne 1998).

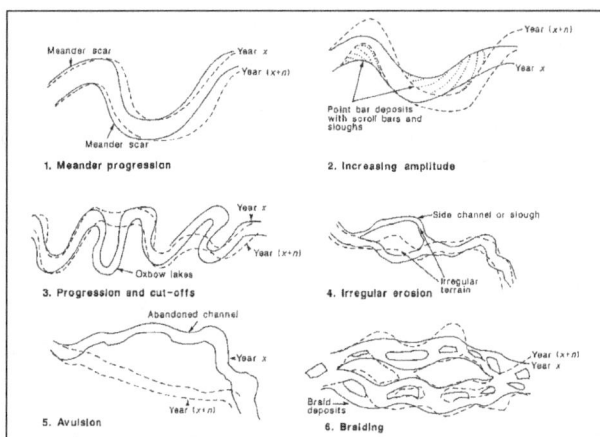

Figure C.11. Types of lateral activity and typical associated floodplain features (Thorne 1998).

**C.4. SECTION 3 – CHANNEL DESCRIPTION (PARTS 6 AND 7).**

General channel dimensions, flow type, geological and artificial controls on vertical and lateral movement, bed sediment characteristics, and the presence of sedimentary forms and features within the channel are documented in this section.

**In Part 6 (Figure C.12)**, the channel is characterized relative to dimensions, flow regime, and possible controls on bed scour and bank retreat. This information can be used to describe and define the channel hydraulics, boundary conditions, and potential instability. *Dimension* measurements need not be detailed or precise but should be representative of the channel in the study reach and based on a reach-average of several measurements. Estimates of Manning "n" and mean velocity may be somewhat subjective.

*Flow type* describes the flow regime based on the principles of free surface flow. Uniform/tranquil flow is fully turbulent, sub-critical with approximately uniform flow velocity. Uniform/shooting flow is super-critical flow with uniform flow velocity. Pools and riffles are alternating deeps and shallows with slower, flatter and faster, steeper flow, respectively. Tumbling flow represents high gradient streams ($\geq 1\%$) with coarse bed material that disrupts the water surface creating locally super-critical flow. Step/pool flow occurs in very steep gradient channels (>10%) with boulder steps and plunge pools. *Bed and width controls* and *types* define the nature and extent of any bed and bank controls that limit vertical incision or lateral migration and/or widening due to local geology, bed, bank, and floodplain materials (including clay plugs and backswamp deposits), large woody debris jams, or engineered structures.

| PART 6 CHANNEL DESCRIPTION | | | | | |
|---|---|---|---|---|---|
| Dimensions | Flow Type | Bed Controls | Control Types | Width Controls | Control Types |
| Av. top bank width (m) ___ | None ___ | None ___ | None ___ | None ___ | None ___ |
| Av. channel depth (m) ___ | Uniform/Tranquil ___ | Occasional ___ | Solid bedrock ___ | Occasional ___ | Bedrock ___ |
| Av. water width (m) ___ | Uniform/Rapid ___ | Frequent ___ | Weathered bedrock ___ | Frequent ___ | Boulders ___ |
| Av. water depth (m) ___ | Pool+Riffle ___ | Confined ___ | Boulders ___ | Confined ___ | Gravel armor ___ |
| Reach slope ___ | Steep+Tumbling ___ | Number of controls ___ | Gravel armor ___ | Number of controls ___ | Revetment ___ |
| Mean velocity (m/s) ___ | Steep+Steppool ___ | | Cohesive materials ___ | | Cohesive materials ___ |
| Manning ___ | (Note: Flow type on day of observation) | | Bridge protection ___ | | Bridge abutments ___ |
| | | | Grade control structures ___ | | Dikes or groins ___ |

Figure C.12. Reconnaissance sheet - channel description.

**In Part 7 (Figure C.13)**, the morphology of the bed and bar sediments are characterized in order to estimate the bed material mobility, transport rate, and to gauge the potential for bed instability. *Bed material* is characterized qualitatively based on the predominant sediment size. *Bed armor* is coarse surface sediment that is identified as either static or mobile. Static armor (pavement) is immobile under all but catastrophic flood flow conditions. Mobile armor is mobilized during events of moderate magnitude and recurrence. *Sediment depth* in the channel defines the amount of sediment available for transport and can be indicative of vertical instability. *Bedforms* record form roughness elements on the channel bed and islands or bars record macro bed features and divided flow. *Bar types* define the shape of bars within the channel (Figure C.14). *Bed and bar surface* and *substrate* (below any armor layer) data represent quantitative data based on sieve-by-weight (bulk sample) or size-by-number (pebble count sample) of sediment samples taken at representative bed or bar locations. *Channel sketch map* and *representative cross-section* are visual representations of the plan view and profile of the channel that includes all pertinent features of the channel and floodplain and all survey and sampling sites. *Photography* of important features can be an important tool. Location and orientation of photos should also be noted.

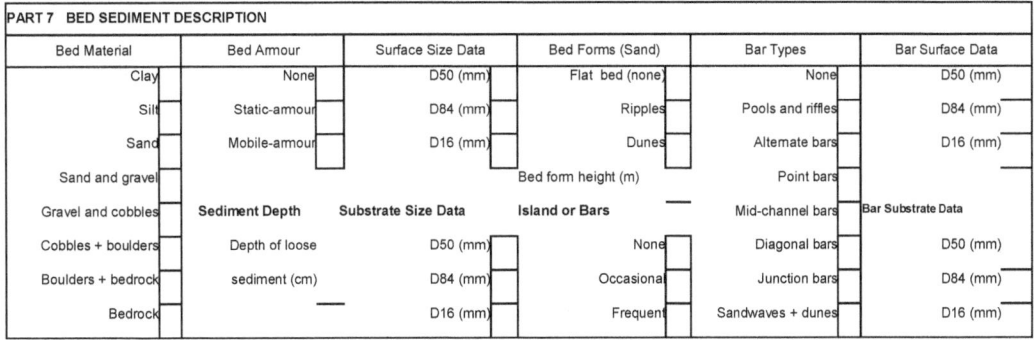

Figure C.13. Reconnaissance sheet - bed sediment description.

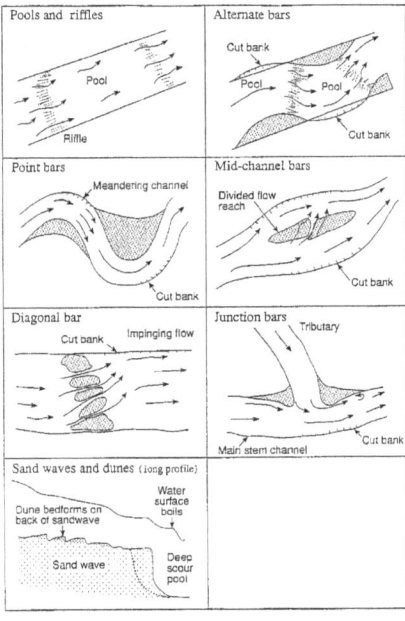

Figure C.14. Channel bar classification and bedforms (Thorne 1998).

## C.5  SECTION 4 – (LEFT AND RIGHT) BANK SURVEY (PARTS 8-12)

The morphology of the banks is documented in detail and includes the geometry, sedimentary characteristics, vegetation, erosional processes, geotechnical failure mechanisms, and the extent of toe sediment buildup. A comprehensive and detailed evaluation of the banks and their dynamics is at the heart of the reconnaissance. It is the basis for planform evolution and bankline migration, and supplies information in support of the selection of appropriate approaches to modeling bank processes and appropriate bank management strategies.

**In Part 8 (Figure C.15)**, the bank characteristics are described in detail relative to type, materials, countermeasures, dimensions, shape, and degree of cracking. *Type* classifies the bank based on being composed of non-cohesive, cohesive, composite, or layered bank materials. *Protection status* records the presence or absence and type of bank protection. *Bank materials* define the composition of the bank including any major stratigraphic layering and the thickness of stratigraphic layers within the bank are recorded. The material composition and location of the stratigraphic horizons are defined under the *distribution and description* of bank materials topic. *Average bank height* and *slope* record overall representative height and steepness of the bank. *Bank profile shape* defines the form of the bank profile as shown in Figure C.16. *Tension cracks*, indicative of inevitable failure, may be present along the top of the bank and their depth is recorded.

| PART 8  LEFT (OR RIGHT) BANK CHARACTERISTICS | | | | | |
|---|---|---|---|---|---|
| Type | Bank Materials | Layer Thickness | Average Bank Height | Bank Profile Shape | Tension Cracks |
| Noncohesive | Silt/clay | Material 1 (m) | Average height (m) | (Figure 4.16) | None |
| Cohesive | Sand/silt/clay | Material 2 (m) | | | Occasional |
| Composite | Sand/silt | Material 3 (m) | Average Bank Slope | | Frequent |
| Layered | Sand | Material 4 (m) | angle (degrees) | | Crack Depth |
| Even Layers | Sand/gravel | | | | Proportion of |
| Thick+thin layers | Gravel | Distribution and Description of Bank Materials in Bank Profile | | | bank height |
| Number of layers | Gravel/cobbles | Material Type 1 | Material Type 2 | Material Type 3 | Material Type 4 |
| Protection Status | Cobbles | Toe | Toe | Toe | Toe |
| Unprotected | Cobbles/boulders | Mid-bank | Mid-bank | Mid-bank | Mid-bank |
| Hard points | Boulders/bedrock | Upper bank | Upper bank | Upper bank | Upper bank |
| Toe protection | | Whole bank | Whole bank | Whole bank | Whole bank |
| Revetments | | D50 (mm) | D50 (mm) | D50 (mm) | D50 (mm) |
| Dykefields | | Sorting coefficient | Sorting coefficient | Sorting coefficient | Sorting coefficient |

Figure C.15.  Reconnaissance sheet - left (or right) bank characteristics.

**In Part 9 (Figure C.17)**, vegetation characteristics along the bank face are described because of the effects on bank morphology, erodibility, and stability. The general types of *vegetation* along the bank face and the orientation of their trunks (vertical or degree of leaning) relative to the bank and channel are described. Depth of rooting, which affects bank cohesiveness, and susceptibility to environmental hazards, which can affect bank stability, are indicated by the *types* and *species* of trees present along the bank. *Density + spacing* define the degree of vegetative cover and, consequently, indicates the amount of erosion protection afforded by the vegetation. Exposed or adventitious *roots* define the relationship with the bank surface relative to aggradation, degradation, or erosion. The location of vegetation on the bank also affects the stability of the bank. *Diversity, health, age, height*, and *lateral extent* of vegetation all affect bank stability and can be indicators of past channel instability. The profile of the bank, important details on bank morphology and vegetation characteristics, and any photographic documentation are recorded in the *bank profile sketch*.

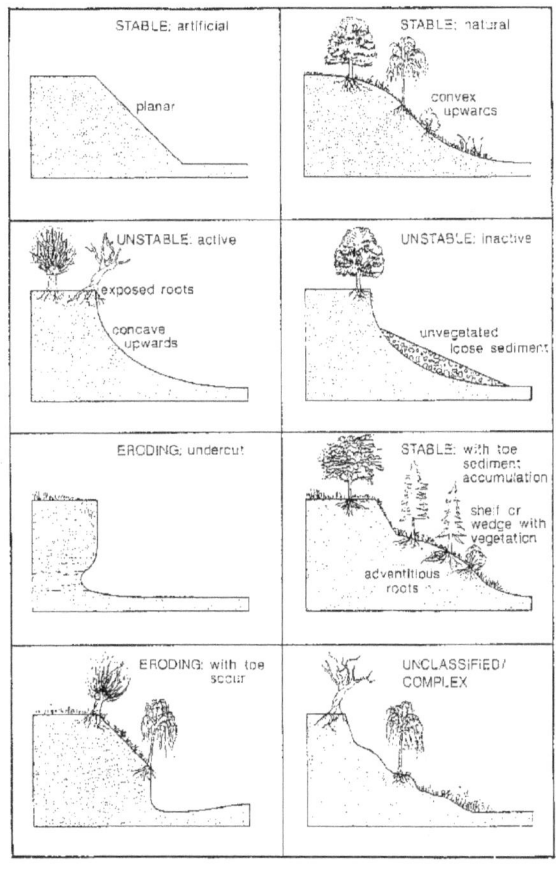

Figure C.16. Classification and morphology of typical bank profiles (Thorne 1998).

| PART 9 | LEFT (OR RIGHT) BANK-FACE VEGETATION | | | | |
|---|---|---|---|---|---|
| Vegetation | Tree Types | Density + Spacing | Location | Health | Height |
| None/fallow | None | None | Whole bank | Healthy | Short |
| Artificially cleared | Deciduous | Sparse/clumps | Upper bank | Fair | Medium |
| Grass and flora | Coniferous | Dense/clumps | Mid-bank | Poor | Tall |
| Reeds and sedges | Mixed | Sparce/continuous | Lower bank | Dead | Height (m) |
| Shrubs | **Tree Species** | Dense/continuous | | | |
| Saplings | if known) | **Roots** | **Diversity** | **Age** | **Lateral Extent** |
| Trees | | Normal | Mono-stand | Imature | Wide belt |
| **Orientation** | | Exposed | Mixed stand | Mature | Narrow belt |
| Angle of leaning (0) | | Adventitious | Climax-vegetation | Old | Single row |

Figure C.17. Reconnaissance sheet - left (or right) bank vegetation.

**In Part 10 (Figure C.18)**, details pertinent to a good understanding of the erosional processes and their distribution along the bank are recorded. The *erosion location* is documented relative to major channel features. The *present status* and *rate* of active or inactive bank erosion (retreat) or advancement (accretion) is also documented where possible. The *processes* responsible for bank erosion and the distribution of the different processes that may be operating on different parts of the bank are identified. Parallel or impinging flow erosion results in detachment and removal of intact grains or aggregates of grains from the bank face by flow that is either parallel or oblique to the long-stream direction. Piping is caused by seepage of groundwater (sapping) from the bank face, especially in high banks, and detaches and entrains grains from the bank. Freeze/thaw processes can remove individual grains to blocks of bank material from the bank or weakens the bank by destroying bank cohesion, increasing its susceptibility to erosion. Sheet erosion removes surface material by non-channelized surface runoff over the bank edge and down the bank face. Rills and gullies form when sufficient uncontrolled surface runoff forms small channels. Wind and vessels cause erosion of the bank through the generation of waves, which increase near-bank velocities and shear stresses, detaches and erodes bank particles, and causes dramatic fluctuations in bank pore water pressures through rapid fluctuations in water level. Ice rafting can mechanically damage the bank making it susceptible to erosion. Other types of erosion processes include trampling by animals and man or by off-road vehicles. Where different erosion processes are operating on different parts of the bank, they should be identified and recorded.

| PART 10  LEFT (OR RIGHT) BANK EROSION | | | | | | | |
|---|---|---|---|---|---|---|---|
| Erosion Location | | Present Status | Rate of Retreat | | Dominant Process | | |
| General | Opposite a structure | Intact | m/yr (if applicable and known) | Parallel flow | | Rilling + gullying | |
| Outside meander | Adjacent to structure | Eroding: dormant | | Impinging flow | | Wind waves | |
| Inside meander | D/S of structure | Eroding: active | | Piping | | Vessel forces | |
| Opposite a bar | U/S of structure | Advancing: dormant | **Rate of Advance** | Freeze/thaw | | Ice rafting | |
| Behind a bar | Other (write in) | Advancing: active | m/yr (if applicable and known) | Sheet erosion | | Other (write in) | |

Figure C.18. Reconnaissance sheet - left (or right) bank erosion.

**In Part 11 (Figure C.19)**, the geotechnical characteristics of bank failures are identified and recorded. Failures may or may not coincide with bank erosion, so the *location* is identified relative to major channel features and the stability *status* of the bank is classified. Failure scars and blocks are prominent features of bank failure and their presence and appearance should be noted. The *failure mode* identifies the type of failure resulting from bank instability and is dependent on bank material composition as described in Chapter 2 and Appendix B. The distribution on the bank of all modes of failure are identified and recorded. A brief description of each of the major failure modes follows:

<u>Soil/rock fall</u> – failure of grains, grain assemblages, or blocks from often undercut, steep, eroding banks with little cohesion.

<u>Shallow slide</u> – is a shallow-seated failure in bank material with little cohesion that occurs along a plane parallel to the ground surface.

<u>Rotational slip</u> – is a deep-seated mass failure in cohesive material that occurs along a curved surface leaving an arcuate scar in the bank and back-tilting toward the bank of the failed mass.

<u>Slab-type block</u> – is a failure in cohesive material formed from sliding and forward toppling of a deep-seated mass into the channel. Tension cracks are often present in the bank.

Cantilever failure – forms in composite and layered banks where undermining of a less cohesive lower layer causes the collapse of the resultant overhanging block of cohesive material into the channel.

Pop-out failure – is formed by a piece of the lower bank falling out as a result of saturation and seepage flow in the lower part of a steep, cohesive bank.

Piping failure – forms when part of the bank collapses as a result of high groundwater seepage pressures and rates of flow.

Dry granular flow – also known as dry ravel and soil avalanches, is a flow-type failure of non-cohesive, dry, granular bank material in an oversteepened bank.

Wet earth flow – is a viscous failure of the bank due to the increased weight of the bank and the loss of cohesion resulting from saturation.

| PART 11 LEFT (OR RIGHT) BANK GEOTECH FAILURES | | | | | | | | | |
|---|---|---|---|---|---|---|---|---|---|
| Failure Location | | | | Present Status | | Failure Scars + Blocks | | Apparent Failure Mode | |
| General | | Opposite a structure | | Stable | | None | | Soil/rock fall | Pop-out failure |
| Outside meander | | Adjacent to structure | | Unreliable | | Old | | Shallow slide | Piping failure |
| Inside meander | | D/S of structure | | Unstable: dormant | | Recent | | Rotational slip | Dry granular flow |
| Opposite a bar | | U/S of structure | | Unstable: active | | Fresh | | Slab-type block | Wet earth flow |
| Behind a bar | | Other (write in) | | | | Contemporary | | Cantilever failure | Other (write in) |

Figure C.19. Reconnaissance sheet - left (or right) bank geotechnical failures.

**In Part 12 (Figure C.20)**, the supply of toe sediment and the degree of removal or accumulation are evaluated relative to basal endpoint control and stability of the bank (see Appendix B). *Stored bank debris* records the presence of material derived from erosion or failure of the bank that is stored along the toe of the bank. *Vegetation* is classified since it can accelerate the accumulation of sediment along the bank toe. *Age* relates to the maturity of the accumulated sediment based on the maturity of vegetation established on the deposit. *Tree species* and *health* provide information on physiology, growth patterns, biomass, and physiography relative to the toe deposit maturity and stability. As noted previously, roots define whether a channel and the toe deposits are undergoing aggradation, degradation or are stable. *Existing debris storage* is an estimate of the total volume of the bank-derived debris that has accumulated at the bank toe.

Stream reconnaissance and the record sheets should not be used as a substitute for conventional hydrographic, hydraulic, and geotechnical site surveys, but rather, should precede and complement such surveys.

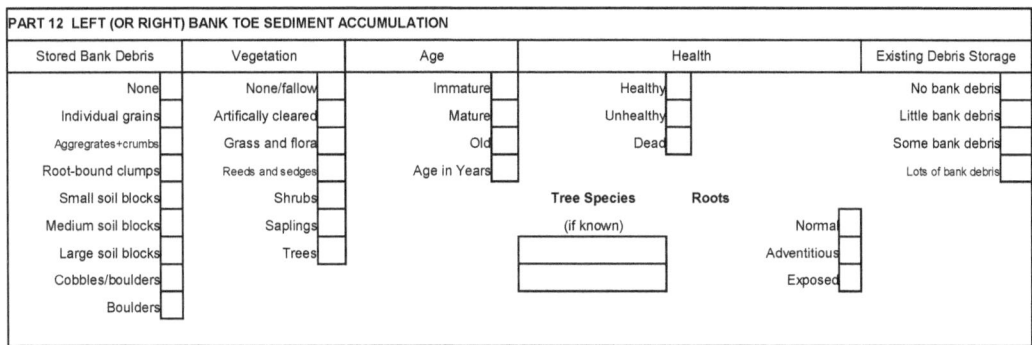

Figure C.20. Reconnaissance sheet - left (or right) bank toe sediment accumulation.

| Channel Sketch Map |||| 
|---|---|---|---|
| Map Symbols<br>(to be determined by field crew) ||||
| Study Reach Limits<br>Cross-Section<br>Bank Profile | North Point<br>Flow Direction<br>Impinging Flow | Cut Bank<br>Exposed Island/Bar<br>Structure | Photo Point<br>Sediment Sampling Point<br>Significant Vegetation |

**Representative Cross-Section**

| Bank Profile Sketches |||
|---|---|---|
| Profile Symbols<br>(to be determined by field crew) |||
| Bank Top Edge<br>Bank Toe<br>Water's Edge | Failed Debris<br>Attached Bar<br>Undercutting | Engineered Structure<br>Significant Vegetation<br>Vegetation Limit |
|  |  |  |

## C.6 REFERENCES

Chorley, R.J., S.A. Schumm, and D.E. Sugden, 1984, "Geomorphology," Methuen & Co. Ltd., London.

Howard, A.D., 1967, "Drainage analysis in geological interpretation: a summation," Bulletin of the American Association of Petroleum Geologists, Vol. 56, p. 275-370.

Thorne, C.R., 1998, "Stream Reconnaissance Handbook," John Wiley & Sons, Chichester.

Varnes, D.J., 1958, "Landslide types and processes," in E.B. Eckel (ed), Landslides and Engineering Practice, Special Report 29, NAS-NRC Publication 544, Highway Research Board, Washington, D.C., p. 20-47.

(page intentionally left blank)

# APPENDIX D

## DATA ITEMS FROM APPENDIX C RECONNAISSANCE SHEETS RELATED TO STREAM STABILITY INDICATORS

(page intentionally left blank)

| Section and Part Number | Reconnaissance Parameter | Relationship to Stability |
|---|---|---|
| Section 1 | general information | basic information on project; should include bridge number |
| Section 2, Part 1. Area around river valley | terrain | important in some classification methods |
| | drainage pattern | descriptive of setting |
| | surface geology | sediment source information, but difficult to identify |
| | rock type | affects erosion rate to some extent, generally difficult to identify during short visit |
| | land use | important to hydrologic response and erosion rates |
| | vegetation | important to hydrologic response and erosion rates |
| Section 2, Part 2. River valley and sides | location of river | indicative of large scale channel behavior |
| | valley shape | large scale descriptor |
| | valley height | along with angle, may be important for indicating sediment source, such as land sliding into river |
| | side slope angle | along with height, may be important for indicating sediment source, such as land sliding into river |
| | valley side failures | sediment source |
| | failure locations | indicates whether potential sediment loadings are upstream or downstream of the bridge |
| Section 2, Part 3. Floodplain | valley floor type and width | indicative of confinement and lateral stability |
| | surface geology | indicates erosion rates |
| | land use | ongoing changes in land use critical to stability |
| | vegetation | important to whether sediment sources are protected |
| | buffer strip and width | important to lateral stability and erosion rates |
| | left and right overbank n | roughness used to assess hydraulics of flow |
| Section 2, Part 4. Vertical relation of channel to valley | terraces | identifies previous incision |
| | trash lines | identifies high water surface elevations |
| | overbank deposits | provides sense of median size of sediment discharge |
| | levees | increases shear stress along bottom; prohibits floodplain activity |
| Section 2, Part 5. Lateral relation of channel to valley | planform | type and dimensions of meanders indicate relative rate of lateral moving |
| | floodplain features | provide evidence of prior meander movement |
| Section 3, Part 6. Channel Description | dimensions | width and depth can indicate entrenchment and stability; slope indicates streampower |
| | flow type | indicates flow energy |
| | bed controls and types | effects vertical stability or where erosion might occur |
| | width controls and types | effects lateral stability or where erosion might occur |
| Section 3, Part 7. Bed Sediment | bed material | overall size category indicates energy of stream to transport |
| | bed armor | a type of vertical control |

| Section and Part Number | Reconnaissance Parameter | Relationship to Stability |
|---|---|---|
| Description | $D_{50, 84, 16}$ | use to compute critical shear stress |
| | substrate size | indicates material available to movement after armor removed |
| | sediment depth | depth of mobile sediment |
| | bed forms | can significantly effect flow resistance |
| | bar types and sediment size | number, size, location, vegetation, and overall sediment size indicative of vertical changes |
| Section 4 and 5, Parts 8 and 13. Left bank characteristics | type | layering and cohesiveness key to bank stability |
| | bank materials | level of cohesiveness controls stability |
| | protection | important to stabilization of bank materials |
| | layer thickness | along with layer materials important to stability |
| | bank height | along with angle, height indicates mass wasting potential |
| | bank slope | along with height, angle indicates mass wasting potential |
| | profile shape | indicates potential for geotech failures |
| | tension cracks | indicates potential for geotech failures |
| Sections 4 and 5, Parts 9 and 14. Left bank-face vegetation | vegetation | plays key role in stabilizing banks and slowing lateral erosion |
| | orientation | indicates rate of bank movement |
| | tree types | roots of different trees better for holding soil in place and providing drainage |
| | density and spacing | important to how much erosion control exerted |
| | location, health, diversity, height | indicates growing rates, and therefore, erosion rates |
| Sections 4 and 5, Parts 10 and 15. Bank erosion | fluvial erosion location | indicates whether erosion activity is occurring in specific locations indicating problems |
| | erosion status and rate | difficult to assess on short site visit, but indicates stabilizing versus destabilizing conditions |
| Sections 4 and 5, Parts 11 and 16. Bank geotech failures | failure location | indicates areas that are contributing to destabilization |
| | present status | difficult to assess on short site visit, but indicates stabilizing versus destabilizing conditions |
| | failure scars | indicates prior mass wasting |
| Sections 4 and 5, Parts 12 and 17. Left bank toe sediment accumulation | stored bank debris | indicative of bank material and failure mechanisms |
| | vegetation | indicative of stabilizing toe |
| | age, health, and type of vegetation | important to showing stability |

# APPENDIX E

## SIMPLIFIED AND REVISED DATA COLLECTION SHEETS BASED ON THORNE (1998) AND JOHNSON (2006)

(page intentionally left blank)

# MODIFIED STREAM RECONNAISSANCE
Based on Thorne (1998)

## SECTION 1 - SITE DESCRIPTION

ROAD NAME/NUMBER                        DATE

BRIDGE NUMBER

STREAM NAME

GPS COORDINATES

## SECTION 2 - REGION AND VALLEY DESCRIPTION

**PART 1: WATERSHED**

| Land Use | Vegetation |
|---|---|
| O Natural | O None |
| O Agricultural | O Grass |
| O Urban | O Pasture |
| O Suburban | O Crops |
| O Rural | O Shrubs |
| O Industrial | O Deciduous Forest/trees |
| O Cattle grazing | O Coniferous Forest/trees |

**PART 2: RIVER VALLEY CONDITION**

| Valley Side Failures | Failure Locations |
|---|---|
| O None | O None |
| O Occasional | O Away from river |
| O Frequent | O Along river |

**PART 3: FLOODPLAIN**

| Floodplain Width | Land Use | Vegetation | Riparian Buffer Strip |
|---|---|---|---|
| O None | O Natural | O None | O None |
| O <1 river width | O Agricultural | O Grass | O <1 river width |
| O 1-5 river widths | O Urban | O Pasture | O 1-5 river widths |
| O 5-10 river widths | O Suburban | O Orchards | O >5 river widths |
| O >10 river widths | O Rural | O Crops | |
| | O Industrial | O Shrubs | |
| | O Mining | O Deciduous Forest/trees | |
| | O Cattle grazing | O Coniferous Forest/trees | |

## PART 4: VERTICAL CONFINEMENT

**Terraces**
O None
O Left bank
O Right bank

**Levees**
O None
O Natural
O Constructed

**Levee Location**
O Along channel bank
O Set back <1 river width
O Set back >1 river width

## PART 5: LATERAL RELATION OF CHANNEL TO VALLEY

**Planform**
O Straight
O Meandering
O Braided
O Anastomosed
O Engineered

**Meander Characteristics**
O Mild bends
O Moderate bends
O Tight bends

# SECTION 3 - CHANNEL DESCRIPTION

## PART 6: CHANNEL DESCRIPTION (select all that apply)

**Bed Controls**
O None
O Occasional
O Frequent
O Confined

**Control Types**
O None
O Bedrock
O Boulders
O Gravel armor
O Bridge protection
O Grade control
O Debris
O Dams (beaver, engineered)

**Width Controls**
O None
O Occasional
O Frequent
O Confined

**Control Types**
O None
O Bedrock
O Boulders
O Gravel armor
O Bridge protection
O Bridge abutments
O Bank stabilization
O Debris

**Other**
O Debris
O Mining
O Reservoir
O Knickpoint

**Flow Habit**
O Perennial
O Flashy perennial
O Intermittent
O Ephemeral

**Channel width = _____**

**M-B classification**
O Cascade or step-pool
O Plane, pool-riffle, dune-ripple
O Braided

**Corps classification (other)**
O Modified (engineered)
O Regulated
O Arroyo

## PART 7: BED SEDIMENT DESCRIPTION (select all that apply)

**Bed Material**
O Clay
O Silt
O Sand
O Gravel
O Cobbles
O Boulders
O Bedrock

**Bar Types**
O None
O Alternate bars
O Point bars
O Mid-channel bars
O Diagonal bars
O Irregular/combination
O Braided

**Bar Material**
O Silt
O Sand
O Gravel
O Cobbles

**Bar Vegetation**
O None
O Grasses
O Reeds/shrubs
O Trees

**Bar Width**
O None
O Narrow
O Moderate
O Wide

**Percent sand in bed = _____ %**

| Bank Characteristic | Left Bank | Right Bank |
|---|---|---|
| Bank material | O clay<br>O silt<br>O sand<br>O gravel<br>O cobbles<br>O boulders<br>O bedrock | O clay<br>O silt<br>O sand<br>O gravel<br>O cobbles<br>O boulders<br>O bedrock |
| Layer material | O no layers<br>O cohesive<br>O sand<br>O oravel<br>O cobbles<br>O boulders | O no layers<br>O cohesive<br>O sand<br>O gravel<br>O cobbles<br>O boulders |
| Bank height | | |
| Bank slope | O steep<br>O moderate<br>O shallow | O steep<br>O moderate<br>O shallow |
| Bank vegetation | O none<br>O grasses/annuals<br>O reeds/shrubs<br>O trees:<br>falling trees? O yes O no<br>tree density: O sparse O dense<br>tree health: O good O poor<br>tree ages: Oyoung Omature Oold<br>tree diversity? O yes O no | O none<br>O grasses/annuals<br>O reeds/shrubs<br>O trees:<br>falling trees? O yes O no<br>tree density: O sparse O dense<br>tree health: O good O poor<br>tree ages: Oyoung Omature Oold<br>tree diversity? O yes O no |
| Bank erosion and failure location | location of erosion:<br>  O outside meander bend<br>  O inside meander bend<br>  O opposite bar or obstruction<br>  O general<br>type of erosion:<br>  O fluvial<br>  O geotechnical | location of erosion:<br>  O outside meander bend<br>  O inside meander bend<br>  O opposite bar or obstruction<br>  O general<br>type of erosion:<br>  O fluvial<br>  O geotechnical |

(page intentionally left blank)